Multi-sensor Fusion in Medical Imaging, Diagnosis and Therapy

Multi-sensor Fusion in Medical Imaging, Diagnosis and Therapy

Guest Editor

Xuming Zhang

MDPI

Basel • Beijing • Wuhan • Barcelona • Belgrade • Novi Sad • Cluj • Manchester

Guest Editor
Xuming Zhang
College of Life Science &
Technology
Huazhong University of
Science and Technology
Wuhan
China

Editorial Office
MDPI AG
Grosspeteranlage 5
4052 Basel, Switzerland

This is a reprint of the Special Issue, published open access by the journal *Sensors* (ISSN 1424-8220), freely accessible at: https://www.mdpi.com/journal/sensors/special_issues/4RG930S0UB.

For citation purposes, cite each article independently as indicated on the article page online and as indicated below:

Lastname, A.A.; Lastname, B.B. Article Title. *Journal Name* **Year**, *Volume Number*, Page Range.

ISBN 978-3-7258-8243-4 (Hbk)
ISBN 978-3-7258-8244-1 (PDF)
https://doi.org/10.3390/books978-3-7258-8244-1

Contents

sensors

MDPI

Article

Rehabilitation Assessment System for Stroke Patients Based on Fusion-Type Optoelectronic Plethysmography Device and Multi-Modality Fusion Model: Design and Validation

Liangwen Yan [1], Ze Long [1], Jie Qian [2], Jianhua Lin [3], Sheng Quan Xie [4] and Bo Sheng [1,*]

[1] School of Mechatronic Engineering and Automation, Shanghai University, Shanghai 200444, China; lw_yan@shu.edu.cn (L.Y.)
[2] Second Affiliated Hospital, School of Medicine, Zhejiang University, Hangzhou 310009, China
[3] Department of Rehabilitation Therapy, Yangzhi Affiliated Rehabilitation Hospital of Tongji University, Shanghai 201619, China
[4] School of Electronic and Electrical Engineering, University of Leeds, Leeds LS2 9JT, UK; s.q.xie@leeds.ac.uk
* Correspondence: shengbo@shu.edu.cn

Abstract: This study aimed to propose a portable and intelligent rehabilitation evaluation system for digital stroke-patient rehabilitation assessment. Specifically, the study designed and developed a fusion device capable of emitting red, green, and infrared lights simultaneously for photoplethysmography (PPG) acquisition. Leveraging the different penetration depths and tissue reflection characteristics of these light wavelengths, the device can provide richer and more comprehensive physiological information. Furthermore, a Multi-Channel Convolutional Neural Network–Long Short-Term Memory–Attention (MCNN-LSTM-Attention) evaluation model was developed. This model, constructed based on multiple convolutional channels, facilitates the feature extraction and fusion of collected multi-modality data. Additionally, it incorporated an attention mechanism module capable of dynamically adjusting the importance weights of input information, thereby enhancing the accuracy of rehabilitation assessment. To validate the effectiveness of the proposed system, sixteen volunteers were recruited for clinical data collection and validation, comprising eight stroke patients and eight healthy subjects. Experimental results demonstrated the system's promising performance metrics (accuracy: 0.9125, precision: 0.8980, recall: 0.8970, F1 score: 0.8949, and loss function: 0.1261). This rehabilitation evaluation system holds the potential for stroke diagnosis and identification, laying a solid foundation for wearable-based stroke risk assessment and stroke rehabilitation assistance.

Keywords: rehabilitation assessment; stroke; photoplethysmography; multi-modality fusion; MCNN-LSTM-Attention

1. Introduction

According to a study published in the international journal *The Lancet*, which focused on 369 common diseases, stroke, ischemic heart disease, and diabetes were identified as the three major threats to human health. Among these diseases, China bears the most severe burden of stroke [1]. The China Stroke Prevention and Control Report (2023) states that China has the highest number of stroke patients globally, and the incidence is increasing among younger populations [2]. Stroke is a neurological symptom and disease caused by damage to specific areas of the brain [3,4]. Stroke is considered one of the most severe diseases in modern society as it can lead to death in severe cases and cause physical and mental disabilities. Rehabilitation assessment after stroke has been proven to be beneficial and essential for most stroke patients [5,6]. Julie Bernhardt et al. [7] highlighted the crucial role of rehabilitation in stroke recovery during the First Stroke Recovery and Rehabilitation Roundtable. Rehabilitation assessment not only improves patients' functional abilities but also enhances their quality of life and social participation.

Commonly used assessment scales include the Brunnstrom Assessment [8], the National Institutes of Health Stroke Scale (NIHSS) [9,10], and the Modified Rankin Scale [11]. The Brunnstrom staging is widely used in clinical practice to classify the rehabilitation progress of stroke patients due to its simplicity and effectiveness. The Brunnstrom staging divides stroke recovery patients into stages of flaccidity, spasticity, voluntary movement, and recovery. By combining changes in neuro-reflex muscle strength and muscle tone at different stages, specific training methods can be applied to achieve the goal of restoring patients' limb and cognitive functions [12–14]. However, most traditional assessment methods are time-consuming, require significant effort, and lack diversity in approaches. Moreover, rehabilitation assessment by healthcare professionals heavily relies on personal experience and subjective judgment, making it difficult to obtain objective and accurate quantitative assessment results that intuitively reflect the patient's rehabilitation status.

To address these issues, numerous researchers have proposed digital methods for assessing the rehabilitation statuses of patients. Common biomedical sensing technologies include surface electromyography (sEMG), motion analysis systems, electrocardiography, and photoplethysmography (PPG). For example, Bo Sheng et al. [15] utilized the Kinect v2 depth sensor to capture motion data and applied singular spectrum analysis and the multi-ReliefF method to assess upper limb function in stroke hemiparetic patients. Hsin-Ta Li [16] used inertial measurement units (IMUs) and surface electromyography (sEMG) to acquire lower limb motion signals and employed the Support Vector Machine (SVM) algorithm to assess lower limb function in hemiparetic patients. However, the primary goal of stroke rehabilitation is to restore overall functional abilities, and assessing the functions of specific muscles or movements may not comprehensively reflect a patient's overall motor capabilities. Additionally, in Hsin-Ta Li's study [16], when participants performed finger movements or movements such as dorsiflexion and knee extension, the signals for each action were manually trimmed through a user interface, introducing subjectivity to the data collection. On the other hand, some researchers have focused on evaluating stroke patients from the perspective of blood signals. For instance, Pei-Wen Huang et al. [17] proposed a multi-modality analysis method incorporating electrocardiography, arterial blood pressure (ABP), and PPG to predict the functional status of stroke patients, achieving an accuracy of 82.7%. However, these studies faced challenges of low identification accuracy and precision.

In recent years, the continuous development of biosensing technology has led to widespread attention to photoplethysmography (PPG) as a convenient and non-invasive monitoring method. The PPG signal waveform reflects changes in blood volume within the measured area. It provides important insights into cardiac ejection capacity and hidden pathological information during blood propagation along the vascular tree [18–20]. However, different wavelengths of PPG signals exhibit varying sensitivities to pathological responses, leading to the need for research on multi-sensor PPG systems to obtain comprehensive and accurate physiological information [21]. Previous studies have shown a close correlation between green light PPG signals and electrocardiogram (ECG) R–R intervals [22]. In contrast, red and infrared lights have stronger penetration capabilities, allowing for deeper penetration into deeper layers of vascular tissue [23]. For example, Yuka Maeda et al. revealed the advantages of green-light PPG in pulse rate measurement while the performance of infrared-light PPG in this aspect was relatively poorer [24]. Additionally, Revati Shriram et al. utilized the characteristics of red and infrared lights to investigate the relationship between arterial stiffness and PPG waveforms, providing a foundation for the future assessment of cerebrovascular health [25]. However, the application of PPG signals in the rehabilitation assessment of stroke still faces challenges. In light of this, this study aimed to explore a multi-sensor PPG evaluation system that comprehensively considers red, green, and infrared lights, further investigating the application of multi-wavelength fusion PPG signals in the rehabilitation assessment of stroke.

In previous research, Wei et al. [26] designed a PPG acquisition device and proposed a model that combines the Convolutional Neural Network (CNN), Long Short-Term Memory

(LSTM), and an attention mechanism This model achieved a good accuracy rate of 99.1% in the diagnosis and identification of hypertension. However, when using this model for stroke-patient rehabilitation assessment, the accuracy rate was found to be relatively low. To address the aforementioned issues, this paper proposes a new device and system aiming to provide a more convenient data acquisition solution and higher accuracy assessment results. Specifically, the objectives are as follows:

1. To design a fusion-based PPG sampling device named "NeuroPulseGuard" with higher accuracy, safety, reliability, and portability;
2. To propose a multi-modality assessment model (MCNN-LSTM-Attention) based on the fusion of multiple PPG signals. In this study, a total of eight patients and eight healthy individuals were recruited for data collection and clinical experiments. Performance validation and assessment were conducted by comparing the performance of different models on the same dataset, using accuracy rate, precision, recall, F1 score, and computational efficiency as evaluation metrics.

2. Materials and Methods

Figure 1 illustrates the operational workflow of the system developed in this study. The process consists of the following stages:

1. Designing and implementing a secure, reliable, and portable fusion-based PPG sampling device for collecting PPG data from stroke patients with varying degrees of severity and transmitting the data via Wi-Fi;
2. Preprocessing the data from patients with different severity levels and healthy volunteers;
3. Analyzing the data using the proposed MCNN-LSTM-Attention model to provide rehabilitation assessment grades for stroke patients;
4. Physicians can employ these results for more informed clinical interventions, thereby facilitating better rehabilitation outcomes for the patients.

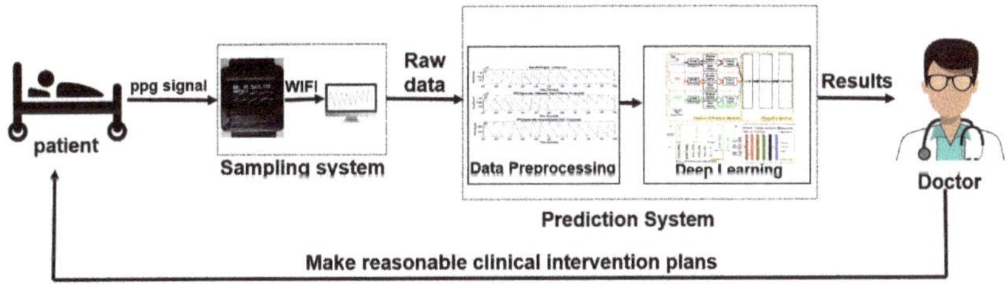

Figure 1. The framework of the stroke assessment system.

2.1. Equipment

In order to obtain high-quality raw PPG signals, a comprehensive signal acquisition process was designed in this study, as shown in Figure 2. The system consists of the following steps: (1) The PPG photoplethysmography (PPG) sensor module collects PPG signals from the patient's finger; (2) The PPG sensor interacts with the microcontroller through the I2C protocol; (3) The microcontroller exchanges data with the PC via Wi-Fi.

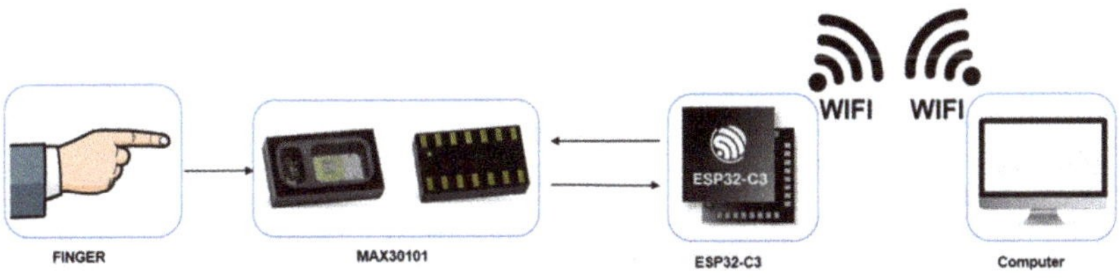

Figure 2. The framework of the PPG acquisition system.

As shown in Figure 3, the NeuroPulseGuard device's primary function is to acquire, transmit, and store PPG signals. The device's dimensions are 35 mm × 35 mm, with the core PCB board measuring only 31 mm × 31 mm. The chosen microcontroller for this device is the ESP32-C3, an IoT chip known for its secure stability, low power consumption, and cost-effectiveness. It is equipped with a 32-bit single-core RISC-V processor clocked at up to 160 MHz. The chip supports 2.4 GHz Wi-Fi and Bluetooth 5 (LE) and has obtained CLS-Ready certification, complying with device network security standards [27]. Additionally, the device utilizes the MAX30101 as the core component of its multi-sensor system. The MAX30101 is a sensor chip integrating green, infrared, and red light sensors, exhibiting exceptional performance in measuring blood volume changes [28]. As depicted in Figure 4, the MAX30101 design integrates various components such as photodetectors, analog-to-digital converters (ADCs), and digital signal processors (DSPs) to achieve high-precision measurement of blood volume changes. The chip's LED light sources include green, red, and infrared lights, with peak wavelengths of 527 nm, 660 nm, and 880 nm, respectively, chosen based on their correlation with blood reflection characteristics. The green light wavelength primarily reflects superficial blood volume changes while the red light wavelength is more sensitive to changes in superficial vessels. The infrared light wavelength can penetrate deeper into vascular tissues, providing more comprehensive information [23]. By analyzing signals from these wavelengths comprehensively, a broader and more accurate assessment of blood volume changes can be obtained, enhancing the system's capability for rehabilitation assessment of stroke.

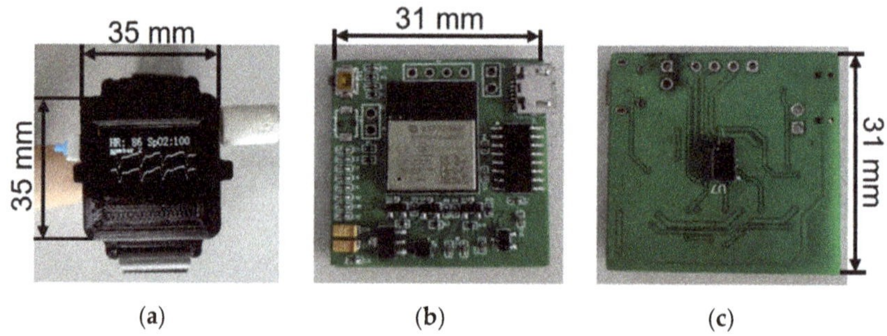

Figure 3. 'NeuroPulseGuard': (**a**) physical diagram of the fusion-based PPG sampling device; (**b**) front view of the internal PCB board of the fusion-based sampling device; (**c**) schematic diagram of the reverse side of the internal PCB board of the fusion-based sampling device.

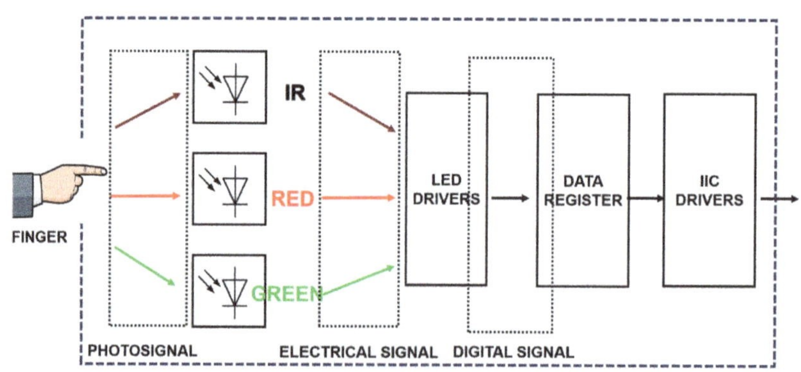

Figure 4. Operational diagram of the multi-functional sensor MAX30101.

To validate the functionality and accuracy of the experimental device, a comparative experiment was conducted involving three participants (including two males and one female). The participants' ages were 45, 52, and 23 years, respectively. During the experiment, the participants were instructed to maintain a supine position on the bed, ensuring a stable physiological state. The PHILIPS DB12 pulse oximeter was used to collect data from the participants' left index fingers while the experimental device was used to capture PPG signals from their right index fingers. Each data collection sequence lasted for 5 s. The obtained blood oxygen saturation and heart rate data were compared with the records from the PHILIPS DB12 pulse oximeter. To ensure reliability, each participant underwent three measurements. The experimental results are presented in Tables 1 and 2. In the experimental device, red and infrared light are highly absorbed by blood and can reflect the oxygenation level of hemoglobin. Therefore, red and infrared light were utilized to calculate blood oxygen saturation. On the other hand, green light with a wavelength that can penetrate the skin and superficial blood vessels is highly sensitive to small blood volume changes caused by cardiac pulsations. Consequently, the green light PPG signal was used to calculate heart rate [22]. The heart rate and blood oxygen saturation data, obtained through the acquisition and processing of the PPG signal, were compared with the corresponding data recorded by the PHILIPS DB12 pulse oximeter. The results demonstrated an error of less than 1%, ensuring the accuracy of data acquisition.

Table 1. Comparison of device blood oxygen values with the commercial instrument.

No. of Volunteer	Experimental Data/%	DB12/%	Error
	98	98	0%
1	98	97	1%
	96	97	1%
	95	95	0%
2	99	100	1%
	97	97	0%
	95	96	1%
3	97	97	0%
	100	100	0%

Table 2. Comparison of device heart rate values with commercial instruments.

No. of Volunteer	Experimental Data/bpm	DB12/bpm	Error
	86	86	0%
1	92	93	1%
	89	89	0%
	89	90	1%
2	99	100	1%
	101	100	1%
	109	108	1%
3	101	101	0%
	86	85	1%

2.2. Data Processing

The objective of preprocessing filtering is to remove or attenuate noise, artifacts, and interference to extract the desired physiological signals. PPG signals of different wavelengths are, fundamentally, intensity signals collected by optical sensors, sharing similar characteristics and spectral content [29]. Therefore, the same filtering approach was applied for preprocessing in this study.

In the process of PPG signal acquisition, device component characteristics, external environment, and unconscious movements can introduce different frequencies of noise. These noise types can be mainly classified into three categories [30]: random noise, low-frequency noise, and high-frequency noise. Random noise originates from interference by light and other electromagnetic signals in the environment, affecting the intensity of the

PPG signal. Low-frequency noise is primarily caused by baseline drift and motion artifacts, including baseline drift due to respiration and short-term tissue deformation caused by motion. High-frequency noise comes from muscle interference and power line interference, appearing as "spike-like" signal noise. Researchers have developed various types of filters to address these issues, such as FIR filters [31,32], Kalman filters [33], Butterworth filters [34,35], moving average filters [36], and Chebyshev Type II filters. According to relevant researchers' studies [37], by comparing nine different filters with 10 different orders, it was found that the Chebyshev Type II filter exhibited superior performance in improving the quality of the PPG signal. This filter demonstrates frequency selectivity, effectively filtering out interference and noise while preserving valuable information in the signal. Additionally, this study also compared the filtering effects of different filters. Based on the results presented in Table 3, the Chebyshev Type II filter performed the best in terms of signal-to-noise (SNR) ratio.

Table 3. Comparison of filtering effects among different filters.

	Chebyshev Type II Filter	Gaussian Filter	Savitzky–Golay Filter	Smooth Filter
SNR (dB)	81.9626	40.845	76.5083	45.2653

The Chebyshev Type II filter is a type of digital filter characterized by having minimal ripple within the passband and maximum attenuation within the stopband. This filter allows for the existence of ripples within the passband to achieve a steeper filter response and provides maximum attenuation within the stopband to suppress unwanted frequency components. To describe the transfer function, we use the following formula(s):

$$\mathrm{H(s)} = 1/\mathrm{sqrt}\left(1 + \varepsilon^2 \times \mathrm{C_2}^2(\mathrm{s})\right) \tag{1}$$

Here, H(s) represents the transfer function of the filter, where s is a complex variable. C_2 denotes the polynomial form of the transfer function for the Chebyshev Type II filter. ε represents the maximum passband ripple parameter. The expression for $C_2(s)$ is given by the following equation:

$$\mathrm{C_2(s)} = (\mathrm{s}^2 - 1)^2/(\mathrm{C_1(s)} \times \mathrm{C_2(s)} \times \ldots \times \mathrm{C_n(s)}) \tag{2}$$

Here, n represents the order of the filter, and $D_1(s)$, $D_2(s)$, ..., $D_n(s)$ are first-order polynomials. They can be expressed using the following formula:

$$\mathrm{D_k(s)} = \mathrm{s} + \alpha_\mathrm{k} \tag{3}$$

Here, α_k represents constants associated with the poles of the filter.

From Figure 5a,b and the 10-min processed graph, it can be observed that the signal, after undergoing Chebyshev Type II processing, removed a significant portion of baseline drift, and the waveform distribution is more concentrated around the central axis. From Figure 6a,b and the 10-s processed detailed graph, it can be seen that the signal curve, after Chebyshev Type II processing, appeared smoother. From Figure 6b,c and the processed detailed graph, it can be observed that the residual offset component in the signal, after the previous processing, was removed using cubic spline interpolation [38].

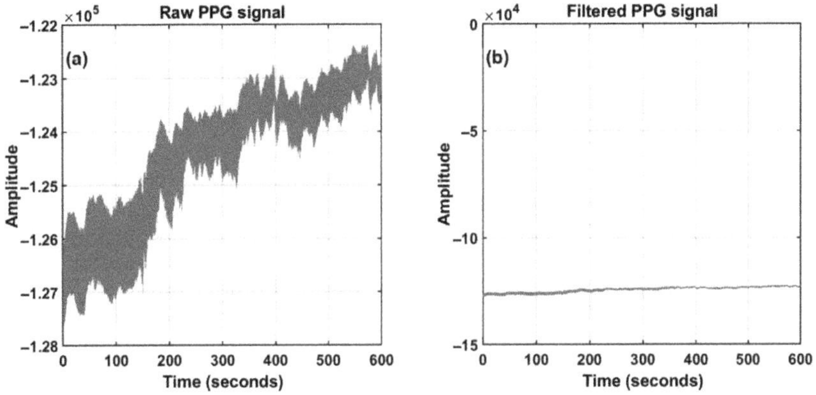

Figure 5. Data processing: (**a**) raw PPG signal (10 min); (**b**) filtered PPG signal (10 min).

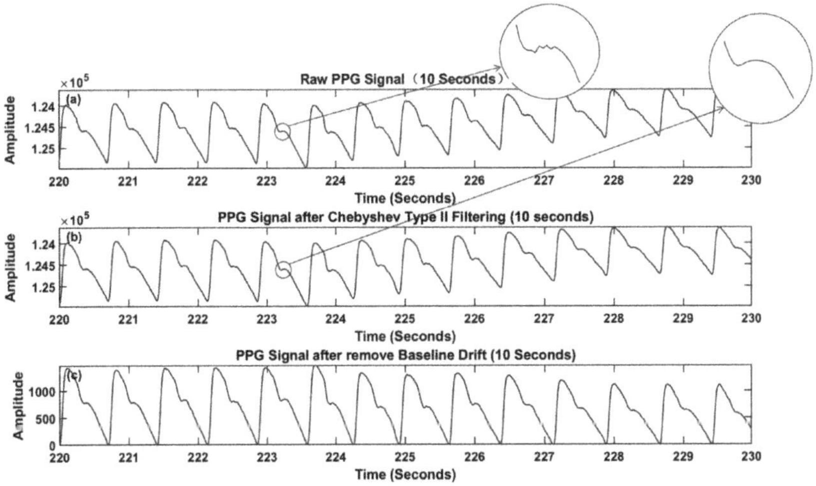

Figure 6. Data processing details: (**a**) raw PPG signal (10 s); (**b**) PPG signal after Chebyshev Type II filtering (10 s); (**c**) PPG signal after removing baseline drift (10 s). The enlarged view of a specific detail in the PPG signal is shown within the green circle.

2.3. Model Architecture

This study aimed to achieve automated classification of stroke patients to provide convenient and accurate rehabilitation assessment. To this end, we have designed an end-to-end deep learning model for extracting features and classifying strokes from preprocessed raw signals. Figure 7a illustrates the overall architecture of the model, consisting of a feature extractor module, depicted in Figure 7b, and a classifier module, shown in Figure 7c. Initially, the preprocessed green, red, and infrared light signals are fed into a Multi-Channel Convolutional Neural Network (MCNN) module, which extracts both shallow and deep features from each signal and subsequently combines them. The fused features from different PPG signals are then input into a Long Short-Term Memory (LSTM) module for further fusion learning. The output of the LSTM module is subsequently passed through an attention module. Finally, the features from the attention module are fed into a classifier for rehabilitation assessment of stroke.

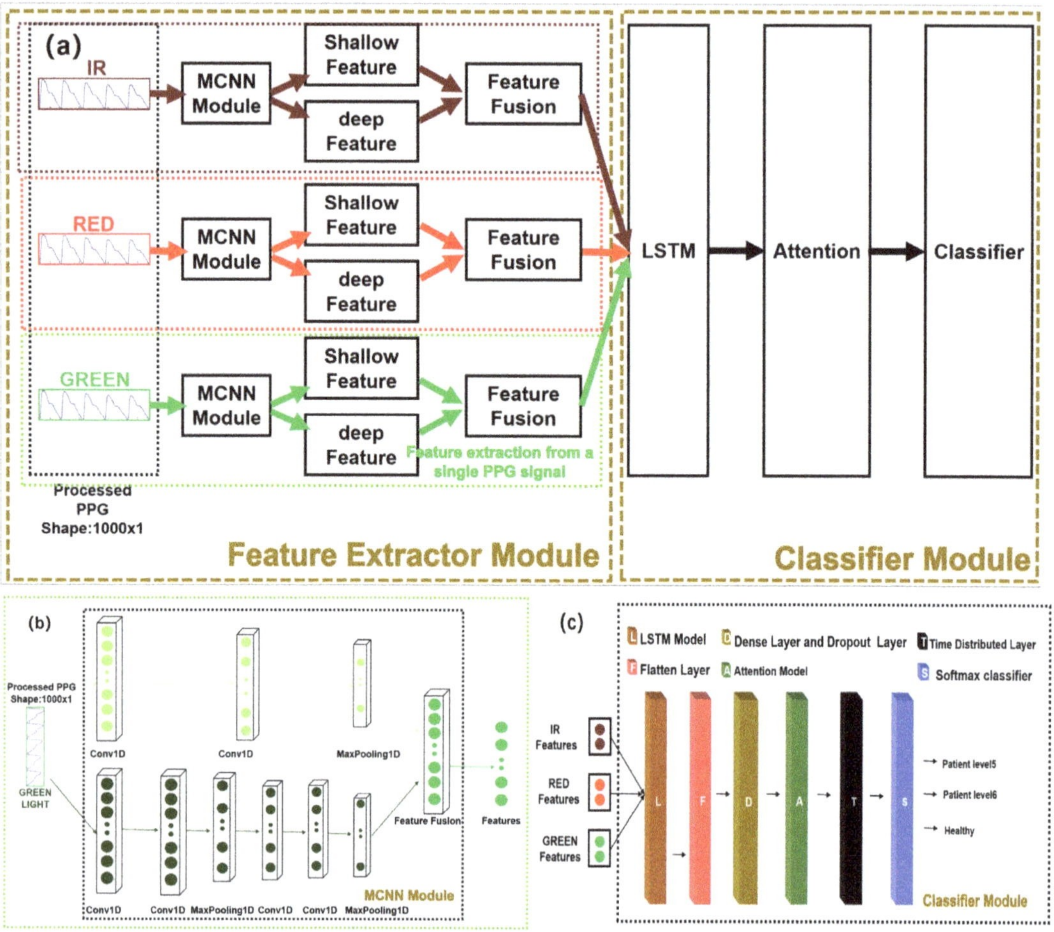

Figure 7. (**a**) MCNN-LSTM-Attention model; (**b**) MCNN module; (**c**) classifier module.

Figure 7b showcases the feature extractor module using the preprocessed green light signal as an example. It comprises two one-dimensional Convolutional Neural Networks (CNNs) with different kernel sizes and depths. These CNNs simultaneously process the preprocessed green light signal as input and extract features in different frequency domains. The larger receptive field is employed for extracting shallow features while the smaller receptive field is utilized for extracting deep features. Using one-dimensional CNNs as feature extractors offers several advantages: firstly, they can automatically learn the underlying meanings of different PPG signals; secondly, weight-sharing strategies significantly reduce the number of parameters in high-dimensional input vectors. The classifier module depicted in Figure 7c consists of hierarchical Long Short-Term Memory (LSTM) networks, an attention layer, and fully connected layers. These components construct a complex nonlinear model that captures the relationship between inputs and outputs. In this architecture, the LSTM module further integrates the features extracted from different PPG signals and serves as an input to the attention layer. The attention mechanism identifies and emphasizes the most important features as determined by the model, thereby enhancing classification performance. The output of the attention layer is then passed to the fully connected layers for classification. Finally, the softmax function is applied to obtain classification results at different levels. Dropout strategies and 10-fold cross-validation were employed to mitigate overfitting during model training.

2.3.1. Single-Signal Fusion Module

Convolutional Neural Networks (CNNs) are feedforward neural networks with convolutional computations and deep structures that are widely used in image processing and natural language processing [39]. They extract features from input data by performing convolutional computations and deep structures, providing effective feature representations. Multi-Channel Convolutional Neural Networks (MCNNs) have been applied in various fields. Liang et al. [40] utilized a multi-channel CNN structure where each channel received a different pre-trained word vector as input. These multiple channels captured additional semantic information from input sentences, enabling the model to learn more discriminative semantic features and have stronger representation capability in natural language processing. X. Chen et al. [41] employed two CNNs with different kernel sizes to extract signal features of different frequencies from raw data and used Long Short-Term Memory (LSTM) networks to classify fault types based on these features. When applying MCNNs to process one-dimensional pulse waveforms (PPG) signals, the following advantages can be obtained:

1. Capturing multi-modal features: PPG signals are measurements of changes in blood volume caused by heartbeats obtained through optical sensors. PPG signals contain components with different frequencies and amplitudes that are related to physiological parameters such as heart rate and blood pressure. By using multiple channels in convolutional layers, MCNN can capture features at different scales simultaneously. For example, lower-frequency channels can capture the overall shape and fluctuations of heartbeats while higher-frequency channels can capture subtle variations in heartbeats [42–44].
2. Multi-modal fusion: PPG signals can be obtained from three different light sources, each providing slightly different characteristics in PPG signals. MCNN can process PPG signals from different lights simultaneously and extract feature representations for light sources through multi-channel convolutional layers. By applying multi-channel convolution and pooling operations, MCNN can fuse the information from different light sources into a unified feature representation, enhancing the model's understanding of PPG signals.
3. Hierarchical feature extraction: Hierarchical feature extraction is an important characteristic of MCNN, typically comprising multiple convolutional and pooling layers. This hierarchical structure enables the progressive extraction of features at different levels of abstraction from PPG signals. As shown in Figure 8, taking the green light PPG signal as an example, MCNN utilizes CNNs to extract deep and shallow features separately. These features originate from different layers and exhibit distinct characteristics. The shallow convolutional layers aim to capture low-level features of PPG signals, such as the shape and fluctuations of heartbeats, to preserve the local information of the signals better. As the network layers deepen, the deep feature encoder can capture more abstract and complex patterns in the signal, such as patterns of heart rate variations or cardiac pathologies [45,46]. By combining these two types of features from different levels, MCNN can fully leverage both global and local information in the signal, resulting in more comprehensive and accurate feature representations. The fused features are then input into the Long Short-Term Memory (LSTM) network for further temporal modeling and processing. Consequently, hierarchical feature extraction enables MCNN to represent and classify PPG signals better, thereby improving the model's performance and robustness.

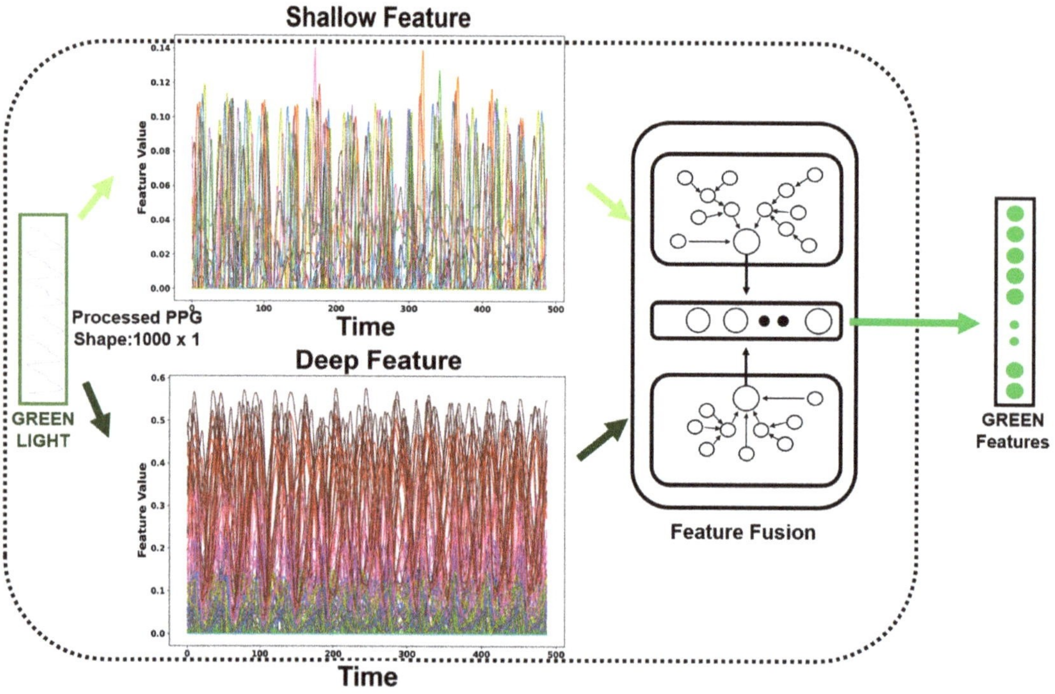

Figure 8. Fusion diagram of MCNN. In the "shallow feature" graph and the "deep feature" graph, different colors represent different features.

One-dimensional convolution is used for feature extraction for one-dimensional time series data such as PPG signals.

$$y_i^{k+1}(j) = w_i^k \times X^k(j) + b_i^k \tag{4}$$

Here, w and b are the weights and biases of the i-th layer filter and the k-th layer, respectively, and X represents the j-th local input in the k-th layer.

The i-th layer of the k + 1 channel after pooling can be described as follows:

$$P_i^{k+1}(j) = \max_{(j-1)W+1 \leq t \leq jW}\left\{q_i^k(t)\right\} \tag{5}$$

q represents the t-th neuron in the i-th channel of the k-th layer and W is the width of the pooling kernel.

In PPG signal processing, one-dimensional convolutional neural networks can be used to extract temporal features from the signals. The convolutional layer learns filters suitable for PPG signals to capture important features while the pooling layer further reduces the dimensionality of the features and retains the main signal patterns. These features can be utilized in various applications such as heart rate detection, emotion analysis, and disease diagnosis.

2.3.2. Multi-Signal Fusion Module

Pulse wave data belong to the category of one-dimensional temporal data, and the recurrent neural network (RNN) structure in deep learning algorithms has shown promising results in processing time series data [47]. However, RNNs are prone to gradient vanishing or exploding problems. Therefore, in this paper, the Long Short-Term Memory (LSTM) network structure is employed as a substitute for the RNN model.

As shown in Figure 9, the LSTM consists of three gating mechanisms: the forget gate, the input gate, and the output gate. The PPG signal features extracted by the Multi-Channel Convolutional Neural Network (MCNN) are fed into the input gate. The forget gate determines the amount of information to retain in the previous time step's state. It computes the activation value of the forget gate by considering the current input, the previous output, and the state information, followed by applying the sigmoid function. The activation value of the forget gate is multiplied by the state of the previous time step, controlling which information should be forgotten from the previous state.

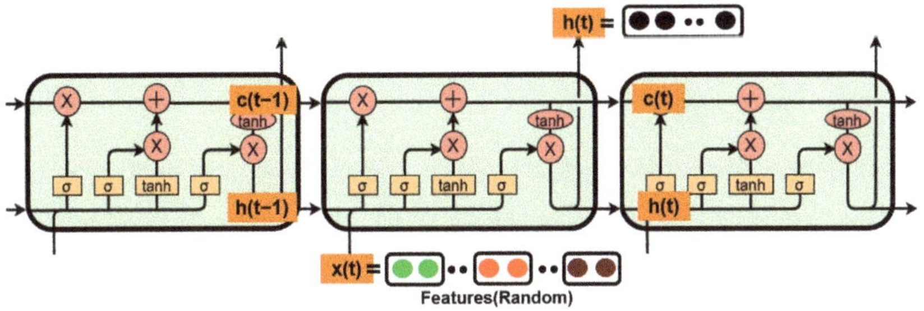

Figure 9. Diagram of LSTM multi-signal fusion structure.

The specific formulas are as follows; the information received by the forget gate includes the current input x_t, the previous node's output h_{t-1}, and the previous node's state C_{t-1}. The activation calculation formula for the forget gate (where b and W are bias and weight vectors, and σ denotes the sigmoid function) is presented below:

$$f_t = \sigma(W_f \times [x_t, h_{t-1}, C_{t-1}] + b_f) * C_{t-1} \tag{6}$$

$$i_t = \sigma(W_i \times [x_t, h_{t-1}, C_{t-1}] + b_i) \tag{7}$$

$$C_t = f_t + i_t \times \tanh(W_c \times [x_t, h_{t-1}, C_{t-1}] + b_c) \tag{8}$$

Based on the new state C_t of this neuron node, the LSTM structure can output the current neuron's output by using the output gate based on h_{t-1}, x_t, and the new state C_t.

$$o_t = \sigma(W_o \times [x_t, h_{t-1}, C_t] + b_o) \tag{9}$$

$$h_t = \tanh(C_t) \times o_t \tag{10}$$

2.3.3. Accuracy Improvement Module

The attention mechanism simulates the behavior of human vision or attention, allowing the model to focus more on important parts of the input sequence [48,49]. General attention mechanisms are based on the Encoder–Decoder framework, as shown in Figure 10. The purpose of this model is to address the task of mapping variable-length input sequences $X = (x_1, x_2, \ldots, x_n)$ to variable-length output sequences $Y = (y_1, y_2, \ldots, y_m)$. The Encoder is responsible for receiving the input sequence X and transforming it into an intermediate abstract representation C through nonlinear transformations: $C = f(x_1, x_2, \ldots, x_n)$. The primary role of the Encoder is to encode the input sequence into a fixed-dimensional representation, capturing the semantic and contextual information of the input sequence. The Decoder's task is to predict and generate the output at time step i, y_i, based on the intermediate abstract representation C of the input sequence X and the previously generated partial outputs $y_1, y_2, \ldots, y_{i-1} : y_i = g(y_1, y_2, \ldots, y_{i-1}, C)$. The Decoder utilizes the intermediate representation C and the already generated partial outputs to infer the next output, gradually constructing the output sequence.

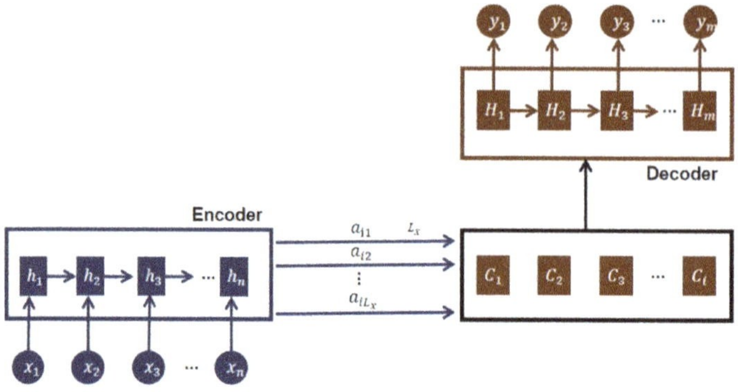

Figure 10. Abstract Encoder–Decoder framework.

H_{m-1} represents the hidden state of the Decoder at time step m−1, y_m is the target value, and C_i is the context vector. Therefore, the hidden state at time step m can be defined as follows.

$$y_m = f(H_{m-1}, y_{m-1}, C_i) \tag{11}$$

C_i depends on the hidden vectors of the input sequence on the Encoder side and can be represented after weighted processing, as shown in Equation (12).

$$C_i = \sum_{j=1}^{L_x} \alpha_{ij} h_j \tag{12}$$

h_j represents the hidden vector of the j-th value on the Encoder side, which contains information from the entire input sequence but focuses on the surrounding portion of the j-th value. L_x is the length of the input side, α_{ij} represents the attention allocation coefficient of the j-th value on the Encoder side to the i-th value on the Decoder side, and the sum of α_{ij} probabilities is 1. The calculation formula for α_{ij} is shown in Equation (13).

$$\alpha_{ij} = \frac{\exp(\alpha_{ij})}{\sum_{j=1}^{L_x} \exp(\alpha_{ij})} \tag{13}$$

$$\alpha_{ij} = a(y_{m-1}, h_j) \tag{14}$$

α_{ij} represents an alignment model that measures the alignment degree between the value at position j on the Encoder side and the value at position i on the Decoder side. Introducing an attention mechanism in the model can help capture the temporal correlation in the signal better and enable the model to adapt to the signal characteristics of each patient individually, improving the personalization effect of classification.

In summary, this model uses MCNN for feature extraction to learn signal features in different frequency domains automatically. Then, LSTM–Attention networks are used for temporal modeling and internal feature extraction, followed by fully connected layers for classification. This structure effectively handles PPG signals and converts them into classification results for different levels of stroke patients.

3. Experiments

3.1. Participants

The subjects were divided into two groups: the stroke patient group and the healthy group. The stroke patient group consisted of 8 individuals diagnosed with ischemic stroke at the Sunshine Rehabilitation Center in Shanghai, China. Correspondingly, during the same period, 8 male volunteers of similar age were selected as the healthy group, with an age range of 50 ± 10 years (Table 4). The average age of the healthy control group was 54.125 years, with a standard deviation of 4.0510, while the average age of the stroke patient group was 49.875 years, with a standard deviation of 8.9512. Previous studies have shown that age is a key factor influencing pulse wave characteristics [50], and many researchers are exploring methods for assessing vascular aging [51,52]. Parameters such as pulse transit time [53] and pulse wave velocity [54] are considered standards for evaluating vascular stiffness. Some researchers have noted a significant correlation between PPG features obtained from the finger and age [55]. This feature resembles the age-related progression of arterial stiffness assessed through pulse wave velocity (PWV). Therefore, age is a factor that influences the evaluation of the model. It is worth noting that both groups included only male participants, ensuring gender consistency in the study.

Table 4. Comparison of basic information of volunteers in group 2.

Group	Gender (Male %)	Age (Years)	Mean Age (Years)	Standard Deviation of Age (Years)	Hypertension (%)	Hemiplegia/ Right (%)	Right-Handedness (%)
Healthy	8 (100)	40-60	54.125	4.0510	0	-	8(100)
Patient	8 (100)	40-60	49.875	8.9512	62.5	75	8(100)

The Brunnstrom Assessment Scale is a grading system used to describe the ex-tent of limb movement recovery in patients. This system consists of six stages, numbered from 1 to 6, representing different levels of limb movement recovery. However, in this paper, the seventh stage is defined as representing normal activity exhibited by individuals who have fully recovered and are in a healthy state. The specific stages are illustrated in Table 5. The stroke patients all experienced their first ischemic stroke within three months. Among these individuals, 6 patients exhibited right-sided hemiplegia and 2 patients exhibited left-sided hemiplegia. Of these individuals, 4 patients were classified as being in stage V and 4 patients as being in stage VI. The selection of participants in stages V and VI was primarily based on the feasibility of data collection and the practical applicability of the study. Firstly, patients in stage I are in the acute phase, during which their neurological and physiological states undergo significant changes and adjustments [56–59]. Therefore, collecting data during this stage may cause secondary harm to the patients. Secondly, patients in stages II, III, and IV typically exhibit more pronounced spasms [12,14], and the significant fluctuations caused by spasms introduce excessive noise, making the collected data less accurate and reliable. Please note that the study design was intended for individuals aged 40 to 60, in stages V and VI, and for healthy individuals. Participants outside this range may not be applicable to this model.

Table 5. Brunnstrom assessment scale [8].

Stage	Description
I Flaccid Stage	No movement was initiated or elicited.
II Spasticity Appears	Synergies or components are first appearing. Spasticity is developing.
III Increased Spasticity	Synergies or components are initiated voluntarily. Spasticity is marked.
IV Decreased Spasticity	Movements are deviating from basic synergies. Spasticity is decreasing
V Complex Movement Combinations	There is relative independence of basic synergies. Spasticity is waning
VI Spasticity Disappears	Movement coordination is near-normal. Spasticity is minimal.

The basic information on all volunteers, including age, gender, type of hemiplegia, and alcohol history, was collected. Simultaneously, the patients' PPG signals were recorded. The control room temperature was maintained at 23 °C while the finger temperature was set at 32 °C. Participants were instructed to remain in a stationary state. The PPG signals from the right index finger of each volunteer were collected continuously for 30 min during the sampling period, which took place between 17:00 and 19:00 in the afternoon. The sampling frequency was set at 100 Hz. The data collection experiment is depicted in Figure 11.

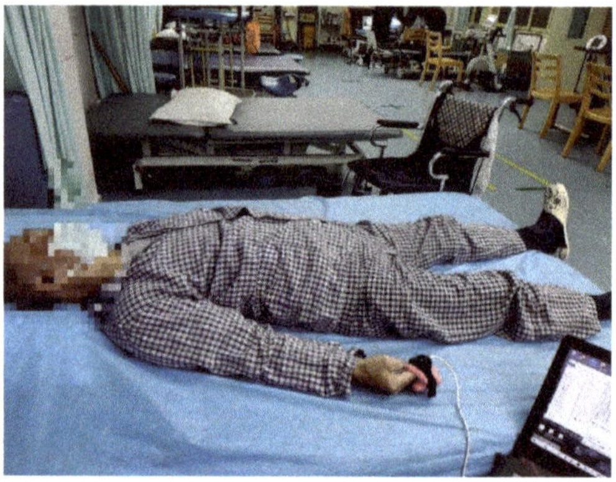

Figure 11. Photo of PPG signal acquisition of a patient in the hospital.

3.2. Testing Protocol

The engagement of stroke patients in this study's voluntary activities underwent a stringent ethical review process, adhering to the principles outlined in the Helsinki Declaration [60]. The primary objective was to safeguard the participants' rights and ensure ethical compliance throughout the activities. A comprehensive set of measures was implemented to protect the participants' welfare. Firstly, prior to commencing the experiments, detailed procedural instructions were provided to each participating subject, accompanied by comprehensive disclosure regarding the data collection requirements and the research objectives, processes, and objectives for which the data would be utilized. Secondly, this study placed utmost importance on preserving the privacy and confidentiality of the participants, ensuring the proper handling and protection of their personal information and medical records. The volunteers were given the autonomy to participate in the activities willingly and possessed the right to withdraw at any point without facing any detrimental repercussions. Stringent adherence to ethical principles was maintained throughout the study, ensuring that the activities were conducted in alignment with the highest moral standards while safeguarding the interests and safety of the participants to the greatest extent possible. The volunteers provided informed consent and signed consent forms while the research protocol underwent thorough deliberation and approval by the Ethics Committee of Shanghai University.

3.3. Validation

3.3.1. K-Fold Cross-Validation

As shown in Figure 12, K-fold cross-validation is a commonly used technique in machine learning to evaluate and validate model performance [61]. Firstly, the original dataset is divided into 10 equally sized subsets, and then, 10 iterations are performed. In each iteration, one of the subsets is selected as the validation set while the remaining 9 subsets are used as the training set. Each subset is used as the validation set once, allowing for multiple rounds of training and validation, which provides a reliable estimate of the model's performance. Ten-fold cross-validation helps detect whether a model is overfitting the training data. If the model performs well on each validation set, it indicates a strong generalization capability.

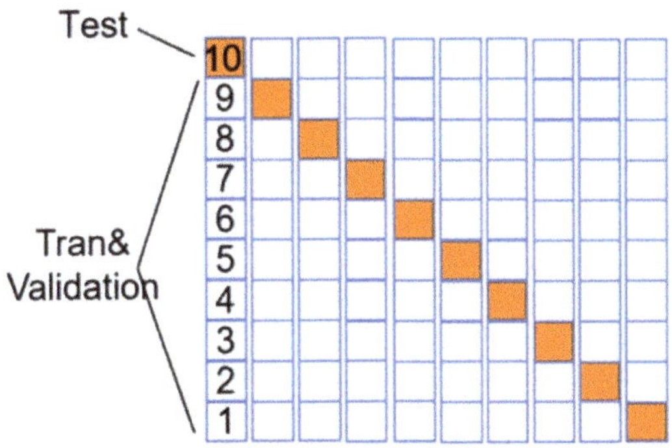

Figure 12. Ten-fold cross-validation.

3.3.2. Performance Evaluation

The model's performance was comprehensively evaluated by calculating the probability distribution of the dataset being stroke-related and the overall accuracy, precision, recall, and F1 score.

The accuracy was computed using Formula (15):

$$\text{Accuracy} = \frac{\text{TP} + \text{TN}}{\text{TP} + \text{TN} + \text{FP} + \text{FN}} \tag{15}$$

The precision was calculated as shown in Formula (16):

$$\text{Precision} = \frac{\text{TP}}{\text{TP} + \text{FP}} \tag{16}$$

The recall was calculated as shown in Formula (17):

$$\text{Recall} = \frac{\text{TP}}{\text{TP} + \text{FN}} \tag{17}$$

The F1 score was calculated as shown in Formula (18):

$$\text{F1} = \frac{2 \times \text{Precision} \times \text{Recall}}{\text{Precision} + \text{Recall}} \tag{18}$$

The relationships between TP, TN, FP, and FN are summarized in Table 6.

Table 6. Correspondence of TP, TN, FP, and FN.

Confusion Matrix		**Ground Truth**	
		Positive	**Negative**
Predicted value	**Positive**	TP	FP
	Negative	FN	TN

4. Results

The training process is illustrated in Figure 13, showing the accuracy and loss curves during a training iteration. The figures indicate that as the number of iterations increases, the model's accuracy gradually improves and reaches its peak while the loss value gradually decreases and eventually stabilizes. The validation accuracy is slightly lower than the training accuracy.

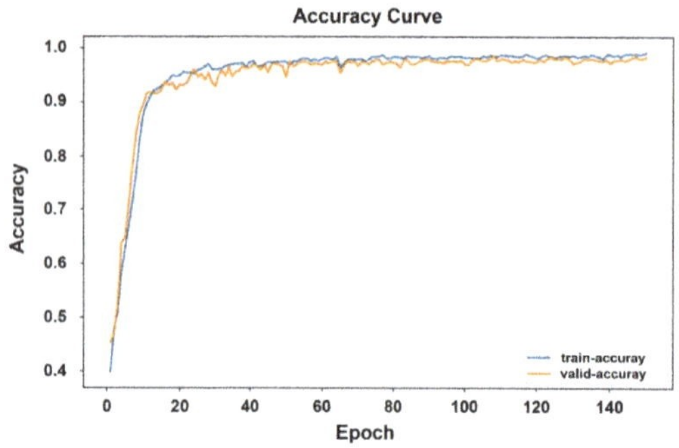

(a)

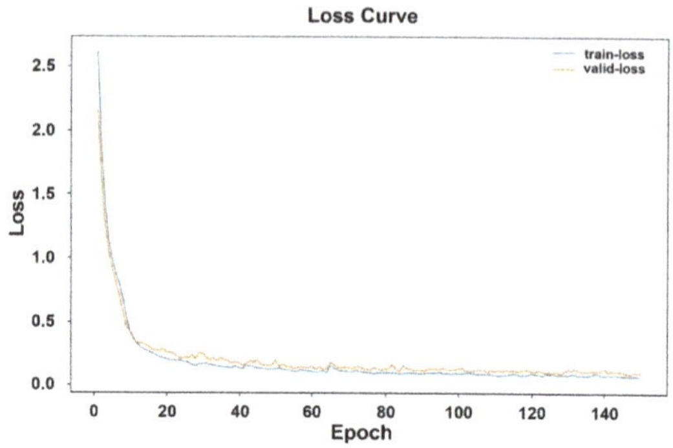

(b)

Figure 13. (**a**) Training: accuracy curve. (**b**) Training: loss curve.

Based on the confusion matrix obtained from the test set (as shown in Figure 14), the model's recognition accuracy can be calculated as 91.25%, precision as 89.80%, recall as 89.70%, and F1 score as 89.49%. It is important to note that in Figure 14, "Level VII" represents healthy individuals.

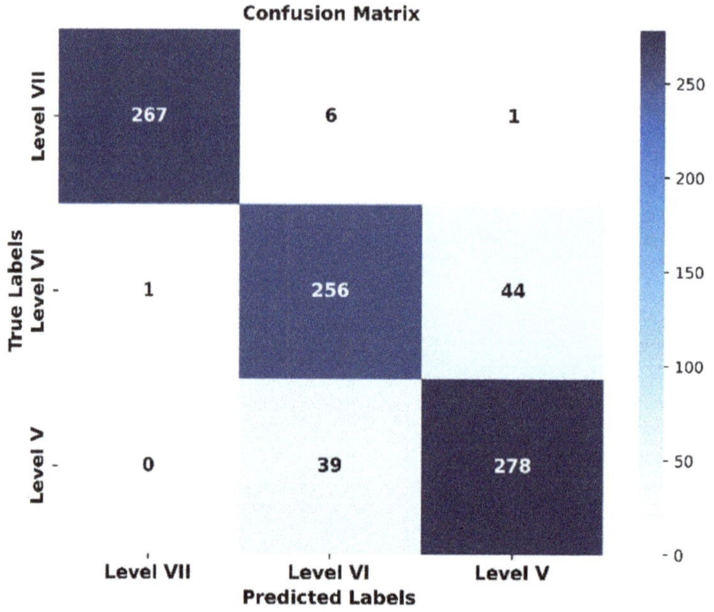

Figure 14. Confusion matrix.

The classification results of different models on the same dataset are presented in Table 7, representing the average values from 10-fold cross-validation. The MCNN-LSTM-Attention model achieves the highest accuracy of 91.25% while the CNN–LSTM model has the lowest accuracy at only 69.04%. The MCNN-LSTM-Attention model outperforms the CNN–LSTM–Attention model by 25.4% in terms of accuracy, and the MCNN–LSTM model surpasses the CNN–LSTM model by 24.8%. The MCNN-LSTM-Attention model achieves a 10.5% higher accuracy than the MCNN–LSTM model, and the CNN–LSTM–Attention model achieves a 5.1% higher accuracy than the CNN–LSTM model.

Table 7. Classification results of different models on the dataset.

Method	Accuracy	Precision	Recall	F1 Score	Loss
CNN–LSTM	0.6904	0.6560	0.6508	0.6534	0.5128
CNN–LSTM–Attention	0.7273	0.6946	0.6362	0.6928	0.4907
MCNN–LSTM	0.8679	0.8861	0.8419	0.8634	0.3677
MCNN-LSTM-Attention	0.9125	0.8980	0.8970	0.8949	0.1261

Table 8 presents the training times required for different algorithm models in a single training iteration. It is worth noting that models with attention mechanisms significantly increase the training time compared to those without attention mechanisms. The difference in training time between the multi-scale CNN model and the single-scale CNN model is minimal.

Table 8. Training times for different models (one epoch).

Model	Time/s
CNN–LSTM–Attention	7.3
CNN–LSTM	1.3
MCNN–LSTM	1.5
MCNN-LSTM-Attention	7.6

5. Discussion

The main findings of this study were as follows: (1) The fusion-based PPG signal acquisition system designed and investigated using ESP32C3 meets the experimental requirements and exhibits advantages such as portability, low cost, and good accuracy. (2) The proposed multi-modality approach significantly improves the accuracy compared to the single-modal approach, with the CNN–LSTM–Attention stroke recognition model achieving an accuracy of 91.94%. Compared to feature extraction from a single modality, the proposed model better captures the interdependencies among types of PPG sequence information, enhancing the feature representation capability and resulting in performance improvement.

5.1. Fusion-Type Device

This paper presents the design of a precision Pulse Plethysmography (PPG) signal acquisition device. The device utilizes an ESP32-C3 microcontroller as its core, offering several advantages over traditional approaches. Firstly, the ESP32-C3 microcontroller enables the collection of PPG signals from the fusion sensor MAX30101, including green, red, and infrared light, facilitating comprehensive signal acquisition for more accurate data in the rehabilitation assessment of stroke. Secondly, the device ensures stable data transmission, high accuracy, low power consumption, and a compact form factor, measuring only 31 mm × 31 mm. It also has a relatively low cost, making it portable and cost-effective.

As shown in Figure 15, in comparison to Wei's previously designed device, which required multi-level data transmission through sensors, STM32, ESP8266, and Wi-Fi to the PC, the 'NeuroPulseGuard' microcontroller ESP32-C3 directly acquires data from the sensor and transmits it via Wi-Fi, simplifying the system architecture and improving system stability and robustness. The device has a smaller form factor, lower power consumption, and lower cost. Table 1 demonstrates that the 'NeuroPulseGuard' device, after collecting and processing PPG signals, achieves a heart rate measurement with an error within 1% compared to the heart rate data recorded by the PHILIPS DB12 pulse oximeter, further enhancing accuracy compared to Wei's device with 2% precision.

Figure 15. Comparison of the internal structures of the devices: (**a**) Wei's equipment; (**b**) 'NeuroPulse-Guard' device.

Additionally, the 'NeuroPulseGuard' device, when compared to those developed by other researchers such as the wrist-worn PPG signal monitoring devices designed by Zhang et al. [62] and Lee et al. [63], offers advantages. Despite sharing the same portability, the 'NeuroPulseGuard' device does not suffer from the weaker PPG signal strength that can result from the small and shallow blood vessels at the wrist. Compared to PPG signals obtained from the fingertip, wrist-worn devices may result in higher levels of signal noise. These issues limit their suitability and performance in real-world applications.

5.2. Multi-Modality Approach

The proposed multi-modality approach in this paper primarily included feature extraction fusion for a single PPG signal and feature fusion for multiple PPG signals. Specifically, the following statements can be made: (1) The feature extraction fusion for a single PPG signal utilized a MCNN, which introduced multiple convolutional channels compared to a traditional single-channel CNN. This design allowed the MCNN to comprehensively capture features at different levels (deep and shallow), thereby enhancing the model's feature extraction capability. By utilizing PPG signals from green, red, and infrared light as inputs, the MCNN could extract comprehensive physiological data, providing richer feature information for the rehabilitation assessment of stroke. (2) In terms of feature fusion for multiple PPG signals, previous research has indicated that LSTM exhibited higher accuracy in the field of multi-signal fusion [64,65]. In this study, by integrating PPG signal features from green, red, and infrared light, a temporal dependency was established. This temporal dependency effectively captured the correlation between features and further improved the model's performance. Additionally, an attention mechanism was introduced to enhance the model's performance. The attention mechanism dynamically adjusted weights based on the importance of input data, enabling the model to focus more on crucial parts and improve its perception of key features [66].

From the perspective of overall model performance, the proposed MCNN-LSTM-Attention model achieved an accuracy of 91.25% in rehabilitation assessment, surpassing the accuracy of 72.73% for the CNN–LSTM–Attention model and 86.79% for the MCNN–LSTM model. Furthermore, compared to other models, the MCNN-LSTM-Attention model demonstrated good precision (0.8980), recall (0.8910), F1 score (0.8949), and loss function (0.1261) values. These findings indicated that the model exhibited good accuracy and reliability in the rehabilitation assessment of stroke. To fully consider the utilized dataset in the experiments, 10-fold cross-validation was employed to effectively reduce biases arising from uneven dataset partitioning; improve the reliability, stability, and generalization ability of model evaluation; and ensure consistent performance across different data subsets, thus enhancing the credibility of the experimental results. In terms of overall model efficiency, the MCNN-LSTM-Attention model required 7.6 s per training cycle. In comparison, the CNN–LSTM–Attention model required 7.3 s, the MCNN–LSTM model required 1.6 s, and the CNN–LSTM model required 1.3 s. The introduction of the MCNN feature extraction module only increased the training time by 0.2–0.3 s but achieved performance improvement.

In the field of pulse wave analysis using photoplethysmography (PPG), traditional machine learning has been a significant topic. However, previous research has shown that deep learning networks outperform traditional machine learning algorithms in PPG signal processing tasks [18,67,68]. Therefore, in this paper, the deep learning algorithm for PPG patient data is not compared with traditional machine learning algorithms. During the feature extraction process of machine learning, researchers found that the waveform in stroke patients is influenced by vascular wall elasticity, making it difficult to locate the dicrotic wave and even resulting in the disappearance of the dicrotic wave. This leads to missing feature points and limited dimensions of manually extracted features, making it challenging to discover deep-level features and correlations between features.

In comparison with other studies in various multi-signal research fields, this study compared the proposed method with other research results, as shown in Table 9. These studies all involved the rehabilitation assessment of stroke. In the first study [26], Wei et al.'s model (CNN–LSTM–Attention) achieved only 72.73% accuracy. In the second study [17], based on the NIHSS scale, 234 features were extracted from EKG, ABP, and PPG signals, and a linear kernel SVM classifier was employed, resulting in an accuracy of 82.7%. Furthermore, in the third study by Zhe Zhang et al. [69], patients' levels were evaluated based on the Brunnstrom scale using a fuzzy inference system, achieving an accuracy of 87.5%. In the fourth study, the C4.5 decision tree algorithm was used for patient-level recognition based on the HINSS scale [70], and an accuracy of 91.11% was achieved. The classification accuracies in these four studies were lower than those of the proposed model in this paper. In comparison, the MCNN-LSTM-Attention rehabilitation assessment of the stroke model proposed in this study can effectively extract deep-level features from PPG signals and improve accuracy through multi-modality feature extraction. In the fifth study [16], based on the Brunnstrom scale, the lower limb movement signals of patients were obtained through inertial measurement units (IMUs) and surface electromyography (sEMG). A total of 480 features were extracted, and an SVM classifier was utilized, resulting in an accuracy of 95.2%. However, in the assessment process, the signals for each movement were manually trimmed from a user interface. The experimental process was complex and cumbersome, and subjective factors were involved, limiting its applicability.

Table 9. Classification performance comparison.

Method	Feature Numbers	Stroke Scale	Classifier	Accuracy
PPG	-	Brunnstrom	MCNN-LSTM-Attention	91.25%
PPG [26]	-	Brunnstrom	CNN–LSTM–Attention	72.73%
EKG-ABP-PPG [17]	234 features	NIHSS	Linear kernel SVM	82.7%
Motion data samples [69]	27 features	Brunnstrom	Fuzzy inference system	87.5%
HINSS [70]	13 features	HINSS	C4.5 decision trees	91.11%
IMU and ECG feature [16]	480 features	Brunnstrom	SVM	95.2%

Furthermore, as observed in Table 9, traditional machine learning algorithms seemed to outperform deep learning algorithms in the rehabilitation assessment of stroke. However, the relevant literature suggests that deep learning networks outperform traditional machine learning algorithms in many cases. The reason for this apparent contradiction is that traditional machine learning algorithms make use of multiple signal sources, such as combinations of IMU-ECG, IMU-ECG, EKG-ABP-PPG, and others. These combinations of different signal types provide richer feature information, resulting in improved classification accuracy. Therefore, the high accuracy achieved by traditional machine learning methods is closely tied to the comprehensive utilization of multi-signal data. In contrast, this study focused on the classification of multi-modal PPG signals for the rehabilitation assessment of stroke. The MCNN-LSTM-Attention rehabilitation assessment of the stroke model proposed in this study achieved high accuracy despite using only multi-modal

PPG signals and not combining multiple signal sources. This suggests that deep learning methods have an advantage in handling multi-modal data from a single signal source. If multiple signal sources are combined in the future, it is likely that further improvements in accuracy can be achieved. This indicates the potential of the proposed deep learning method in the future.

5.3. Limitations and Future Work

The study had certain limitations that could impact the results. Firstly, similar to many studies, the number of volunteers in this experiment was limited. However, the proposed MCNN-LSTM-Attention model focused on the raw data rather than the traditional feature-based algorithms that require feature extraction from raw data. In this preliminary study, a total of 8,640,000 data points (three wavelengths $\times$ 100 Hz $\times$ 60 s $\times$ 30 min $\times$ 16 participants) were collected, and 91.25% accuracy was achieved. Secondly, age is a crucial factor influencing PPG features, and this study only targeted participants aged 40 to 60. The model might not be applicable to individuals outside this age range. Lastly, this study only involved stroke patients in stages V and VI, as well as healthy individuals, limiting the generalizability of the model to all stroke patients undergoing rehabilitation.

In the future, the further validation and improvement of the model, as well as the enhancement of its accuracy, generalization, and robustness, can be achieved by collecting more patient data and increasing the sample size. Additionally, seeking higher-quality sensors to enhance hardware and continuously optimizing the deep learning model used in this study will broaden its applicability and scope. Finally, the NeuroPulseGuard data acquisition device, which enables the convenient collection of multi-modal PPG signals, holds the potential to assist post-recovery stroke patients and is suitable for community-wide adoption.

6. Conclusions

This paper presents the design and investigation of a fusion-based photoplethysmography (PPG) and capacitive volume pulse acquisition device for the rehabilitation assessment of stroke. Subsequently, a recognition model based on MCNN-LSTM-Attention has been proposed. The incorporation of a multi-modality fusion mechanism and attention mechanism enhances the model's capability to represent features, leading to improved recognition accuracy. Experimental results demonstrate that the MCNN-LSTM-Attention rehabilitation assessment of the stroke model achieves an average accuracy of 91.9% in 10-fold cross-validation. The proposed model outperforms other deep learning models in terms of precision, accuracy, and F1 score metrics.

Author Contributions: Conceptualization, L.Y. and Z.L.; methodology, Z.L.; software, Z.L.; validation, L.Y., B.S. and S.Q.X.; formal analysis, Z.L.; investigation, Z.L.; resources, J.Q., J.L. and B.S.; data curation, Z.L., J.Q. and J.L.; writing—original draft preparation, Z.L.; writing—review and editing, B.S. and L.Y.; visualization, Z.L.; supervision, B.S., L.Y. and S.Q.X.; project administration, L.Y., B.S., J.Q. and J.L.; funding acquisition, B.S. All authors have read and agreed to the published version of the manuscript.

Funding: This research was supported by the Shanghai Pujiang Program, under Grant 21PJ1404000, and the National Natural Science Foundation of China, under Grant 62103252.

Institutional Review Board Statement: The study was conducted in accordance with the Declaration of Helsinki and approved by the Science and Technology Ethics Committee of Shanghai University (No. ECSHU 2021-102).

Informed Consent Statement: Informed consent was obtained from all subjects involved in the study. Written informed consent has been obtained from the patients to publish this paper.

Data Availability Statement: Data are unavailable due to privacy and ethical restrictions.

Acknowledgments: The authors would like to thank Jennifer Qiao for English improvement, Mingsen Wei for the equipment development, and Jianyu Zhao and Jiadong Ni for the data collection.

Conflicts of Interest: The authors declare no conflicts of interest.

References

1. Vos, T.; Lim, S.S.; Abbafati, C.; Abbas, K.M.; Abbasi, M.; Abbasifard, M.; Abbasi-Kangevari, M.; Abbastabar, H.; Abd-Allah, F.; Abdelalim, A.; et al. Global burden of 369 diseases and injuries in 204 countries and territories, 1990–2019: A systematic analysis for the Global Burden of Disease Study 2019. *Lancet* **2020**, *396*, 1204–1222. [CrossRef] [PubMed]
2. Tu, W.-J.; Wang, L.-D. China stroke surveillance report 2021. *Mil. Med. Res.* **2023**, *10*, 33. [CrossRef] [PubMed]
3. Warlow, C.P. Epidemiology of stroke. *Lancet* **1998**, *352*, S1–S4. [CrossRef]
4. Musuka, T.D.; Wilton, S.B.; Traboulsi, M.; Hill, M.D. Diagnosis and management of acute ischemic stroke: Speed is critical. *Can. Med. Assoc. J.* **2015**, *187*, 887–893. [CrossRef]
5. Langhorne, P.; Bernhardt, J.; Kwakkel, G. Stroke rehabilitation. *Lancet* **2011**, *377*, 1693–1702. [CrossRef] [PubMed]
6. Stinear, C.M.; Lang, C.E.; Zeiler, S.; Byblow, W.D. Advances and challenges in stroke rehabilitation. *Lancet Neurol.* **2020**, *19*, 348–360. [CrossRef]
7. Bernhardt, J.; Hayward, K.S.; Kwakkel, G.; Ward, N.S.; Wolf, S.L.; Borschmann, K.; Krakauer, J.W.; Boyd, L.A.; Carmichael, S.T.; Corbett, D.; et al. Agreed definitions and a shared vision for new standards in stroke recovery research: The Stroke Recovery and Rehabilitation Roundtable taskforce. *Int. J. Stroke* **2017**, *12*, 444–450. [CrossRef]
8. Brunnstrom, S. Motor Testing Procedures in Hemiplegia: Based on Sequential Recovery Stages. *Phys. Ther.* **1966**, *46*, 357–375. [CrossRef]
9. Shrestha, S.; Poudel, R.S.; Khatiwada, D.; Thapa, L. Stroke subtype, age, and baseline NIHSS score predict ischemic stroke outcomes at 3 months: A preliminary study from Central Nepal. *J. Multidiscip. Healthc.* **2015**, *8*, 443–448. [CrossRef]
10. Powers, D.W. Assessment of the stroke patient using the NIH stroke scale. *Emerg. Med. Serv.* **2001**, *30*, 52–56.
11. Liu, F.; Tsang, R.C.C.; Zhou, J.; Zhou, M.; Zha, F.; Long, J.; Wang, Y. Relationship of Barthel Index and its Short Form with the Modified Rankin Scale in acute stroke patients. *J. Stroke Cerebrovasc. Dis.* **2020**, *29*, 105033. [CrossRef] [PubMed]
12. Safaz, I.; Ylmaz, B.; Yasar, E.; Alaca, R. Brunnstrom recovery stage and motricity index for the evaluation of upper extremity in stroke: Analysis for correlation and responsiveness. *Int. J. Rehabil. Res.* **2009**, *32*, 228–231. [CrossRef] [PubMed]
13. Kawahira, K.; Shimodozono, M.; Ogata, A.; Tanaka, N. Addition of intensive repetition of facilitation exercise to multidisciplinary rehabilitation promotes motor functional recovery of the hemiplegic lower limb. *J. Rehabil. Med.* **2004**, *36*, 159–164. [CrossRef]
14. Brunnstrom, S. *Movement Therapy for Stroke Patients: A Neurophysiologic Approach*; Harper & Row: Manhattan, NY, USA, 1965.
15. Sheng, B.; Chen, X.; Cheng, J.; Zhang, Y.; Tao, J.; Duan, C. A novel scoring approach for the Wolf Motor Function Test in stroke survivors using motion-sensing technology and machine learning: A preliminary study. *Comput. Methods Programs Biomed.* **2024**, *243*, 107887. [CrossRef] [PubMed]
16. Li, H.-T.; Han, S.-L.; Pan, M.-C. Lower-limb motion classification for hemiparetic patients through IMU and EMG signal processing. In Proceedings of the 2016 International Conference on Biomedical Engineering (BME-HUST), Hanoi, Vietnam, 5–6 October 2016; pp. 113–118.
17. Huang, P.-W.; Tang, S.-C.; Lin, Y.-M.; Liu, Y.-C.; Jou, W.-J.; Jen, H.-I.; Lai, D.-M.; Wu, A.-Y. Predicting stroke outcomes based on multi-modal analysis of physiological signals. In Proceedings of the 2015 IEEE International Conference on Digital Signal Processing (DSP), Singapore, 21–24 July 2015; pp. 454–457.
18. Georgieva-Tsaneva, G.; Gospodinova, E.; Cheshmedzhiev, K. Cardiodiagnostics Based on Photoplethysmographic Signals. *Diagnostics* **2022**, *12*, 412. [CrossRef]
19. Uluç, N.; Glasl, S.; Gasparin, F.; Yuan, T.; He, H.; Jüstel, D.; Pleitez, M.A.; Ntziachristos, V. Non-invasive measurements of blood glucose levels by time-gating mid-infrared optoacoustic signals. *Nat. Metab.* **2024**, *6*, 678–686. [CrossRef] [PubMed]
20. Al Ashry, H.S.; Ni, Y.; Thomas, R.J. Cardiopulmonary Sleep Spectrograms Open a Novel Window Into Sleep Biology—Implications for Health and Disease. *Front. Neurosci.* **2021**, *15*, 755464. [CrossRef] [PubMed]
21. Ray, D.; Collins, T.; Woolley, S.I.; Ponnapalli, P.V.S. A review of wearable multi-wavelength photoplethysmography. *IEEE Rev. Biomed. Eng.* **2021**, *16*, 136–151. [CrossRef]
22. Lee, J.; Matsumura, K.; Yamakoshi, K.-I.; Rolfe, P.; Tanaka, S.; Yamakoshi, T. Comparison between red, green and blue light reflection photoplethysmography for heart rate monitoring during motion. In Proceedings of the 2013 35th Annual International Conference of the IEEE Engineering in Medicine and Biology Society (EMBC), Osaka, Japan, 3–7 July 2013; pp. 1724–1727.
23. Moço, A.V.; Stuijk, S.; de Haan, G. New insights into the origin of remote PPG signals in visible light and infrared. *Sci. Rep.* **2018**, *8*, 8501. [CrossRef]
24. Maeda, Y.; Sekine, M.; Tamura, T. The advantages of wearable green reflected photoplethysmography. *J. Med. Syst.* **2011**, *35*, 829–834. [CrossRef]
25. Shriram, R.; Martin, B.; Sundhararajan, M.; Daimiwal, N. Effect of source wavelength on second derivative of finger photoplethysmogram in healthy young female volunteers. *Biomed. Res.-India* **2016**, *27*, S454–S459.
26. Yan, L.; Wei, M.; Hu, S.; Sheng, B. Photoplethysmography Driven Hypertension Identification: A Pilot Study. *Sensors* **2023**, *23*, 3359. [CrossRef] [PubMed]
27. Espressif Systems. *ESP32-C3 Series Datasheet*; Espressif Systems: Shanghai, China, 2024.
28. Maxim Integrated. *MAX30101: High-Sensitivity Pulse Oximeter and Heart-Rate Sensor for Wearable Health Data Sheet*; Maxim Integrated: San Jose, CA, USA, 2020.

29. Zhang, Q.; Yang, L.C. The Photoplethysmography Signal Analysis Based on Wavelet Transform. In Proceedings of the 2015 8th International Symposium on Computational Intelligence and Design (ISCID), Hangzhou, China, 12–13 December 2015; Volume 1, pp. 228–231.
30. Mishra, B.; Nirala, N.S. A survey on denoising techniques of PPG signal. In Proceedings of the 2020 IEEE International Conference for Innovation in Technology (INOCON), Bangluru, India, 6–8 November 2020; pp. 1–8.
31. Becerra-Luna, B.; Sánchez, J.C.; Martinez-Memije, R.; Infante, O. Photoplethysmographic fingertip with miokinetic noise attenuation using an accelerometer and adaptive filtering techniques. In Proceedings of the 2013 Pan American Health Care Exchanges (PAHCE), Medellin, Colombia, 29 April–4 May 2013; pp. 1–5.
32. Mejía-Mejía, E.; Kyriacou, P.A. Effects of noise and filtering strategies on the extraction of pulse rate variability from photoplethysmograms. *Biomed. Signal Process. Control.* **2023**, *80*, 104291. [CrossRef]
33. Lee, J.-W. Design of kalman filter to estimate heart rate variability from ppg signal for mobile healthcare. *J. Inf. Commun. Converg. Eng.* **2010**, *8*, 201–204. [CrossRef]
34. Elgendi, M.; Norton, I.; Brearley, M.; Abbott, D.; Schuurmans, D. Detection of a and b waves in the acceleration photoplethysmogram. *Biomed. Eng. Online* **2014**, *13*, 139. [CrossRef] [PubMed]
35. Elgendi, M.; Norton, I.; Brearley, M.; Abbott, D.; Schuurmans, D. Systolic peak detection in acceleration photoplethysmograms measured from emergency responders in tropical conditions. *PLoS ONE* **2013**, *8*, e76585. [CrossRef] [PubMed]
36. Lee, J. Motion artifacts reduction from PPG using cyclic moving average filter. *Technol. Health Care* **2014**, *22*, 409–417. [CrossRef] [PubMed]
37. Liang, Y.; Elgendi, M.; Chen, Z.; Ward, R. An optimal filter for short photoplethysmogram signals. *Sci. Data* **2018**, *5*, 180076. [CrossRef] [PubMed]
38. Wang, H.; Wang, X.; Deller, J.R.; Fu, J. Shape-preserving preprocessing for human pulse signals based on adaptive parameter determination. *IEEE Trans. Biomed. Circuits Syst.* **2013**, *8*, 594–604. [CrossRef]
39. Bhatt, D.; Patel, C.; Talsania, H.; Patel, J.; Vaghela, R.; Pandya, S.; Modi, K.; Ghayvat, H. CNN variants for computer vision: History, architecture, application, challenges and future scope. *Electronics* **2021**, *10*, 2470. [CrossRef]
40. Liang, Y.; Ailian, F. Chinese text relation extraction based on a multi-channel convolutional neural network. *J. East China Norm. Univ. (Nat. Sci.)* **2021**, *2021*, 96.
41. Chen, X.; Zhang, B.; Gao, D. Bearing fault diagnosis base on multi-scale CNN and LSTM model. *J. Intell. Manuf.* **2021**, *32*, 971–987. [CrossRef]
42. Han, X.; Zhai, Q.; Zhang, N.; Zhang, X.; He, L.; Pan, M.; Zhang, B.; Liu, T. A real-time evaluation algorithm for noncontact heart rate variability monitoring. *Sensors* **2023**, *23*, 6681. [CrossRef]
43. Alugubelli, N.; Abuissa, H.; Roka, A. Wearable devices for remote monitoring of heart rate and heart rate variability—What we know and what is coming. *Sensors* **2022**, *22*, 8903. [CrossRef]
44. Pereira, T.; Tran, N.; Gadhoumi, K.; Pelter, M.M.; Do, D.H.; Lee, R.J.; Colorado, R.; Meisel, K.; Hu, X. Photoplethysmography based atrial fibrillation detection: A review. *NPJ Digit. Med.* **2020**, *3*, 3. [CrossRef] [PubMed]
45. Park, J.; Seok, H.S.; Kim, S.-S.; Shin, H. Photoplethysmogram analysis and applications: An integrative review. *Front. Physiol.* **2022**, *12*, 808451. [CrossRef] [PubMed]
46. Sahoo, R.R.; Ghosh, S.; Lala, A.; Kundu, P.K. Classification of Cardiac Abnormality using PPG Fiducial Parameters and Probabilistic Neural Network. In Proceedings of the 2022 IEEE Calcutta Conference (CALCON), Kolkata, India, 10–11 December 2022; pp. 49–54.
47. Alessandrini, M.; Biagetti, G.; Crippa, P.; Falaschetti, L.; Turchetti, C. Recurrent neural network for human activity recognition in embedded systems using ppg and accelerometer data. *Electronics* **2021**, *10*, 1715. [CrossRef]
48. Brauwers, G.; Frasincar, F. A general survey on attention mechanisms in deep learning. *IEEE Trans. Knowl. Data Eng.* **2021**, *35*, 3279–3298. [CrossRef]
49. Niu, Z.; Zhong, G.; Yu, H. A review on the attention mechanism of deep learning. *Neurocomputing* **2021**, *452*, 48–62. [CrossRef]
50. Qawqzeh, Y.; Reaz, M.B.I.; Maskon, O.; Chellappan, K. Photoplethysmogram Reflection Index and Aging. In Proceedings of the International Conference on Graphic and Image Processing (ICGIP 2011), Cairo, Egypt, 1–3 October 2011; Volume 8285.
51. Shin, H.; Noh, G.; Choi, B.-M. Photoplethysmogram based vascular aging assessment using the deep convolutional neural network. *Sci. Rep.* **2022**, *12*, 11377. [CrossRef]
52. Djurić, B.; Žikić, K.; Nestorović, Z.; Lepojević-Stefanović, D.; Milošević, N.; Žikić, D. Using the photoplethysmography method to monitor age-related changes in the cardiovascular system. *Front. Physiol.* **2023**, *14*, 1191272. [CrossRef] [PubMed]
53. Jannasz, I.; Sondej, T.; Targowski, T.; Mańczak, M.; Obiała, K.; Dobrowolski, A.P.; Olszewski, R. Relationship between the Central and Regional Pulse Wave Velocity in the Assessment of Arterial Stiffness Depending on Gender in the Geriatric Population. *Sensors* **2023**, *23*, 5823. [CrossRef] [PubMed]
54. Liao, S.; Liu, H.; Lin, W.-H.; Zheng, D.; Chen, F. Filtering-induced changes of pulse transmit time across different ages: A neglected concern in photoplethysmography-based cuffless blood pressure measurement. *Front. Physiol.* **2023**, *14*, 1172150. [CrossRef] [PubMed]
55. Lin, W.H.; Zheng, D.; Li, G. Age-Related Changes in Blood Volume Pulse Wave at Fingers and Ears. *IEEE J. Biomed. Health Inform.* **2023**. [CrossRef] [PubMed]

56. Aguilar-Salas, E.; Rodríguez-Aquino, G.; García-Domínguez, K.; Garfias-Guzmán, C.; Hernández-Camarillo, E.; Oropeza-Bustos, N.; Arguelles-Castro, R.; Mitre-Salazar, A.; García-Torres, G.; Reynoso-Marenco, M.; et al. Acute Stroke Care in Mexico City: The Hospital Phase of a Stroke Surveillance Study. *Brain Sci.* **2022**, *12*, 865. [CrossRef]
57. Duloquin, G.; Graber, M.; Baptiste, L.; Mohr, S.; Garnier, L.; Ndiaye, M.; Blanc-Labarre, C.; Hervieu-Bègue, M.; Osseby, G.V.; Giroud, M.; et al. Prise en charge de l'infarctus cérébral à la phase initiale. *Rev. Méd. Interne* **2022**, *43*, 286–292. [CrossRef]
58. Dziedzic, T. Clinical significance of acute phase reaction in stroke patients. *Front. Biosci.* **2008**, *13*, 2922–2927. [CrossRef]
59. Finley Caulfield, A.; Wijman, C.A.C. Critical Care of Acute Ischemic Stroke. *Crit. Care Clin.* **2006**, *22*, 581–606. [CrossRef] [PubMed]
60. Association, W.M. World Medical Association Declaration of Helsinki. Ethical principles for medical research involving human subjects. *Bull. World Health Organ.* **2001**, *79*, 373.
61. Wong, T.-T. Performance evaluation of classification algorithms by k-fold and leave-one-out cross validation. *Pattern Recognit.* **2015**, *48*, 2839–2846. [CrossRef]
62. Zhang, Z.; Pi, Z.; Liu, B. TROIKA: A general framework for heart rate monitoring using wrist-type photoplethysmographic signals during intensive physical exercise. *IEEE Trans. Biomed. Eng.* **2014**, *62*, 522–531. [CrossRef] [PubMed]
63. Lee, E.; Lee, C.-Y. PPG-based smart wearable device with energy-efficient computing for mobile health-care applications. *IEEE Sens. J.* **2021**, *21*, 13564–13573. [CrossRef]
64. Mellouk, W.; Handouzi, W. CNN-LSTM for automatic emotion recognition using contactless photoplythesmographic signals. *Biomed. Signal Process. Control.* **2023**, *85*, 104907. [CrossRef]
65. Kumar, A.K.; Ritam, M.; Han, L.; Guo, S.; Chandra, R. Deep learning for predicting respiratory rate from biosignals. *Comput. Biol. Med.* **2022**, *144*, 105338. [CrossRef] [PubMed]
66. Park, S.-A.; Lee, H.-C.; Jung, C.-W.; Yang, H.-L. Attention mechanisms for physiological signal deep learning: Which attention should we take? In Proceedings of the International Conference on Medical Image Computing and Computer-Assisted Intervention, Singapore, 18–22 September 2022; pp. 613–622.
67. Pereira, T.; Ding, C.; Gadhoumi, K.; Tran, N.; Colorado, R.A.; Meisel, K.; Hu, X. Deep learning approaches for plethysmography signal quality assessment in the presence of atrial fibrillation. *Physiol. Meas.* **2019**, *40*, 125002. [CrossRef] [PubMed]
68. Ortiz, B.L.; Gupta, V.; Chong, J.W.; Jung, K.; Dallas, T. User Authentication Recognition Process Using Long Short-Term Memory Model. *Multimodal Technol. Interact.* **2022**, *6*, 107. [CrossRef]
69. Zhang, Z.; Fang, Q.; Gu, X. Fuzzy inference system based automatic Brunnstrom stage classification for upper-extremity rehabilitation. *Expert Syst. Appl.* **2014**, *41*, 1973–1980. [CrossRef]
70. Yu, J.; Park, S.; Lee, H.; Pyo, C.-S.; Lee, Y.S. An elderly health monitoring system using machine learning and in-depth analysis techniques on the NIH stroke scale. *Mathematics* **2020**, *8*, 1115. [CrossRef]

sensors

MDPI

Article

CIRF: Coupled Image Reconstruction and Fusion Strategy for Deep Learning Based Multi-Modal Image Fusion

Junze Zheng †, Junyan Xiao †, Yaowei Wang † and Xuming Zhang *

Department of Biomedical Engineering, College of Life Science and Technology, Huazhong University of Science and Technology, Wuhan 430074, China

* Correspondence: zxmboshi@hust.edu.cn

† These authors contributed equally to this work.

Abstract: Multi-modal medical image fusion (MMIF) is crucial for disease diagnosis and treatment because the images reconstructed from signals collected by different sensors can provide complementary information. In recent years, deep learning (DL) based methods have been widely used in MMIF. However, these methods often adopt a serial fusion strategy without feature decomposition, causing error accumulation and confusion of characteristics across different scales. To address these issues, we have proposed the Coupled Image Reconstruction and Fusion (CIRF) strategy. Our method parallels the image fusion and reconstruction branches which are linked by a common encoder. Firstly, CIRF uses the lightweight encoder to extract base and detail features, respectively, through the Vision Transformer (ViT) and the Convolutional Neural Network (CNN) branches, where the two branches interact to supplement information. Then, two types of features are fused separately via different blocks and finally decoded into fusion results. In the loss function, both the supervised loss from the reconstruction branch and the unsupervised loss from the fusion branch are included. As a whole, CIRF increases its expressivity by adding multi-task learning and feature decomposition. Additionally, we have also explored the impact of image masking on the network's feature extraction ability and validated the generalization capability of the model. Through experiments on three datasets, it has been demonstrated both subjectively and objectively, that the images fused by CIRF exhibit appropriate brightness and smooth edge transition with more competitive evaluation metrics than those fused by several other traditional and DL-based methods.

Keywords: multi-modal medical image fusion; multi-task learning; deep learning; vision transformer; convolutional neural network

1. Introduction

With the development of medical imaging technology, a variety of imaging modalities have emerged, such as magnetic resonance imaging (MRI) [1], computed tomography (CT) [2], positron emission tomography (PET) [3] and single-photon emission computed tomography (SPECT) [4]. They all have unique information and characteristics [5]. MR images have better soft tissue definition and higher spatial resolution but are often accompanied by motion artifacts. CT images can facilitate the detection of dense structures like bones and implants; however, CT imaging involves a certain level of radiation and is limited in its ability to provide qualitative diagnosis. PET and SPECT images have high sensitivity and are often used for metabolic information gauging, vascular disease diagnosis and tumor detection, but their spatial resolution is relatively low.

From the above discussion, it is clear that each imaging modality has its own scope of application and limitations. Furthermore, information from a single sensor is not enough to handle scene changes effectively, and the information from different modalities is exceedingly significant. Additionally, even when there are multi-modal medical images (MMI), the high requirement of spatial imagination capability for doctors still poses a challenge. Therefore, the multi-modal medical image fusion (MMIF) algorithm is the key to resolving this awkward situation [5]. Generally, MMIF is a process of combining salient and complementary information into images with high visual perceptual quality, thereby benefiting more comprehensive and accurate disease diagnosis and treatment.

Currently, MMIF methods are mainly divided into traditional and deep learning (DL) based fusion methods. The former consists of three parts: image decomposition and reconstruction, image fusion rules, and image quality assessment [6]. The traditional methods do not require model training but need to fix specific fusion strategies in advance. However, manually designed complex image decomposition methods are usually ineffective in retaining important information from the source images and may produce artifacts in the fused image. In addition, feature extraction methods are usually designed for specific tasks, leading to the poor generalization ability and robustness of the fusion methods.

As for existing DL-based [7] methods, they have improved the fusion quality to some extent, but their fusion effect is greatly influenced by the lack of gold standards, the limitations of the adopted network structure and improper loss function. Besides, unlike many traditional methods, previous DL-based fusion methods have rarely used feature decomposition. Recently, Zhao et al. [8] have proposed the Correlation-Driven Dual-Branch Feature Decomposition based fusion (CDDFuse) method, which combines Convolutional Neural Network (CNN) with Vision Transformer (ViT). In CDDFuse, the distinction between cross-modal features and shared features is facilitated to increase the correlation between the low-frequency features and decrease the correlation between the high-frequency features. However, when handling low-resolution MMI, rich detailed textures and blurred edges, CDDFuse cannot always work well. One example is that the CDDFuse does not perform well in CT-MR fusion in RIRE dataset [9] with large amounts of low-frequency monochromatic smearing, i.e., a pure grey background of the image covers the detailed textures of the MR image. Therefore, the following drawbacks can not be ignored. Firstly, the network uses a two-stage training strategy in which the cascading structure of image fusion and image reconstruction modules is trained in a serial way, leading to the accumulation of errors. Moreover, the feature decomposition network involves insufficient feature interaction, resulting in the deterioration of complementary information. Finally, the loss function used in CDDFuse cannot ensure the preservation of smooth boundary transition and high-quality visual fidelity.

To address the above-mentioned problems of the CDDFuse, we have proposed the Coupled Image Reconstruction and Fusion (CIRF) strategy. In this strategy, we have optimized the network structure and applied a new loss function. Our contributions can be briefly summarized as:

- We have proposed a novel fusion network with parallel image fusion and image reconstruction modules that share the same encoder and use the image masking strategy to enhance the feature learning ability of the encoder, thereby reducing error accumulation.
- The base-detail feature decomposition is optimized by adopting a concise parallel ViT-CNN structure, where base-details are processed separately but interact with each other to facilitate producing the complementary information, making the feature decomposition more effective.
- A new loss function combination is applied, i.e., the weighted sum of the reconstruction loss and the fusion loss. The former takes into account detail recovery, structural fidelity, and edge preservation. The latter utilizes a powerful unsupervised evaluation function.
- The performance of our method has been evaluated on three datasets with five types of multi-modal samples, and it demonstrates superior fusion performance to several traditional and DL-based fusion algorithms.

2. Related Works

2.1. The Traditional Fusion Methods

Image fusion has been extensively studied before the prevalence of DL. The traditional fusion methods have used relevant mathematical transformations to manually analyze the activity level and design fusion rules in the spatial or transform domain [10].

Spatial domain based fusion methods typically compute a weighted average of the local or pixel-level saliency of the two source images to obtain a fused image. However, these methods usually have problems in pseudo-color-image decomposition, i.e., the base and detail images obtained after decomposition are in grayscale. To tackle this problem, Du et al. [11] have come up with the Adaptive Two-scale Image Fusion (ATF) method, which uses Otsu's method [12,13] to decompose the pseudo-color input image into a base image and a detailed image, thereby obtaining an adaptive threshold for two-scale image fusion [14].

Transform domain-based fusion methods usually start by transforming the source images into the transform domain (e.g., wavelet domain [15]) to obtain different frequency components. For instance, Yin et al. [16] have proposed a medical image fusion method in the Nonsubsampled Shearlet Transform (NSST) domain [17]. Firstly the high frequency bands and low frequency bands are obtained by performing NSST decomposition of the input image. Then, the high-frequency bands are fused by the PAPCNN model [18]. As for the low-frequency bands, two new measures of activity level are introduced, namely the Weighted Local Energy (WLE) and the Weighted Sum of Eight-neighborhood-based Modified Laplacian (WSEML). WLE is utilized to address the issue of energy loss that arises due to the average-based conventional low-frequency fusion rule, and WSEML is fully employed to extract the detailed information present in the low-frequency band. The fused high-frequency and low-frequency bands are passed through inverse NSST to generate the final fused image. Besides, Li et al. [19] have proposed the Laplacian Redecomposition (LRD) framework. Here, the source images are processed by Gradient-domain Image Enhancement (GDIE) which is used for increasing the LRD ability of detail extraction through mapping gradient information adaptively. Then, the enhanced image undergoes Laplacian pyramid (LP) transform [20] to decompose it into the High-frequency Subband Image (HSI) with edges and details and the Low-frequency Subband Image (LSI) with background information. Through the pre-set fusion rules, image fusion is performed both on HSI and LSI to generate the high and low-frequency components of the fused image, respectively. Eventually, these components are subjected to inverse LP to produce the final fused image.

2.2. The DL-Based Fusion Methods

At present, the two most commonly used models in image fusion are CNN and Transformer. However, due to the giant computational overhead, pure Transformer methods are rare and CNN-Transformer hybrid networks are often used for image fusion.

2.2.1. The CNN Based Image Fusion

The most popular DL network in image processing is CNN. By training a CNN model, it is capable of recognizing and extracting different features for image fusion. Usually, in a CNN with multiple layers, each network layer produces several feature maps which are calculated through convolution, spatial pooling, and non-linear activation [21]. Besides, the CNN network can model the local area quite well by selecting an appropriate window size. However, it needs to stack very deep CNN layers to meet the requirement of the global perspective.

Some fusion methods usually contain CNN layers to extract multi-scale information. For example, Zhang et al. [22] have come up with a general image fusion framework based on a convolutional neural network (IFCNN). The most remarkable characteristic of this model is that it is fully convolutional so that it can be trained in an end-to-end manner without any post-processing procedures. To avoid the loss of fusion capabilities when training a single model for different scenes sequentially, Xu et al. [23] have presented a unified unsupervised image fusion network, termed U2Fusion, to solve multiple-territory fusion problems. In addition, some fusion methods which are initially proposed for infrared-visible image fusion are also inspiring for MMIF. For example, Li and Wu have proposed a DL architecture named DenseFuse [24] which consists of an encoder, a fusion layer, and a decoder. To extract salient features from source images effectively, the encoder is constructed with convolutional layers and dense blocks where the output of each layer is used as the input of all the subsequent layers. This prevents excessive information loss within the encoder. Li et al. have introduced an image fusion architecture, i.e., NestFuse [25], by developing a nest connection network and spatial/channel attention models. To begin with, they use pooling-assisted convolution to extract the multi-scale features. Then, several proposed spatial/channel attention models are utilized to fuse these multi-scale deep features in each scale. Li et al. [26] have also proposed a residual fusion network (RFN) based on a residual architecture to replace the traditional fusion approach. The learning of model parameters is accomplished by a novel two-stage training strategy. In the first stage, an auto-encoder network based on Nest connection is trained for better feature extraction and image reconstruction ability. Next, the RFN is trained using a specially designed loss function for fusion.

2.2.2. The CNN-Transformer-Based Image Fusion

Another widely used paradigm is the Transformer [27]. As an architecture initially proposed for natural language processing (NLP), the Transformer works by using stacked layers of self-attention and feed-forward networks to deal with data sequences. In the field of computer vision (CV), the Vision Transformer (ViT) [28] has been proposed to extend the application of the attention mechanism. Its basic principle is to treat images as sequence data and use self-attention mechanisms to capture their spatial and temporal information. Firstly, the input images are divided into multiple patches (e.g., with the size of 16×16), flattened and concatenated with positional encoding, and projected into the Transformer encoder. Then, by calculating the correlation between embedded patches, attention weight distribution is obtained to enable the model to focus on different positions in the image, thereby facilitating better global information transmission.

Although the cascaded self-attention modules can capture global representations, the ViT still cannot perform well in extracting the positional-encoded features with low computational consumption. Hence, the idea of using the convolution operators to extract local features and the self-attention mechanisms to capture global representations has been presented. For the MMIF, Tang et al. [29] have proposed an adaptive Transformer to capture long-range dependencies, which improves the global semantic extraction ability. They also make use of adaptive convolution instead of vanilla convolution to modulate the convolutional kernel automatically based on the wide-field context information. Zhang et al. [30] have introduced the Transformer as the fusion block, and applied multi-scale CNN as encoders and decoders. By interacting across fusion Transformers at multiple scales, the global contextual information from different modalities is incorporated more effectively. Zhou et al. [31] have proposed a novel architecture that combines a densely connected high-resolution network (DHRNet) with a hybrid transformer. Specifically, the hybrid transformer employs the fine-grained attention module to generate global features by exploring long-range dependencies, while the DHRNet is responsible for local information processing. Liu et al. [32] have used a CNN and Transformer module to build the extraction network and the decoder network. Besides, they have designed a self-adaptive weighted rule for image fusion.

3. Proposed Method

In this section, we present the architecture of CIRF and explain how each component works. Then, we introduce the entire model workflow and the loss function.

3.1. Framework of CIRF

Our CIRF consists of two parallel branches. The fusion branch adopts an encoder-decoder architecture with feature decomposition, fusing the base and detail features separately. The reconstruction branch, as a multi-task branch, assists in training a more powerful encoder and contributes to the reduction in the overall loss. The two branches share one common encoder so that ViT and CNN are parallel while the subsequent branch modules are different, and they complete the reconstruction and fusion tasks, respectively. In each epoch, the weighted summation of the reconstruction loss and the fusion loss is performed.

As shown in Figure 1, the framework of CIRF contains a Parallel Decomposition Encoder (PDE), Decoupling Reconstruction Decoder (DRD), Base Fusion Block (BFB), Detail Fusion Block (DFB), and Decoupling Fusion Decoder (DFD). In the following, these modules will be referred to by abbreviation for simplicity and clarity.

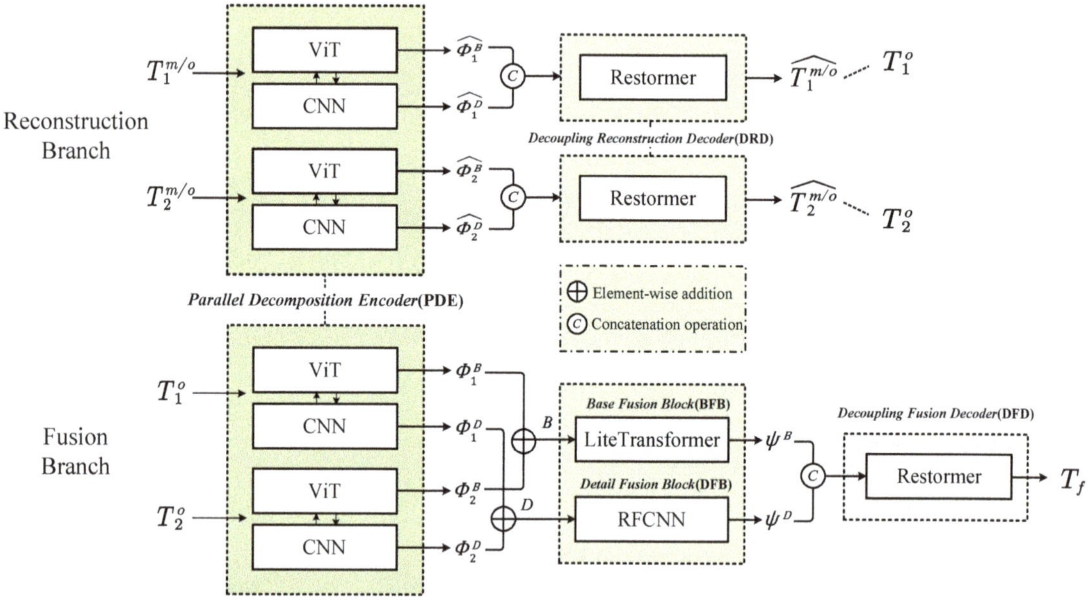

Figure 1. The brief workflow of the CIRF network. The architecture consists of two branches: the reconstruction branch at the top and the fusion branch at the bottom. During training, both branches are calculated simultaneously, and their total loss is added by an adjustable weight. However, in model inference, only the fusion branch is retained.

Furthermore, to make narration easier, here we agree on some symbols.

- We use o and m to distinguish original and masked images, e.g., T_1^o and T_1^m.
- We use $\widehat{(\cdot)}$ to denote information extracted from masked inputs in the reconstruction branch, e.g., $\widehat{\Phi_1^B}$ and $\widehat{T_1^m}$.
- We use B and D to abbreviate base and detail, r and f to abbreviate reconstruction and fusion, e.g., ϕ^B and ψ^D.
- The outputs of the encoder, two fusion blocks, and two decoders are represented by $E(\cdot)$, $F_b(\cdot)$, $F_d(\cdot)$, $D_r(\cdot)$ and $D_f(\cdot)$, respectively.

3.2. Fusion Branch

3.2.1. Overview

The fusion branch utilizes an encoder-fusion-decoder structure that involves feature decomposition. It has four components: PDE, BFB, DFB and DFD.

The inputs of this branch are two batches of original multi-modal images T_1^o and T_2^o. These images are firstly decomposed into base and detail features through PDE, i.e., a paralleled ViT-CNN encoder, formulated as:

$$\Phi_1^B, \Phi_1^D = E(T_1^o) \tag{1}$$

$$\Phi_2^B, \Phi_2^D = E(T_2^o) \tag{2}$$

Then, two types of features are added for high-frequency and low-frequency information fusion, respectively. For BFB, a Lite Transformer (LT) [33] module with long-short-range attention is chosen. In essence, it is a Transformer that is assisted with the Gated Linear Unit (GLU) and convolution block, and thus it is suitable for long-range information fusion while taking into account the local details. For DFB, we have constructed the Residual Fusion CNN (RFCNN) which is a pure convolutional neural network with various residuals so as to keep more detailed information. This process can be expressed as:

$$\psi^B = F_b\left(\Phi_1^B + \Phi_2^B\right) \tag{3}$$

$$\psi^D = F_d\left(\Phi_1^D + \Phi_2^D\right) \tag{4}$$

Finally, the outputs of fusion blocks are concatenated and sent into DFD (Restormer module) [34] for image restoration until we obtain:

$$T_f = D_f\left(cat\left\{\psi^B, \psi^D\right\}\right) \tag{5}$$

3.2.2. Parallel Decomposition Encoder

When it comes to traditional multi-modal medical image fusion (MMIF) methods, there have been several strategies based on frequency decomposition, but most of them are ineffective and time-consuming. In CDDFuse [8], a dual-branch Transformer-CNN framework that performs cross-modal feature decomposition extraction through a shared encoder is proposed and has obtained relatively good results. However, in the specific scene of MMIF, given the low-resolution input images, the detail loss caused by CDDFuse is more serious, thereby leading to contrast distortion and obvious artifacts. Inspired by [35,36], we have developed a lite encoder that can retain detail representations and base features to the maximum extent, whose framework is shown in Figure 2.

Here, the inputs of the network can be denoted as a four-dimensional matrix $[N, C, H, W]$, which represents the batch size, channel, height and width, respectively. Generally, most medical images are single-channel gray-scale images. While processing an RGB image, we first convert it into YUV space, where the Y channel contains gray-scale information, and then fuse the Y channel with another gray-scale image separately. Finally, we re-stitch the image with UV channels to restore a colored one [16]. Accordingly, after data pre-processing, the input tensor can be unified as $[N, 1, H, W]$.

In PDE, the input tensor is initially processed by a coarse feature extraction module with large convolution kernels (e.g., 7 × 7) and pooling layers, and then it is sent to two parallel branches comprising multi-head Transformer Block and CNN Block. Notably, features input into the Transformer Block need to go through an extra convolution layer before being reshaped into 8 × 8 patches [28]. By doing so, the number of feature channels is increased and the size of the feature maps is reduced, which is more conducive to effective and efficient feature extraction by attention layers. The number of heads in the self-attention layer is set to 4, and the stack depth is set to 6 with a drop rate of 10%.

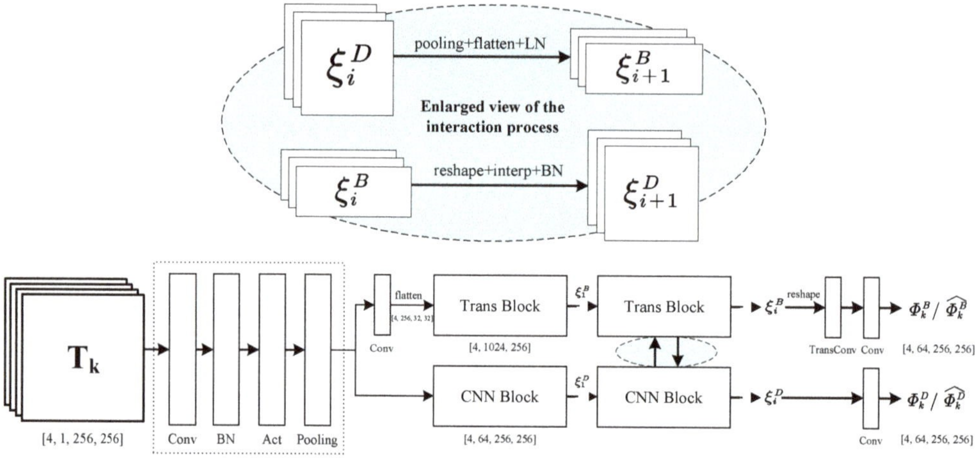

Figure 2. The brief workflow of PDE. The overall architecture comprises two parallel branches for feature decomposition, the ViT branch and the CNN branch, which are also connected with each other to form information complementation. Here, the inputs are two modalities of single-channel gray-scale images denoted as T_k, where $k = 1, 2$. In the experiments, we fix $N = 4$, $H = W = 256$, $C = 1$. Therefore, after data pre-processing, the input tensor dimension can be written as $[4, 1, 256, 256]$ while the output tensor will be $[4, 64, 256, 256]$. The complementary process is illustrated by feature map transformations shown at the top of the figure.

Subsequently, the ViT and CNN Blocks are repeatedly stacked for i times. Considering the complementation of base and detail features [35], we have added information interaction between the multi-head Transformer and CNN Block when $i \geqslant 2$, which contributes to better preserving detailed texture features and protecting image edge contours. During the transformation from detail feature maps (e.g., ξ_i^D) to base ones (e.g., ξ_{i+1}^B,), pooling, flattening, and layer-normalization operations are applied. On the contrary, reshaping, interpolation, and batch-normalization operations are adopted for transforming from base feature maps to detail ones (e.g., from ξ_i^B to ξ_{i+1}^D). Experiments show that setting $i = 2$ is enough to obtain satisfactory outcomes and can help to limit network parameters to a relatively small scale.

Eventually, through reshaping and trans-convolution, we can restore the feature maps back to their original visible sizes. However, the extracted deep-layer information has increased, which can be described as $[N, 64, H, W]$.

3.2.3. Base and Detail Fusion Block

For the MMI, it is still important to pay close attention to the local features when fusing the global information in BFB. Unfortunately, the traditional Transformer architecture can be inefficient due to its large time and space consumption as well as computational redundancy. To tackle this, a CNN-assisted lite Transformer which offers a trade-off between the feed-forward computation for wider attention layers is applied [33]. Here, one group of heads is responsible for the local context modeling via convolution, while the other conducts long-distance relationship modeling via attention.

As for DFB, reducing information loss is the most urgent goal. Therefore, we should not only improve the richness of information (i.e., improve the dynamic range of output representation) but also prevent gradient explosion and model non-convergence. As shown in Figure 3, a simple CNN cell (the yellow box) and a residual line (the yellow line) composed of convolution layers and batch normalization are first defined. Additionally, between two CNN cells comes an Exponential Linear Unit (ELU) [37] activation function, which is unilaterally saturated and outputs tensors with zero-mean distribution, thereby speeding up training and accelerating convergence. Besides, we have utilized convolutional residuals to link the output of the front module to the input of the rear module with a ReLU6 activation function [38] added after post-merger residuals. By doing so, the output is limited to the maximum of 6, thereby preventing gradient explosion, benefiting gradient descent at low precision, and improving decimal expression ability [39]. Under such an architecture of detail feature fusion, detail fidelity will be ensured by continuous optimization.

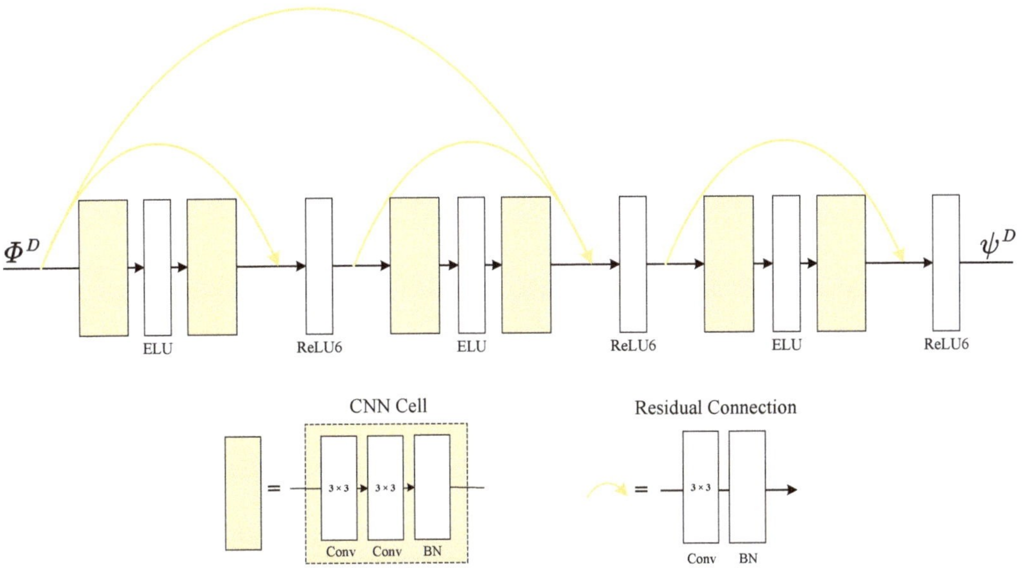

Figure 3. The architecture of RFCNN. In this figure, the yellow box and the yellow line respectively represent different CNN modules with a small kernel size of 3 × 3. Notably, the ELU and ReLU6 activation functions are specifically used to enhance expressivity and prevent gradient explosion. By adding residuals, RFCNN can effectively accomplish detail-feature fusion tasks from Φ^D to ψ^D.

3.2.4. Decoupling Fusion Decoder

To restore noise-disturbed images, Zamir et al. have developed an efficient Transformer model [34] that can output high-resolution images in restoration tasks. This is also used in [8] for fused image decoding. In this paper, we retain this module.

3.3. *Reconstruction Branch*

In RFN-Nest [26], a two-stage training strategy has been presented for the first time. By pre-training the network via reconstruction tasks, the quality of the fused image is greatly improved, which also alleviates the challenge caused by the lack of gold standard to some extent. However, two-stage training can cause error accumulation, raising a stage time allocation problem, and resulting in redundant time overhead and low robustness. Therefore, a multi-task network that couples the reconstruction branch and the fusion branch with one common encoder is proposed. Here, the reconstruction branch aims at training a more powerful feature extraction encoder. By paralleling the two stages, the total loss of the task can better reflect model capability at any time.

Besides, inspired by [40], we have figured out that in some cases (e.g., when given low-quality source images), adding random image masks can enhance the expressivity of the shared encoder. Hence, the encoding process can be characterized by:

$$\widehat{\Phi_1^B}, \widehat{\Phi_1^D} = E\left(T_1^{m/o}\right) \tag{6}$$

$$\widehat{\Phi_2^B}, \widehat{\Phi_2^D} = E\left(T_2^{m/o}\right) \tag{7}$$

Then, features derived from the same image are concatenated and fed into the shared PDE that will be discarded later. Since the reconstruction branch mainly contributes to the encoder, the Restormer module used in Section 3.2.4 is again selected here as DRD for convenience. Here, it can be any simple decoding structure. The function of DRD can be expressed as:

$$\widehat{T_1^{m/o}} = D_r\left(cat\left\{\widehat{\Phi_1^B}, \widehat{\Phi_1^D}\right\}\right) \tag{8}$$

$$\widehat{T_2^{m/o}} = D_r\left(cat\left\{\widehat{\Phi_2^B}, \widehat{\Phi_2^D}\right\}\right) \tag{9}$$

Here, it is worth mentioning that in the inference process, the reconstruction branch will be cut off.

3.4. Loss Function

The workflow of the reconstruction branch is a supervised process with a given ground truth (i.e., source images). Therefore, the reconstruction loss is composed of three components: mean square error (MSE), structural similarity (SSIM) [41] and spatial gradient loss (SG) [42,43]. For each source image k, the reconstruction loss $L_{rec,k}$ can be calculated by:

$$L_{rec,k} = L_{MSE}\left(\widehat{T_k^{m/o}}, T_k^o\right) + \alpha \cdot L_{SSIM}\left(\widehat{T_k^{m/o}}, T_k^o\right) + \beta \cdot L_{SG}\left(\widehat{T_k^{m/o}}, T_k^o\right), k = 1, 2. \tag{10}$$

where α, β are adjustable weights; $L_{MSE}, L_{SSIM}, L_{SG}$ protect the local pixel information, regional structure information, and edge contour information, respectively. Meanwhile, the gradient loss can be described as $L_{SG} = \frac{1}{HW}\sum \| |\nabla \widehat{T_k^{m/o}}| - |\nabla T_k^o| \|_1$.

Furthermore, the total reconstruction loss is computed as:

$$Loss1 = \mu \cdot (L_{rec,1} + L_{rec,2}) \tag{11}$$

where μ is a weight for numeral balance, i.e., adjusting the order of magnitude.

On the other hand, the fusion branch lacks ground truth, so the unsupervised loss function should be able to effectively measure the intensity correlation and structural information between the source and fused images. Inspired by [39], we choose mutual information (MI), the sum of the correlations of differences (SCD) [44], structural similarity (SSIM), and edge retentiveness ($Q^{AB/F}$) [45] as our four metrics that make up the fusion loss function. Such a function can be described as:

$$Loss2 = \lambda \cdot (L_{MI} + L_{SCD}) + (1 - \lambda) \cdot \left(L_{SSIM} + L_{Q^{AB/F}}\right) \tag{12}$$

where λ is a hyper-parameter and $L_{MI}, L_{SCD}, L_{SSIM}, L_{Q^{AB/F}}$ reflect the amount of common information, the correlations of image differences, the similarity of luminance, contrast and structure as well as the preservation of edge information, respectively. For each metric in Equation (12), each loss is one minus the normalized average of each metric of the two source images and the fused image.

As a whole, the total loss function is as follows:

$$Loss_{total} = (1 - \sigma) \cdot Loss1 + \sigma \cdot Loss2 \tag{13}$$

where σ is also a hyper-parameter to balance our network's preference for reconstruction and fusion, which will be discussed later in the ablation study.

4. Experimental Settings

In this section, we discuss the settings of our dataset, the compared algorithms and the metrics we have chosen to evaluate the algorithms.

4.1. Dataset

The Whole Brain Atlas (Atlas) [46], the IXI Brain Development Dataset (IXI) [47], and the Retrospective Image Registration Evaluation (RIRE) [9] are used to evaluate our algorithm. The Atlas dataset collected by Harvard Medical School includes CT, MR, PET, and SPECT images from patients with various diseases. The IXI dataset collected by three hospitals in London includes 3D MR images from 600 healthy test-takers, including T1-, T2-, and PD-weighted images. The RIRE dataset collected by the National Institute of Biomedical Imaging and Bioengineering includes CT, MR, and PET images.

In the Atlas dataset, we have acquired three groups of images, including 388 pairs of SPECT-CT/MR images, 590 pairs of multi-modal MR (i.e., T1-T2) images, and 140 pairs of CT-MR images. For multi-modal MR image pairs, we have randomly divided them into training and testing sets in the ratio of 8:1. To increase the number of training images, we have augmented the training set by performing rotations and mirroring operations on the original images, resulting in six times the amount of data. The generated methods are shown in Figure 4. SPECT-CT/MR and CT-MR images in the Atlas dataset are all retained to test images for evaluating model generalization ability.

In the IXI and RIRE datasets, we have, respectively, acquired 3936 multi-modal MR (i.e., PD-T2) image pairs and 476 CT-MR image pairs. We have used similar methods to process the dataset, i.e., dividing them into training and testing sets in a ratio of 8:1. Considering that the IXI images are enough for training and testing, we have only enhanced the training set of the RIRE dataset. It is noteworthy that we have first registered the RIRE dataset using the Elastix algorithm [48,49], and then used these registered image pairs to produce training and testing sets. Specifically, the MR images and CT images are chosen as the fixed and moving images, respectively, for registration. See Table 1.

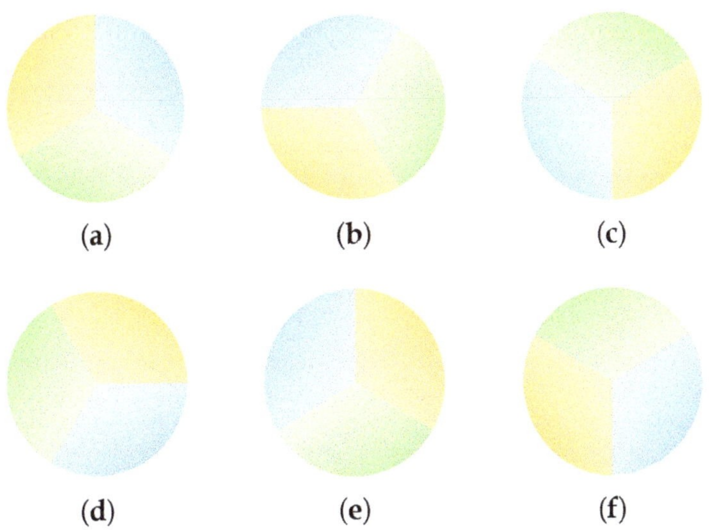

Figure 4. Visualization of training dataset flipping via a three-color-leaf example. (**a**) Original image; (**b**) 90° counterclockwise rotation; (**c**) 180° counterclockwise rotation; (**d**) 90° clockwise rotation; (**e**) Left-right mirror symmetry; (**f**) Up-down mirror symmetry.

Table 1. Details of the three datasets.

Datasets	Image Pairs	Training	Testing
Atlas	Multi-modal MR	2976	62
	CT-MR	-	140
	SPECT-CT/MR	-	388
IXI	Multi-modal MR	3504	432
RIRE	CT-MR	2538	53

4.2. The Comparison Algorithms

Our method will be compared with several traditional algorithms (i.e., PAPCNN [16], TIF [11], and ReLP [19]) and DL-based algorithms (i.e., CDDFuse [8], DenseFuse [24], IFCNN [22], NestFuse [25], U2Fusion [23], and RFN-Nest [26]). For the traditional methods, the fusion results are directly obtained by running the corresponding algorithm. For DL-based methods, we have trained the U2Fusion on the datasets mentioned above. However, for other methods, we have directly used the models trained by their corresponding authors as these models are already sufficiently trained.

4.3. Fusion Metrics

We have used eight metrics to evaluate our algorithm. Standard deviation (SD) measures the contrast of the fused image. Peak-signal-to-noise ratio (PSNR) measures the effective signal intensity of the fused image. For the computation of PSNR, the two mean square errors (MSE) between the source images and the fused image are first averaged to produce the mean MSE. Then, the ratio of the square of the maximum pixel intensity to the mean MSE is computed, and the logarithm (base 10) of the ratio is multiplied by 10 to produce the PSNR according to [50]. The sum of the correlations of differences (SCD) measures distortion and loss of information of the fused image [44]. Mutual information (MI) measures the amount of information from the original images that is captured in the fused image. The structural similarity (SSIM) evaluates the structural similarity between the fused image and the source image, the overall SSIM is calculated by directly averaging the two SSIM values of the two source images and the fused image according to [51]. $Q^{AB/F}$ evaluates the edge information from the original image [45]. The visual information fidelity for fusion (VIFF) evaluates the quality of an image based on the calculation of visual information fidelity [52]. The ratio of spatial frequency error ($|rSFe|$) evaluates the ratio of spatial frequency error calculated from the source image referred to as SF. A value of $|rSFe|$ greater than zero indicates the introduction of noise during image fusion, while a value less than zero indicates the loss of information [53]. In general, the closer the $|rSFe|$ is to 0, the better the fusion effect. However, larger values of other metrics indicate better fusion performance.

5. Ablation Experiments

Our algorithm is realized using Python 3.10, Pytorch 2.0.1 on Ubuntu 22.04.3 LTS, and CUDA 11.8. Meanwhile, it is run on a server with the Intel(R) Xeon(R) Gold 6248R CPU (Intel, Santa Clara, CA, USA) and the NVIDIA RTX A100 with 40 G VRAM (NVIDIA, Santa Clara, CA, USA). Additionally, we use the Adam optimizer to update the model parameters.

In the DL-based MMIF tasks, the loss function is extremely important. Here, our loss function has two adjustable hyper-parameters, λ (in Equation (12)) and σ (in Equation (13)) which will be determined subsequently. Moreover, we have also explored the impact of inputting images with different masking ratios based on the considerations that in some cases masking can enhance the feature extraction capability of PDE and reduce the fusion artifacts. To obtain the optimal values for the above three parameters, we have conducted ablation experiments on three datasets separately.

5.1. Parameter Setting on Atlas Dataset

On the Atlas dataset, we have firstly fixed the values of parameters λ and σ based on our experience, and then found the best value for the masking ratio by increasing it with a step size of 0.1. The results are shown in Table 2. Obviously, the value of MI reaches the maximum when the masking ratio is 0.1, while the values of SCD and $VIFF$ decrease with the increase in the masking ratio. Besides, by setting the masking ratio to 0.1, we observe fewer fusion artifacts in the fused results compared with those produced with the masking ratio = 0. Taking all these into account, we will set the masking ratio = 0.

Table 2. The metric values of CIRF using different masking ratios ($\lambda = 0.3, \sigma = 0.2$) in Atlas Dataset.

Ratio	SD	$PSNR$	SCD	MI	$SSIM$	$Q^{AB/F}$	$VIFF$	$\lvert rSFe \rvert$
0	92.705	24.709	1.769	1.478	0.889	0.549	0.679	0.143
0.1	91.686	24.975	1.761	1.483	0.885	0.539	0.673	0.230
0.2	92.980	26.700	1.758	1.456	0.743	0.502	0.671	0.252
0.3	91.946	26.698	1.749	1.458	0.862	0.500	0.662	0.308
0.4	93.808	27.799	1.736	1.441	0.442	0.460	0.658	0.311
0.5	92.734	29.311	1.733	1.460	0.854	0.494	0.658	0.320

Next, to make the fusion branch achieve the best effect, we have fixed the masking ratio at 0.1 and preset σ at 0.2 while altering the value of λ. According to the metric values in Table 3, we will choose $\lambda = 0.3$ to achieve the trade-off among all evaluation indicators.

Table 3. The metric values of CIRF using different λ (masking ratio = 0.1, $\sigma = 0.2$) in Atlas Dataset.

λ	SD	$PSNR$	SCD	MI	$SSIM$	$Q^{AB/F}$	$VIFF$	$\lvert rSFe \rvert$
0.1	83.998	18.678	1.632	1.489	0.894	0.555	0.638	0.256
0.2	89.120	21.384	1.732	1.466	0.880	0.531	0.660	0.232
0.3	91.686	24.975	1.761	1.483	0.885	0.539	0.673	0.230
0.4	94.165	23.402	1.778	1.443	0.787	0.491	0.674	0.208
0.5	97.534	28.796	1.790	1.458	0.779	0.488	0.684	0.207
0.6	97.995	27.097	1.790	1.438	0.764	0.451	0.682	0.202
0.7	100.142	26.650	1.802	1.437	0.753	0.436	0.686	0.184
0.8	100.008	23.018	1.796	1.418	0.724	0.395	0.675	0.217
0.9	104.269	25.418	1.787	1.422	0.411	0.358	0.677	0.185

Furthermore, to balance the performance of the two model branches, another parameter σ needs to be determined. Thus, we have fixed the masking ratio and λ at their optimal value (i.e., masking ratio = 0.1 and $\lambda = 0.3$), and changed the value of σ. As shown in Table 4, when σ is set to 0.2, relatively high SCD, MI, $SSIM$, $Q^{AB/F}$ and $VIFF$ values can be obtained, so we choose $\sigma = 0.2$. Additionally, the comparison of the metrics when $\sigma = 1.0$ and $\sigma = 0.2$ clearly demonstrates the contribution of the reconstruction branch to the PDE.

Table 4. The metric values of CIRF using different σ (masking ratio = 0.1, $\lambda = 0.3$) in Atlas Dataset.

σ	SD	$PSNR$	SCD	MI	$SSIM$	$Q^{AB/F}$	$VIFF$	$\lvert rSFe \rvert$
0.1	92.022	25.153	1.761	1.461	0.878	0.519	0.670	0.243
0.2	91.686	24.975	1.761	1.483	0.885	0.539	0.673	0.230
0.3	91.863	24.283	1.761	1.476	0.867	0.537	0.676	0.210
0.4	92.272	23.768	1.768	1.466	0.806	0.523	0.675	0.223
0.5	92.088	24.461	1.762	1.473	0.782	0.534	0.674	0.210
0.6	91.952	26.342	1.758	1.493	0.887	0.544	0.673	0.218
0.7	90.754	23.280	1.751	1.490	0.884	0.536	0.670	0.227
0.8	94.179	27.839	1.767	1.465	0.846	0.529	0.677	0.217

Table 4. *Cont.*

σ	SD	$PSNR$	SCD	MI	$SSIM$	$Q^{AB/F}$	$VIFF$	$\|rSFe\|$
0.9	90.745	23.266	1.755	1.481	0.841	0.536	0.667	0.242
1.0	89.141	21.417	1.747	1.487	0.887	0.535	0.667	0.239

5.2. Parameter Setting on IXI Dataset

To ensure the rigor of the experiment and verify the robustness of our method, we have further used two other datasets to compute the metric value with different masking ratios, λ and σ. The results from using two different masking ratios are shown in Table 5. Due to the high quality of source images of the IXI dataset (i.e., rich and clear details, few artifacts), adding masking to the original images will not improve the fusion effect. Therefore, we set the masking ratio to 0 to obtain the optimal results.

Table 5. The metric values of CIRF using different masking ratio ($\lambda = 0.3$, $\sigma = 0.2$) in IXI Dataset.

Ratio	SD	$PSNR$	SCD	MI	$SSIM$	$Q^{AB/F}$	$VIFF$	$\|rSFe\|$
0	72.361	39.814	1.872	2.347	0.744	0.740	0.981	0.047
0.1	72.122	36.206	1.809	2.251	0.709	0.727	0.945	0.160

From the results using different λ in Table 6, we can see that as λ increases, $PSNR$ generally improves firstly but then experiences a decline. Moreover, with increasing λ, SD, SCD and $VIFF$ also increase while $Q^{AB/F}$ decreases. Based on the above analysis, we will set $\lambda = 0.3$.

Table 6. The metric values of CIRF using different λ (masking ratio = 0, $\sigma = 0.2$) in IXI Dataset.

λ	SD	$PSNR$	SCD	MI	$SSIM$	$Q^{AB/F}$	$VIFF$	$\|rSFe\|$
0.1	66.062	28.838	1.738	2.336	0.798	0.761	0.906	0.100
0.2	71.483	36.240	1.844	2.393	0.832	0.738	0.949	0.063
0.3	72.361	39.814	1.872	2.347	0.744	0.740	0.981	0.047
0.4	76.164	35.067	1.890	2.287	0.730	0.717	1.004	0.030
0.5	75.549	38.379	1.888	2.341	0.702	0.718	1.028	0.034
0.6	74.979	35.110	1.878	2.325	0.742	0.697	0.990	0.029
0.7	75.190	30.003	1.907	2.213	0.657	0.693	1.047	0.052
0.8	79.857	31.399	1.895	2.214	0.689	0.659	1.048	0.058
0.9	79.347	29.511	1.904	2.177	0.634	0.643	1.088	0.114

From the results using different σ in Table 7, we can see that all metrics have no obvious changes but minor fluctuations. However, as σ increases, $PSNR$, MI, $SSIM$ and $Q^{AB/F}$ reach their maximum when $\sigma = 0.2$ while other metrics are also competitive. Accordingly, we will fix $\sigma = 0.2$.

Table 7. The metric values of CIRF using different σ (masking ratio = 0.1, $\lambda = 0.3$) in IXI Dataset.

σ	SD	$PSNR$	SCD	MI	$SSIM$	$Q^{AB/F}$	$VIFF$	$\|rSFe\|$
0.1	72.160	34.644	1.889	2.336	0.730	0.733	1.004	0.033
0.2	72.361	39.814	1.872	2.347	0.744	0.740	0.981	0.047
0.3	70.964	32.695	1.870	2.315	0.734	0.735	0.970	0.033
0.4	72.028	30.786	1.876	2.223	0.683	0.728	0.993	0.033
0.5	70.364	31.237	1.854	2.278	0.715	0.731	0.974	0.040
0.6	72.816	33.506	1.881	2.288	0.713	0.735	0.991	0.036
0.7	73.270	33.861	1.877	2.269	0.710	0.728	0.990	0.032
0.8	72.297	29.842	1.873	2.221	0.680	0.733	0.978	0.044

Table 7. *Cont.*

σ	SD	$PSNR$	SCD	MI	$SSIM$	$Q^{AB/F}$	$VIFF$	$\lvert rSFe \rvert$
0.9	69.995	28.954	1.847	2.213	0.671	0.732	0.955	0.056
1.0	73.257	33.528	1.878	2.272	0.694	0.728	0.993	0.031

5.3. Parameter Setting on RIRE Dataset

As for the RIRE dataset, we have computed the metrics using different masking ratios. Table 8 indicates that by inputting non-masked source images, our method works best.

Table 8. The metric values of CIRF using different masking ratios ($\lambda = 0.3$, $\sigma = 0.2$) in RIRE Dataset.

Ratio	SD	$PSNR$	SCD	MI	$SSIM$	$Q^{AB/F}$	$VIFF$	$\lvert rSFe \rvert$
0	72.361	39.814	1.872	2.347	0.744	0.740	0.981	0.047
0.1	72.122	36.206	1.809	2.251	0.709	0.727	0.945	0.160

From the results of λ in Table 9, it can be seen that the value of SD reaches the maximum and the value of $|rSFe|$ reaches the minimum when λ is equal to 0.3. Meanwhile, the values of $PSNR$, MI, and $Q^{AB/F}$ are also relatively high with $\lambda = 0.3$. Based on comprehensive consideration, we fix $\lambda = 0.3$.

Table 9. The metric values of CIRF using different λ (masking ratio = 0.1, $\sigma = 0.2$) in RIRE Dataset.

λ	SD	$PSNR$	SCD	MI	$SSIM$	$Q^{AB/F}$	$VIFF$	$\lvert rSFe \rvert$
0.1	62.308	15.467	1.709	1.462	0.474	0.574	0.593	0.177
0.2	67.920	16.578	1.838	1.466	0.656	0.562	0.627	0.189
0.3	79.581	18.465	1.862	1.514	0.468	0.560	0.630	0.147
0.4	71.692	18.962	1.887	1.485	0.409	0.541	0.649	0.205
0.5	73.452	17.282	1.909	1.485	0.656	0.519	0.648	0.223
0.6	77.384	18.085	1.921	1.518	0.668	0.505	0.664	0.215
0.7	74.868	16.913	1.921	1.462	0.619	0.488	0.644	0.239
0.8	74.641	16.254	1.916	1.446	0.591	0.452	0.631	0.242
0.9	78.625	17.103	1.918	1.456	0.613	0.415	0.649	0.250

The results from using different σ in Table 10 show that when σ equals 0.4, $SSIM$ reaches the maximum while SCD, MI, $Q^{AB/F}$ and $VIFF$ achieve relatively high values. Therefore, we choose $\sigma = 0.4$.

Table 10. The metric values of CIRF using different σ (masking ratio = 0.1, $\lambda = 0.3$) in RIRE Dataset.

σ	SD	$PSNR$	SCD	MI	$SSIM$	$Q^{AB/F}$	$VIFF$	$\lvert rSFe \rvert$
0.1	71.200	17.396	1.888	1.491	0.680	0.551	0.638	0.206
0.2	73.346	19.670	1.897	1.489	0.403	0.552	0.653	0.173
0.3	68.477	16.624	1.863	1.477	0.651	0.548	0.614	0.226
0.4	71.697	18.221	1.885	1.544	0.706	0.552	0.637	0.186
0.5	70.296	18.061	1.878	1.542	0.624	0.552	0.632	0.186
0.6	69.733	19.324	1.871	1.539	0.429	0.557	0.628	0.215
0.7	72.654	18.500	1.894	1.552	0.691	0.552	0.644	0.217
0.8	69.656	16.727	1.873	1.469	0.650	0.555	0.628	0.201
0.9	68.291	16.391	1.856	1.469	0.652	0.550	0.616	0.217
1.0	69.401	16.830	1.870	1.483	0.659	0.558	0.629	0.197

After testing on three datasets, we have found that our algorithm performs best when the parameter λ is set to 0.3. However, the optimal values of masking ratio and σ vary with the characteristics of different images.

6. Qualitative Analysis

6.1. The Results of the Atlas Dataset

For the CT-MR and SPECT-CT/MR image pairs from the Atlas dataset, we have not trained a specific CIRF on them but directly used the model trained on T1-T2 image pairs.

The fusion results of multi-modal MR (i.e., T1 and T2) image pairs from the Atlas dataset are shown in Figure 5. Generally, except for TIF, CDDFuse and CIRF, the brightness and intensity of all other algorithms are insufficient. Specifically, from the areas marked by the red boxes, the upper parts of the brainstem are blurred or missing in the fused results of U2Fusion, DenseFuse, RFN-Nest, PAPCNN, ReLP and CDDFuse. Additionally, as labeled by the green boxes, except for CDDFuse and CIRF, all other methods produce blurry and incomplete boundaries of the occipital lobe. By comparison, the CIRF algorithm performs better than other algorithms in terms of edge preservation.

The fusion results of the CT-MR image pairs are shown in Figure 6. Clearly, ReLP, TIF and CIRF outperform other methods in preserving the white cranium cross-section from the CT image. From the green boxes and the yellow arrows, we can clearly observe that only U2Fusion, DenseFuse and CIRF can simultaneously preserve low-intensity information and retain crucial information from the CT image. However, CIRF produces the fused result with higher contrast than U2Fusion and DenseFuse. The fused results marked with the red boxes show that CIRF can preserve the details from MR images better than other methods. Therefore, it is evident that CIRF simultaneously retains the features derived from both CT and MR images, which indicates its strong feature extraction and fusion capability.

The fusion results of SPECT-CT/MR image pairs are shown in Figure 7. As indicated by the red boxes, U2Fusion, DenseFuse and RFN-Nest cannot maintain the sharpness of details from MR image, and IFCNN, PAPCNN and TIF produce the unwanted ringing artifacts. As shown by the green boxes, NestFuse, ReLP and CDDFuse reduce the contrast of the details from the SPECT image. By comparison, CIRF can not only avoid undesirable artifacts but also preserve the important details from MR and SPECT images effectively. These results also indicate that CIRF has a good generalization ability when applied to different datasets.

6.2. The Results of the IXI Dataset

For the multi-modal MR image pairs in the IXI dataset, the results are shown in Figure 8. From the fusion results marked by the red boxes, we can see that TIF damages the low-intensity information seriously, and ReLP causes the loss of some low gray-scale detail information in that the contour of the ventricular boundary is missing in Figure 8(b7). Compared with U2Fusion, DenseFuse and CDDFuse, CIRF maintains the continuity of the gray line in the green box, which indicates that CIRF can preserve the fine edges better.

6.3. The Results of the RIRE Dataset

For the CT-MR image pairs from the RIRE dataset, the results are shown in Figure 9. It is evident that CIRF produces clearer image details, higher image contrast, and less loss of original information. However, U2Fusion, DenseFuse and RFN-Nest fail to effectively fuse the bright cranium from the CT image as depicted by the green arrows. As shown by the yellow arrows, IFCNN, NestFuse, PAPCNN, ReLP and CDDFuse fail to retain the low-intensity areas in the MR image, and TIF produces blocky artifacts. Besides, NestFuse and CDDFuse result in a serious loss of structural information as pointed out by the red arrows.

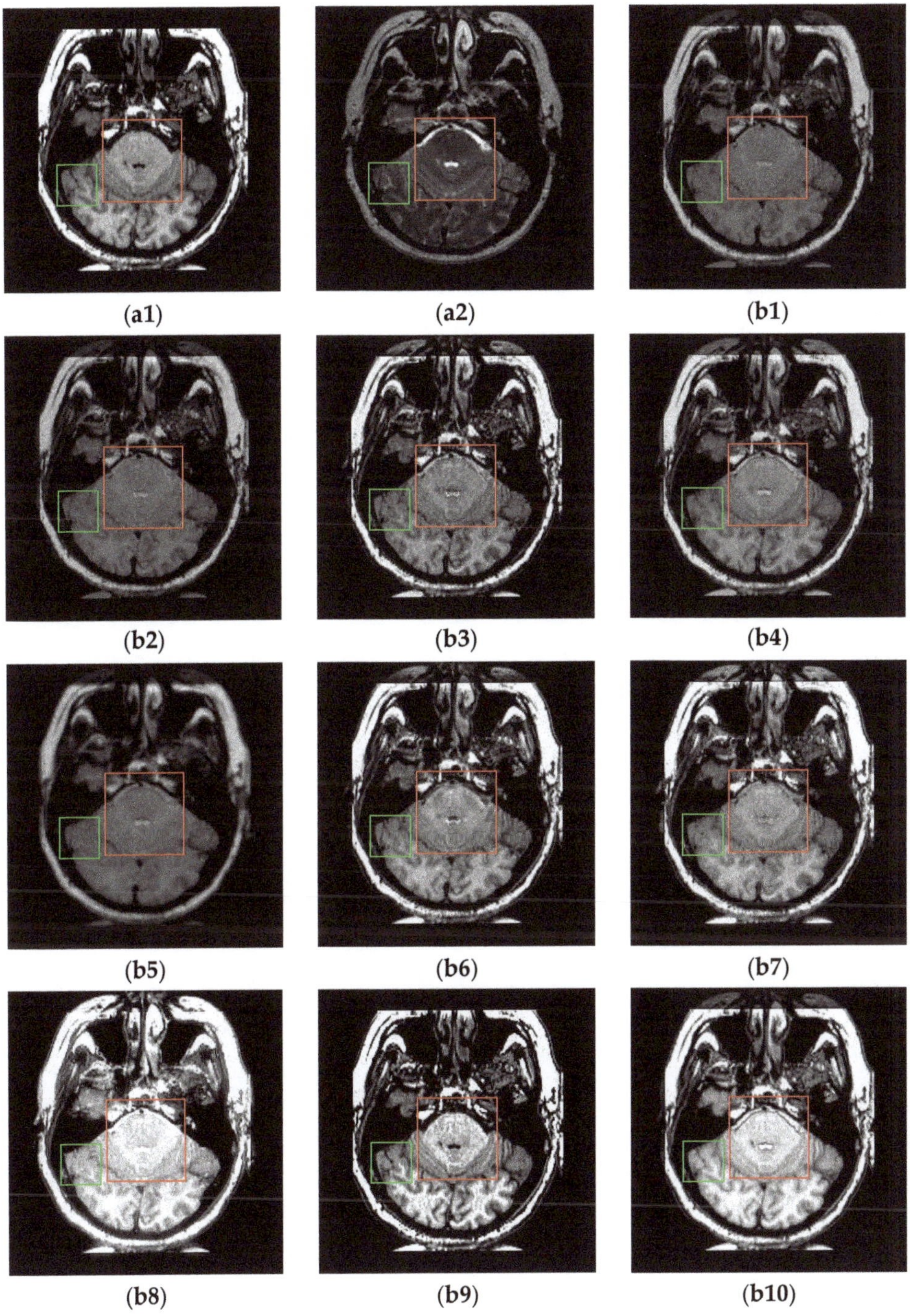

Figure 5. The results of all algorithms on the multi-modal image pairs from Atlas dataset. (**a1**) Source T1 image; (**a2**) Source T2 image; (**b1**) U2Fusion; (**b2**) DenseFuse; (**b3**) IFCNN; (**b4**) NestFuse; (**b5**) RFN-Nest; (**b6**) PAPCNN; (**b7**) ReLP; (**b8**) TIF; (**b9**) CDDFuse; (**b10**) CIRF.

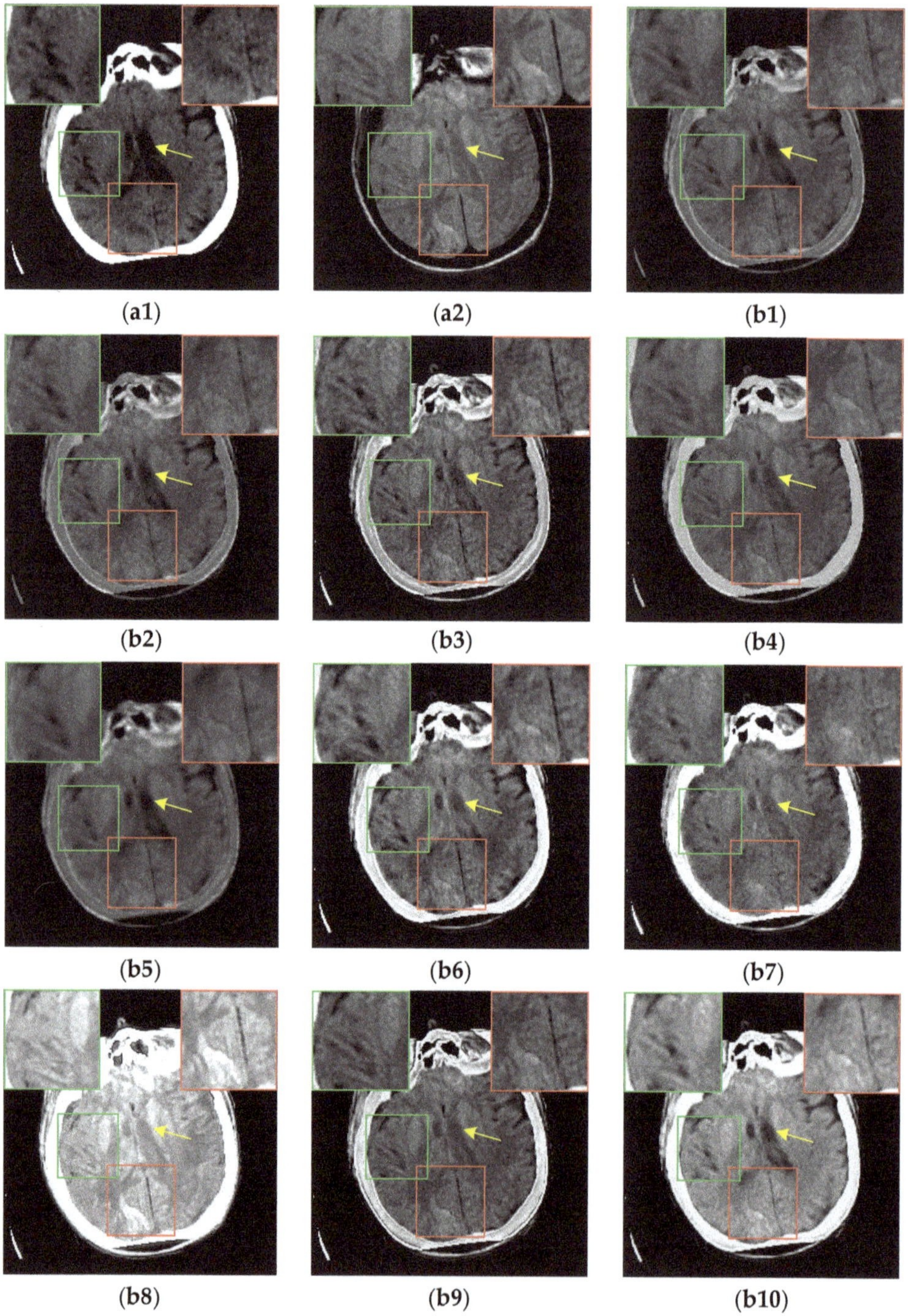

Figure 6. The results of all algorithms on the CT-MR image pairs from Atlas dataset. (**a1**) Source CT image; (**a2**) Source MR image; (**b1**) U2Fusion; (**b2**) DenseFuse; (**b3**) IFCNN; (**b4**) NestFuse; (**b5**) RFN-Nest; (**b6**) PAPCNN; (**b7**) ReLP; (**b8**) TIF; (**b9**) CDDFuse; (**b10**) CIRF.

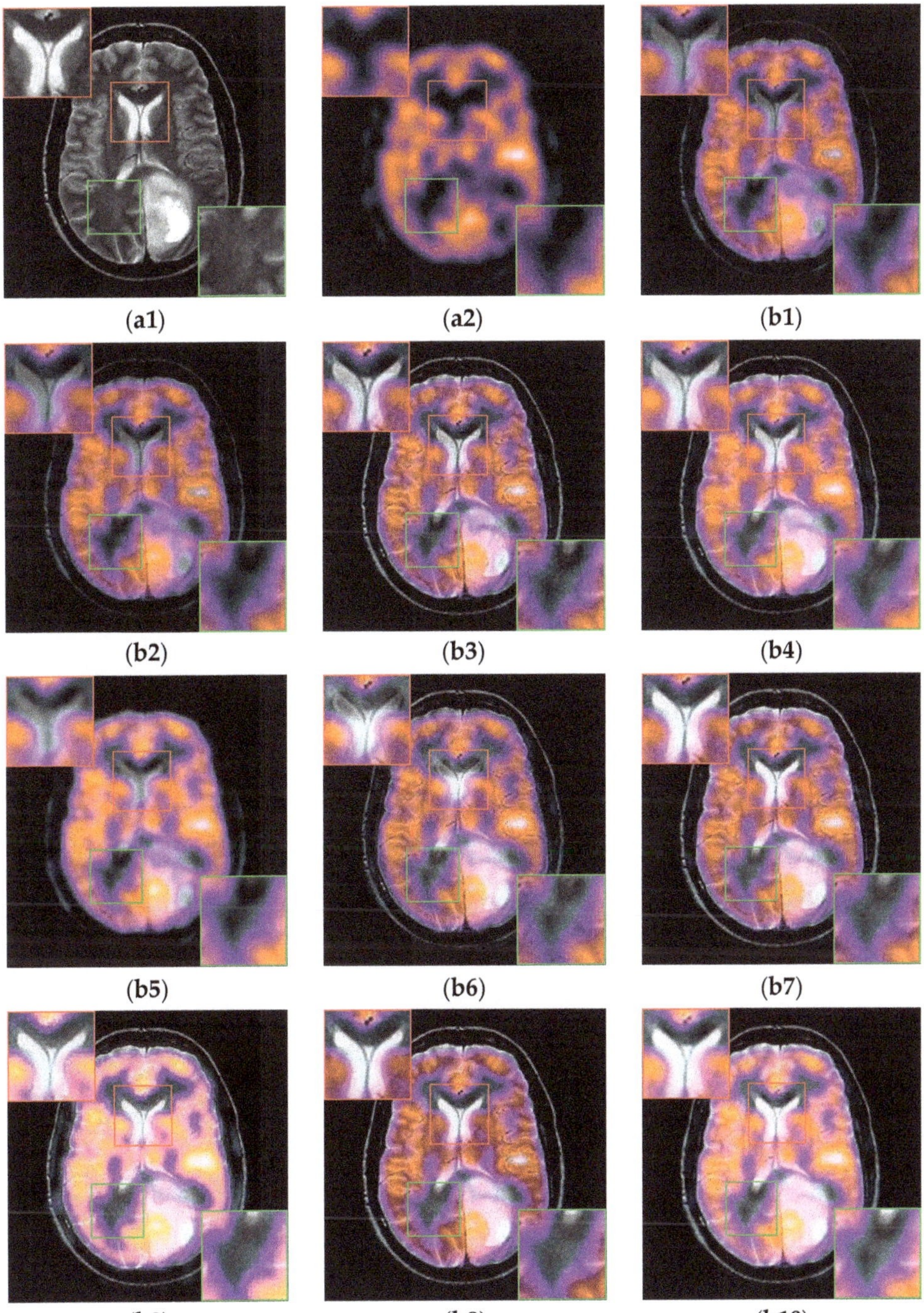

Figure 7. The results of all algorithms on the SPECT-CT/MR image pairs from Atlas dataset. (**a1**) Source CT/MR image; (**a2**) Source SPECT image; (**b1**) U2Fusion; (**b2**) DenseFuse; (**b3**) IFCNN; (**b4**) NestFuse; (**b5**) RFN-Nest; (**b6**) PAPCNN; (**b7**) ReLP; (**b8**) TIF; (**b9**) CDDFuse; (**b10**) CIRF.

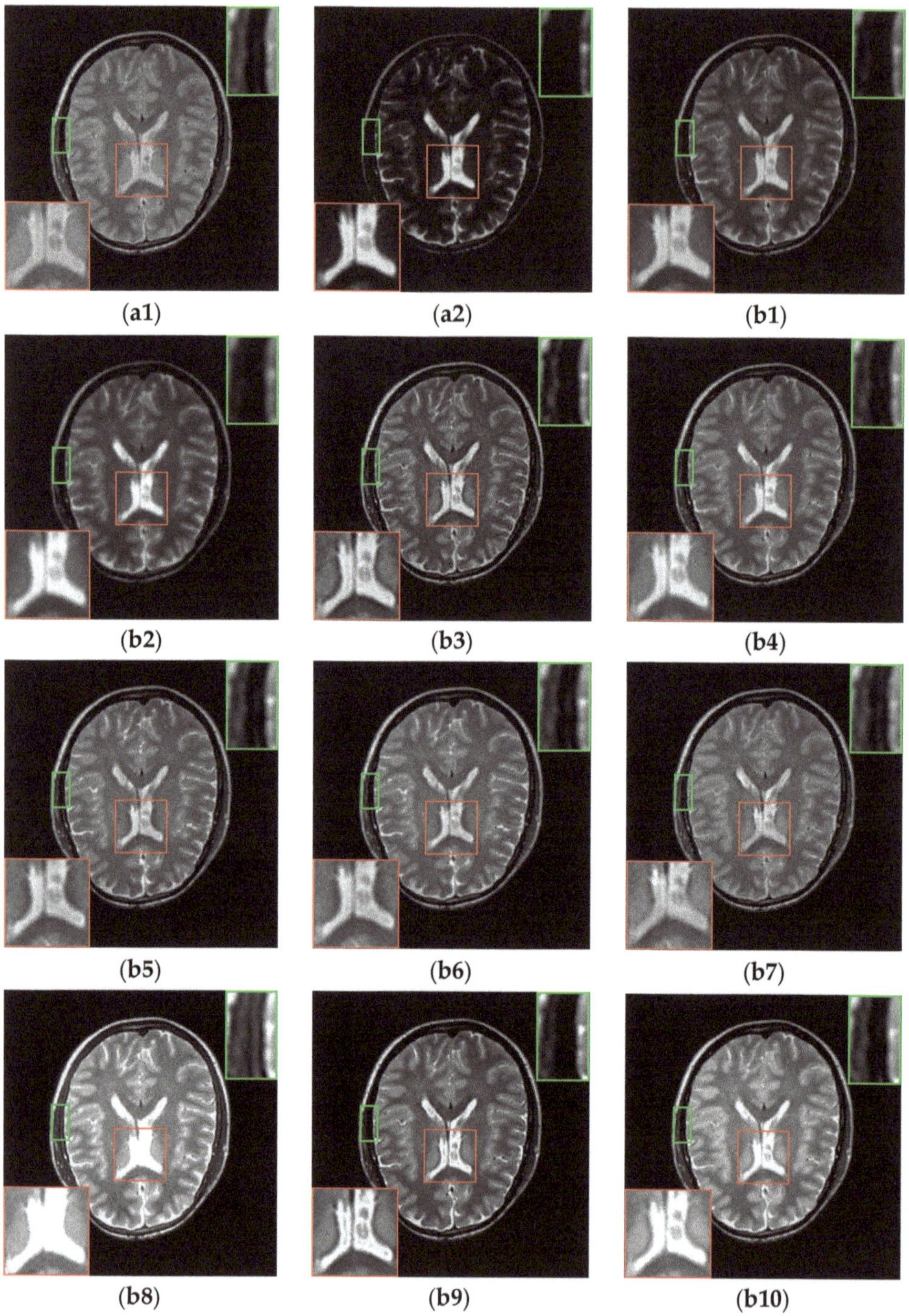

Figure 8. The results of all algorithms on the multi-modal MR image pairs from IXI dataset. (**a1**) Source PD image; (**a2**) Source T2 image; (**b1**) U2Fusion; (**b2**) DenseFuse; (**b3**) IFCNN; (**b4**) NestFuse; (**b5**) RFN-Nest; (**b6**) PAPCNN; (**b7**) ReLP; (**b8**) TIF; (**b9**) CDDFuse; (**b10**) CIRF.

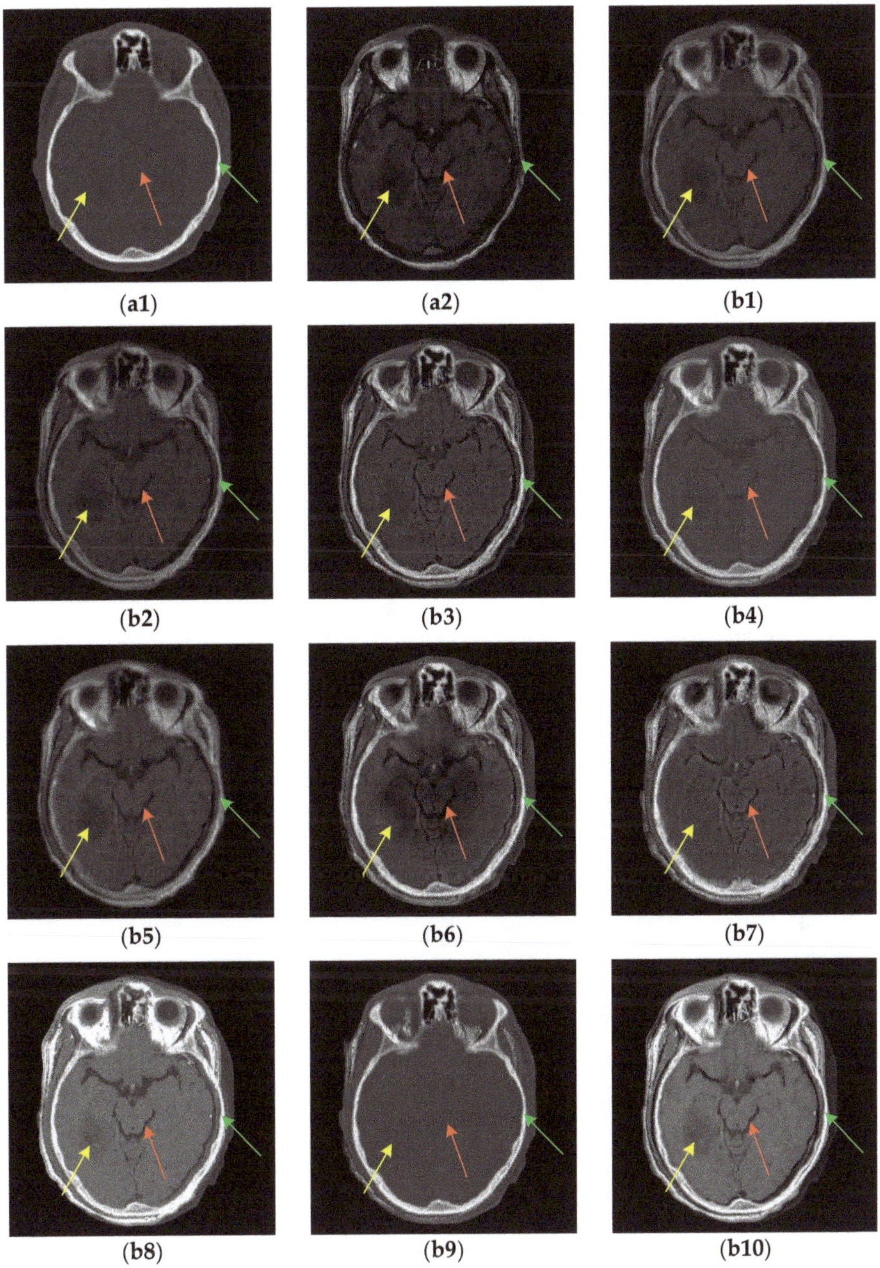

Figure 9. The results of all algorithms on the CT-MR image pairs from RIRE dataset. (**a1**) Source CT image; (**a2**) Source MR image; (**b1**) U2Fusion; (**b2**) DenseFuse; (**b3**) IFCNN; (**b4**) NestFuse; (**b5**) RFN-Nest; (**b6**) PAPCNN; (**b7**) ReLP; (**b8**) TIF; (**b9**) CDDFuse; (**b10**) CIRF.

7. Quantitative Evaluation

To quantitatively evaluate the fusion performance datasets, we have computed eight metrics of ten algorithms on three datasets. Table 11 lists the mean and deviation of each algorithm's values across all datasets, where the deviation refers to the dispersion of all values of each metric from their mean [39].

Table 11. The results of all computed algorithms on three datasets.

Methods	SD	$PSNR$	SCD	MI	$SSIM$	$Q^{AB/F}$	$VIFF$	$\lvert rSFe \rvert$
CIRF	83.566 ± 11.21	25.520 ± 14.29	1.817 ± 0.06	1.628 ± 0.72	0.769 ± 0.30	0.585 ± 0.16	0.683 ± 0.30	0.179 ± 0.13
CDDFuse	71.144 ± 20.81	15.874 ± 10.47	1.412 ± 0.35	1.806 ± 0.61	0.837 ± 0.25	0.631 ± 0.14	0.593 ± 0.32	0.153 ± 0.12
U2Fusion	57.930 ± 27.62	3.913 ± 3.11	1.278 ± 0.13	1.677 ± 0.59	0.718 ± 0.31	0.488 ± 0.30	0.507 ± 0.23	0.389 ± 0.21
DenseFuse	53.216 ± 9.05	3.313 ± 3.27	1.339 ± 0.10	1.891 ± 0.60	0.772 ± 0.12	0.485 ± 0.32	0.499 ± 0.24	0.423 ± 0.23
IFCNN	61.115 ± 15.42	7.147 ± 5.89	1.363 ± 0.12	1.698 ± 0.61	0.857 ± 0.08	0.617 ± 0.19	0.565 ± 0.26	0.146 ± 0.29
NestFuse	66.902 ± 10.90	14.424 ± 8.09	1.480 ± 0.11	1.955 ± 0.68	0.857 ± 0.17	0.630 ± 0.16	0.599 ± 0.25	0.180 ± 0.16
RFN-Nest	63.489 ± 10.25	8.350 ± 7.44	1.558 ± 0.20	1.625 ± 0.53	0.700 ± 0.30	0.424 ± 0.26	0.550 ± 0.26	0.479 ± 0.24
PAPCNN	66.375 ± 19.16	12.104 ± 8.74	1.407 ± 0.14	1.588 ± 0.89	0.776 ± 0.14	0.542 ± 0.25	0.543 ± 0.31	0.251 ± 0.35
ReLP	70.915 ± 16.98	19.235 ± 8.04	1.338 ± 0.23	1.827 ± 0.81	0.869 ± 0.05	0.603 ± 0.14	0.589 ± 0.21	0.157 ± 0.08
TIF	84.423 ± 19.47	43.666 ± 24.94	1.708 ± 0.28	1.796 ± 0.58	0.781 ± 0.08	0.524 ± 0.13	0.672 ± 0.37	0.165 ± 0.11

As can be seen from Table 11, CIRF has significant advantages over all other algorithms in terms of SD, $PSNR$, $VIFF$ and SCD. Meanwhile, CIRF can provide relatively close $Q^{AB/F}$ and $|rSFe|$ to CDDFuse.

Figure 10 shows the values of eight metrics for various algorithms on five kinds of multi-modal image pairs in three datasets. Overall, CIRF outperforms all other algorithms in terms of SCD and its $VIFF$ achieves the highest value across the datasets except for IXI and RIRE. In addition, CIRF is only outperformed by TIF in $PSNR$ and SD. Furthermore, CIRF achieves the most competitive $|rSFe|$ on the IXI dataset, and provides comparable $|rSFe|$ to CDDFuse on other datasets.

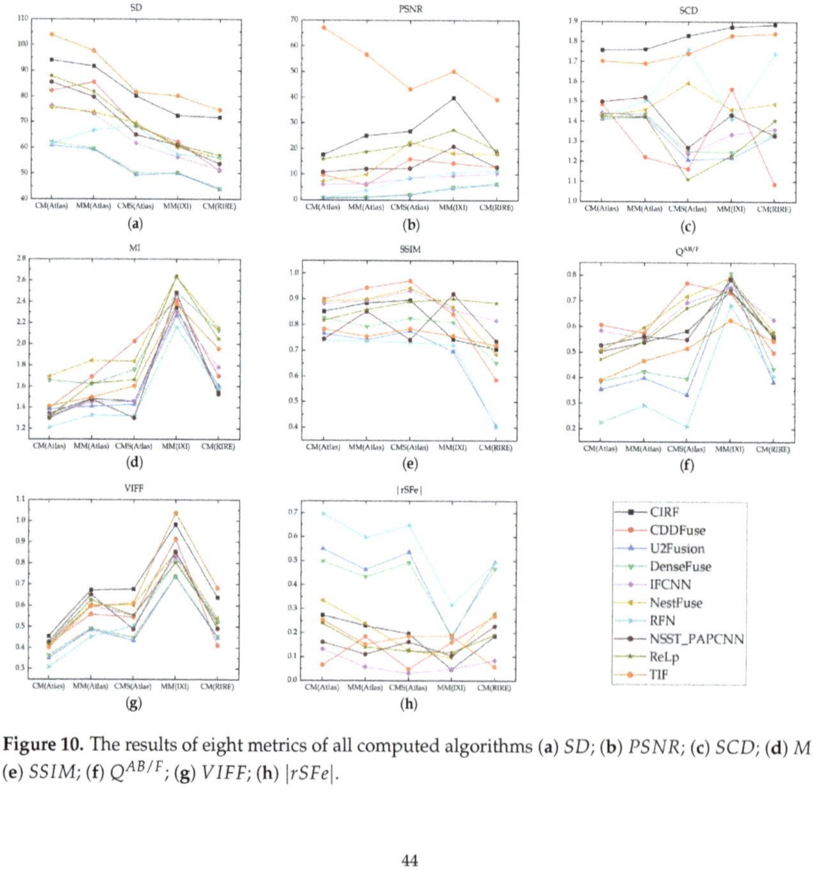

Figure 10. The results of eight metrics of all computed algorithms (**a**) SD; (**b**) $PSNR$; (**c**) SCD; (**d**) MI; (**e**) $SSIM$; (**f**) $Q^{AB/F}$; (**g**) $VIFF$; (**h**) $|rSFe|$.

8. Conclusions

This paper has come up with a coupled reconstruction and fusion network for multi-modal medical image fusion. On the one hand, this architecture parallels the reconstruction branch and the fusion branch which are linked by a shared encoder, thereby reducing error accumulation and improving the network's feature extraction ability by multi-task learning. On the other hand, we have further constructed a feature decomposition network using parallel ViT and CNN modules to fuse base and detail features separately, while adding complementary links of high/low frequency information.

Experiments on three datasets demonstrate that our methods perform better than several typical traditional and DL-based image fusion algorithms in terms of eight fusion metrics and qualitative evaluations. Specifically, on multi-modal MR image fusion, our method produces fused images with excellent retention of bright details and smooth edge transition. For CT-MR image fusion, CIRF provides higher image contrast and better preservation of detail features from the original images. On SPECT-CT/MR image fusion, the fused images generated by CIRF are smoother while still retaining significant edge information. Furthermore, our method also exhibits strong generalization capability. In the future, we hope to extend our method to 3D medical image fusion.

Author Contributions: Conceptualization, X.Z. and J.Z.; methodology, J.Z. and X.Z.; software, Y.W. and J.Z.; validation, Y.W., J.X. and J.Z.; formal analysis, J.X. and J.Z.; investigation, J.Z.; resources, X.Z.; data curation, J.X. and J.Z.; writing—original draft preparation, J.Z., J.X. and Y.W.; writing—review and editing, X.Z., J.Z., J.X. and Y.W.; visualization, J.Z.; supervision, X.Z. and J.Z.; project administration, X.Z. and J.Z.; funding acquisition, X.Z. All authors have read and agreed to the published version of the manuscript.

Funding: This work was supported by the National Key Research and Development Program of China under Grant No. 2018YFE0206900, the Open Project of Key Laboratory for Quality Evaluation of Ultrasound Surgical Equipment of National Medical Products Administration (Grant No: SMDTKL-2023-1-01).

Institutional Review Board Statement: This study used public data instead of private data.

Informed Consent Statement: Not applicable.

Data Availability Statement: This study used open access online datasets [9,46,47].

Conflicts of Interest: The authors declare no conflicts of interest.

References

1. Brown, M.; Semelka, R.; Nishino, T.K. MRI: Basic principles and applications. *Med. Phys.* **1995**, *31*, 170. [CrossRef]
2. Goldman, L.W. Principles of CT and CT technology. *J. Nucl. Med. Technol.* **2007**, *35*, 115–128. [CrossRef] [PubMed]
3. Townsend, D.W.; Carney, J.P.J.; Yap, J.T.; Hall, N.C. PET/CT today and tomorrow. *J. Nucl. Med.* **2004**, *45* (Suppl. S1), 4S–14S. [PubMed]
4. Mariani, G.; Bruselli, L.; Kuwert, T.; Kim, E.E.; Flotats, A.; Israel, O.; Dondi, M.; Watanabe, N. A review on the clinical uses of SPECT/CT. *Eur. J. Nucl. Med. Mol. Imaging* **2010**, *37*, 1959–1985. [CrossRef] [PubMed]
5. Du, J.; Li, W.; Lu, K.; Xiao, B. An overview of multi-modal medical image fusion. *Neurocomputing* **2016**, *215*, 3–20. [CrossRef]
6. Wang, Z.; Cuia, Z.; Zhu, Y. Multi-modal medical image fusion by Laplacian pyramid and adaptive sparse representation. *Comput. Biol. Med.* **2020**, *123*, 103823. [CrossRef] [PubMed]
7. Li, Y.; Zhao, J.; Lv, Z.; Li, J. Medical image fusion method by deep learning. *Int. J. Cogn. Comput. Eng.* **2021**, *2*, 21–29. [CrossRef]
8. Zhao, Z.; Bai, H.; Zhang, J.; Zhang, Y.; Xu, S.; Lin, Z.; Timofte, R.; Van Gool, L. Cddfuse: Correlation-driven dual-branch feature decomposition for multi-modality image fusion. In Proceedings of the IEEE/CVF Conference on Computer Vision and Pattern Recognition, Vancouver, BC, Canada, 17–24 June 2023; pp. 5906–5916.
9. RIRE. Available online: https://rire.insight-journal.org/ (accessed on 25 February 2024).
10. Zhang, H.; Xu, H.; Tian, X.; Jiang, J.; Ma, J. Image fusion meets deep learning: A survey and perspective. *Inf. Fusion* **2021**, *76*, 323–336. [CrossRef]
11. Du, J.; Fang, M.; Yu, Y.; Lu, G. An adaptive two-scale biomedical image fusion method with statistical comparisons. *Comput. Methods Programs Biomed.* **2020**, *196*, 105603. [CrossRef]
12. Huang, D.Y.; Wang, C.H. Optimal multi-level thresholding using a two-stage Otsu optimization approach. *Pattern Recognit. Lett.* **2009**, *30*, 275–284. [CrossRef]

13. Bhandari, A.K.; Kumar, A.; Singh, G.K. Modified artificial bee colony based computationally efficient multilevel thresholding for satellite image segmentation using Kapur's, Otsu and Tsallis functions. *Expert Syst. Appl.* **2015**, *42*, 1573–1601. [CrossRef]
14. Li, S.; Kang, X.; Hu, J. Image fusion with guided filtering. *IEEE Trans. Image Process.* **2013**, *22*, 2864–2875. [PubMed]
15. Kremp, S.; Schaefer-Schuler, A.; Nestle, U.; Sebastian-Welsch, C.; Rube, C.; Kirsch, C. Comparison of CT and CT-PET-fusion based 3D treatment plans in the percutaneous radiotherapy of lung cancer. *Proc. Radiother. Oncol.* **2004**, 73, S447–S448.
16. Yin, M.; Liu, X.; Liu, Y.; Chen, X. Medical image fusion with parameter-adaptive pulse coupled neural network in nonsubsampled shearlet transform domain. *IEEE Trans. Instrum. Meas.* **2018**, *68*, 49–64. [CrossRef]
17. Gao, G.; Xu, L.; Feng, D. Multi-focus image fusion based on non-subsampled shearlet transform. *IET Image Process.* **2013**, *7*, 633–639.
18. Chen, Y.; Park, S.K.; Ma, Y.; Ala, R. A new automatic parameter setting method of a simplified PCNN for image segmentation. *IEEE Trans. Neural Netw.* **2011**, *22*, 880–892. [CrossRef]
19. Li, X.; Guo, X.; Han, P.; Wang, X.; Li, H.; Luo, T. Laplacian redecomposition for multimodal medical image fusion. *IEEE Trans. Instrum. Meas.* **2020**, *69*, 6880–6890. [CrossRef]
20. Burt, P.J.; Adelson, E.H. The laplacian pyramid as a compact image code. *Readings Comput. Vis.* **1987**, *31*, 671–679.
21. Reddy, G.B.; Kumar, H.P. A review on multimodal medical image fusion. *Int. J. Biomed. Eng. Technol.* **2020**, *34*, 119–132. [CrossRef]
22. Zhang, Y.; Liu, Y.; Sun, P.; Yan, H.; Zhao, X.; Zhang, L. IFCNN: A general image fusion framework based on convolutional neural network. *Inf. Fusion* **2020**, *54*, 99–118. [CrossRef]
23. Xu, H.; Ma, J.; Jiang, J.; Guo, X.; Ling, H. U2Fusion: A unified unsupervised image fusion network. *IEEE Trans. Pattern Anal. Mach. Intell.* **2020**, *44*, 502–518. [CrossRef]
24. Li, H.; Wu, X.J. DenseFuse: A fusion approach to infrared and visible images. *IEEE Trans. Image Process.* **2018**, *28*, 2614–2623. [CrossRef] [PubMed]
25. Li, H.; Wu, X.J.; Durrani, T. NestFuse: An infrared and visible image fusion architecture based on nest connection and spatial/channel attention models. *IEEE Trans. Instrum. Meas.* **2020**, *69*, 9645–9656. [CrossRef]
26. Li, H.; Wu, X.J.; Kittler, J. RFN-Nest: An end-to-end residual fusion network for infrared and visible images. *Inf. Fusion* **2021**, *73*, 72–86. [CrossRef]
27. Vaswani, A.; Shazeer, N.; Parmar, N.; Uszkoreit, J.; Jones, L.; Gomez, A.N.; Kaiser, Ł.; Polosukhin, I. Attention is all you need. In Proceedings of the Advances in Neural Information Processing Systems, Long Beach, CA, USA, 4–9 December 2017; pp. 6000–6010.
28. Dosovitskiy, A.; Beyer, L.; Kolesnikov, A.; Weissenborn, D.; Zhai, X.; Unterthiner, T.; Dehghani, M.; Minderer, M.; Heigold, G.; Gelly, S.; et al. An image is worth 16 × 16 words: Transformers for image recognition at scale. *arXiv* **2020**, arXiv:2010.11929.
29. Tang, W.; He, F.; Liu, Y.; Duan, Y. MATR: Multimodal medical image fusion via multiscale adaptive transformer. *IEEE Trans. Image Process.* **2022**, *31*, 5134–5149. [CrossRef] [PubMed]
30. Zhang, J.; Liu, A.; Wang, D.; Liu, Y.; Wang, Z.J.; Chen, X. Transformer-based end-to-end anatomical and functional image fusion. *IEEE Trans. Instrum. Meas.* **2022**, *71*, 5019711. [CrossRef]
31. Zhou, Q.; Ye, S.; Wen, M.; Huang, Z.; Ding, M.; Zhang, X. Multi-modal medical image fusion based on densely-connected high-resolution CNN and hybrid transformer. *Neural Comput. Appl.* **2022**, *34*, 21741–21761. [CrossRef]
32. Liu, Y.; Zang, Y.; Zhou, D.; Cao, J.; Nie, R.; Hou, R.; Ding, Z.; Mei, J. An Improved Hybrid Network With a Transformer Module for Medical Image Fusion. *IEEE J. Biomed. Health Inform.* **2023**, *27*, 3489–3500. [CrossRef] [PubMed]
33. Wu, Z.; Liu, Z.; Lin, J.; Lin, Y.; Han, S. Lite transformer with long-short range attention. *arXiv* **2020**, arXiv:2004.11886.
34. Zamir, S.W.; Arora, A.; Khan, S.; Hayat, M.; Khan, F.S.; Yang, M.H. Restormer: Efficient transformer for high-resolution image restoration. In Proceedings of the IEEE/CVF Conference on Computer Vision and Pattern Recognition, New Orleans, LA, USA, 18–24 June 2022; pp. 5728–5739.
35. Peng, Z.; Huang, W.; Gu, S.; Xie, L.; Wang, Y.; Jiao, J.; Ye, Q. Conformer: Local features coupling global representations for visual recognition. In Proceedings of the IEEE/CVF International Conference on Computer Vision, Montreal, BC, Canada, 11–17 October 2021; pp. 367–376.
36. Fang, J.; Lin, H.; Chen, X.; Zeng, K. A hybrid network of cnn and transformer for lightweight image super-resolution. In Proceedings of the IEEE/CVF Conference on Computer Vision and Pattern Recognition, New Orleans, LA, USA, 18–24 June 2022; pp. 1103–1112.
37. Clevert, D.A.; Unterthiner, T.; Hochreiter, S. Fast and accurate deep network learning by exponential linear units (elus). *arXiv* **2015**, arXiv:1511.07289.
38. Sandler, M.; Howard, A.; Zhu, M.; Zhmoginov, A.; Chen, L.C. Mobilenetv2: Inverted residuals and linear bottlenecks. In Proceedings of the IEEE Conference on Computer Vision and Pattern Recognition, Salt Lake City, UT, USA, 18–23 June 2018; pp. 4510–4520.
39. Ye, S.; Wang, T.; Ding, M.; Zhang, X. F-DARTS: Foveated differentiable architecture search based multimodal medical image fusion. *IEEE Trans. Med. Imaging* **2023**, *42*, 3348–3361. [CrossRef] [PubMed]
40. He, K.; Chen, X.; Xie, S.; Li, Y.; Dollár, P.; Girshick, R. Masked autoencoders are scalable vision learners. In Proceedings of the IEEE/CVF Conference on Computer Vision and Pattern Recognition, New Orleans, LA, USA, 18–24 June 2022; pp. 16000–16009.
41. Wang, Z.; Bovik, A.C.; Sheikh, H.R.; Simoncelli, E.P. Image quality assessment: From error visibility to structural similarity. *IEEE Trans. Image Process.* **2004**, *13*, 600–612. [CrossRef] [PubMed]

42. Sobel, I.; Feldman, G. A 3 × 3 isotropic gradient operator for image processing. *Talk Stanf. Artif. Proj.* **1968**, *1968*, 271–272.
43. Tang, L.; Yuan, J.; Ma, J. Image fusion in the loop of high-level vision tasks: A semantic-aware real-time infrared and visible image fusion network. *Inf. Fusion* **2022**, *82*, 28–42. [CrossRef]
44. Aslantas, V.; Bendes, E. A new image quality metric for image fusion: The sum of the correlations of differences. *Aeu-Int. J. Electron. Commun.* **2015**, *69*, 1890–1896. [CrossRef]
45. Xydeas, C.S.; Petrovic, V. Objective image fusion performance measure. *Electron. Lett.* **2000**, *36*, 308–309. [CrossRef]
46. Atlas. Available online: https://www.med.harvard.edu/aanlib/ (accessed on 25 February 2024).
47. IXI. Available online: https://brain-development.org/ (accessed on 25 February 2024).
48. Klein, S.; Staring, M.; Murphy, K.; Viergever, M.A.; Pluim, J.P. Elastix: A toolbox for intensity-based medical image registration. *IEEE Trans. Med. Imaging* **2009**, *29*, 196–205. [CrossRef] [PubMed]
49. Shamonin, D.P.; Bron, E.E.; Lelieveldt, B.P.; Smits, M.; Klein, S.; Staring, M.; Initiative, A.D.N. Fast parallel image registration on CPU and GPU for diagnostic classification of Alzheimer's disease. *Front. Neuroinform.* **2014**, *7*, 50. [CrossRef]
50. Zhang, X. Deep learning-based multi-focus image fusion: A survey and a comparative study. *IEEE Trans. Pattern Anal. Mach. Intell.* **2021**, *44*, 4819–4838. [CrossRef]
51. Huang, B.; Yang, F.; Yin, M.; Mo, X.; Zhong, C. A review of multimodal medical image fusion techniques. *Comput. Math. Methods Med.* **2020**, *2020*, 1–16. [CrossRef] [PubMed]
52. Han, Y.; Cai, Y.; Cao, Y.; Xu, X. A new image fusion performance metric based on visual information fidelity. *Inf. Fusion* **2013**, *14*, 127–135. [CrossRef]
53. Zheng, Y.; Essock, E.A.; Hansen, B.C.; Haun, A.M. A new metric based on extended spatial frequency and its application to DWT based fusion algorithms. *Inf. Fusion* **2007**, *8*, 177–192. [CrossRef]

sensors

MDPI

Article

Machine Learning-Empowered Real-Time Acoustic Trapping: An Enabling Technique for Increasing MRI-Guided Microbubble Accumulation

Mengjie Wu [1,*] and Wentao Liao [2]

[1] Department of Mechanical Engineering, The University of Hong Kong, Hong Kong 999077, China
[2] Medical Imaging Center, Shenzhen Hospital of Southern Medical University, Shenzhen 518005, China; liaowentao628@i.smu.edu.cn
* Correspondence: mengjiewu@connect.hku.hk

Abstract: Acoustic trap, using ultrasound interference to ensnare bioparticles, has emerged as a versatile tool for life sciences due to its non-invasive nature. Bolstered by magnetic resonance imaging's advances in sensing acoustic interference and tracking drug carriers (e.g., microbubble), acoustic trap holds promise for increasing MRI-guided microbubbles (MBs) accumulation in target microvessels, improving drug carrier concentration. However, accurate trap generation remains challenging due to complex ultrasound propagation in tissues. Moreover, the MBs' short lifetime demands high computation efficiency for trap position adjustments based on real-time MRI-guided carrier monitoring. To this end, we propose a machine learning-based model to modulate the transducer array. Our model delivers accurate prediction of both time-of-flight (ToF) and pressure amplitude, achieving low average prediction errors for ToF (−0.45 µs to 0.67 µs, with only a few isolated outliers) and amplitude (−0.34% to 1.75%). Compared with the existing methods, our model enables rapid prediction (<10 ms), achieving a four-order of magnitude improvement in computational efficiency. Validation results based on different transducer sizes and penetration depths support the model's adaptability and potential for future ultrasound treatments.

Keywords: acoustic trap; microbubble (MB); machine learning; magnetic resonance imaging (MRI); heterogeneous media

1. Introduction

Harnessing acoustic interference within the human body, ultrasound has expanded its applications beyond diagnostic imaging to therapy. Focused ultrasound (FUS) is a typical non-invasive treatment that relies on acoustic interference to generate a focal spot, successfully inducing evident temperature rise for thermal ablation inside organs [1], such as the brain [2], liver [3], and uterus [4]. It can also vibrate target tissues to stimulate mechanical effects in non-thermal applications, among which FUS-mediated blood–brain barrier opening has advanced to clinical Phase II [5]. In addition, inspired by optical tweezers, FUS has also been employed to capture microparticles [6]. This phenomenon arises from the acoustic radiation force (ARF) exerted by the focused ultrasound waves, which creates a stable aggregate at the focal spot [6]. Apart from enabling waves to be focused at a point, acoustic interference can form a "well"-shaped acoustic pressure zone [7,8]. Within the zone, exterior high pressure is converged towards the interior with the lowest amplitude [9,10]. This pressure gradient results in ARF, such that this zone can highly localize solid objects with a diameter of sub-wavelength (<λ) [11], e.g., cells (Ø1 µm–30 µm [12,13]) and bacteria cluster (Ø45.6 µm–228 µm [14]). Therefore, both the focal beam and "well"-shaped pressure zone are generally referred to as "acoustic trap" [7,9,10]. In the latest decade, some laboratory investigations have showcased ARF is sufficient to withstand object trapping against the low-velocity (e.g., 2.5 mm/s) flow in water tanks [11,15]. More recently, the trap successfully located MBs and thereby increased its aggregation in mouse capillaries, showing its applicability for *in vivo* applications [16]. Accredited to these aforementioned advances, the acoustic trap deserves further exploration for the treatment of vascular-rich solid tumors, e.g., hepatocellular carcinoma (HCC) [11,15]. In current preclinical practice, the FDA-approved ultrasound-mediated MBs (Ø1–10 μm [16]) successfully increased local anti-tumor drug (e.g., Doxil) accumulation because the induced microbubble oscillation increases permeation of membranes and enhances drug diffusivity [17,18]. However, the drag forces generated by high-velocity pulsatile (1–4 cm/s [16]) in feeding microvessels limit the MB retention time, leading to heavy drug leakage and reduced absorption [16]. Although optical tweezer [19–21], dielectrophoresis [22], and electrokinetic tweezer [23] have achieved contactless trapping of biomolecules, optical beams are unsuitable for use in opaque medium, and the current of electrokinetic tweezer and dielectrophoresis can induce heating, thereby affecting cell physiology [16]. By contrast, ultrasound waves can penetrate thick tissues and possess excellent biocompatibility [24]. Therefore, the feasibility of acoustic trap generation within human organs, potentially prolonging the retention of MBs in counterflow [16], is highly desired.

This envisioned medical potential of acoustic trapping is tightly bound to the capability of tracking carriers and sensing ultrasound interference. It not only provides spatial guidance for payload release but also monitors if the trapping beams are maintained on the target zone (Figure 1a) [18]. Currently, magnetic resonance imaging (MRI) has enabled real-time carrier tracking in *in vivo* trials attributed to magnetic resonance (MR) imageable materials [25]. Apart from carrier tracking, MRI can sense ultrasound interference by measuring resulting temperature changes or tissue displacements [26]. Specifically, MR thermometry has achieved real-time heat deposition monitor to sense focal spots in FUS [26,27]. Recently, MR acoustic radiation force imaging (MR-ARFI) emerged as an alternative to MR thermometry, addressing imaging accuracy degradation in tumors surrounded by fat via detecting micro-scale tissue displacement (Figure 1b) [28–30]. These advances show that MRI-guided MB tracking and trap sensing are becoming immensely promising for medical scenarios. Nevertheless, the accurate and rapid trap generation at the target remains a formidable challenge, as the heterogeneous tissues complicate the computation of beam propagation [31]. Specifically, as shown in Figure 1a, ultrasound undergoes numerous refractions between tissue layers (e.g., fat, muscle, and bone) with varying thicknesses, resulting in a highly curved propagation pathway within the body Then, considering that each pathway has its unique pressure attenuation coefficient, the beams emitted by multiple transducers exhibit a non-uniform wavefront when arriving at the target. This would weaken the intensity profiles of acoustic traps. Moreover, variations in body tissue structure would induce phase distortion to the propagating waves, thus decreasing the computation accuracy. Furthermore, the computational efficiency becomes an aggravated challenge when fine adjustment to trap position is required to target a more suitable MB seeding site or microvessel, relying on the MRI-guided MB monitor. The short circulation lifetime of most MBs (typically 3–5 min [32]) underscores the urgency of generating a new trap before MBs rupture.

Tracing back the progress in trap generation, researchers primarily focused on hardware and algorithm design. For hardware, the single-side phased array is increasingly favored [33,34]. Compared to the dual-sided [35,36], closed [37,38], or spiral [39] array arrangement, it eliminates spatial constraint so as to fit the parts of the body [33]. Assisted by cumbersome mechanical actuation, the system can relocate the trap in two dimensions (2D) or three dimensions (3D). Yet, such an approach offers restricted dexterity for fine-tuning trap positions [34,35]. Then, a few algorithms for electronically steering elements' phase and amplitude have been developed, permitting the generation of dynamic traps without mechanical actuation [10,14]. Phase is the most critical variable, determining the trap generation and its position [10]. Amplitude attenuates during propagation due to absorption, refraction, etc., influencing the trap's acoustic intensity profiles [15]. Among the existing algorithms, only three, namely, iterative backpropagation (IB), time reversal-based method (TRM), and holographic acoustic element framework (HAEF), have been applied in heterogeneous media [10,14,40]. IB, introduced in 2019 [41], only enabled trap generation in a mouse's skin-fold chamber model (tissue thickness: <1 mm) and cannot solve the complicated acoustic field [14,42]. Its application to thick tissues, phantoms, or the human body remains unachieved. In contrast, TRM has been repeatedly used in large-volume tissues [43]. It converts phase calculation into computing the beam's propagating time between elements and target in the time domain, termed time-of-flight (ToF). Owing to TRM's inherent capability of correcting phase distortion, it can achieve accurate solutions. Moreover, TRM can compute the amplitudes, thus allowing the algorithms to steer elements' emission pressure to correct the non-uniform wavefront. However, TRM primarily relies on numerical computation and still has low efficiency for ToF and amplitude computation. For instance, Yang et al. used the k-Wave toolbox to compute ToF as beams traversed through a macaque skull (65 mm × 68 mm × 38 mm) [43]. This method usually costs an excessive computational expense (e.g., tens of minutes for per target). Regarding HAEF, it diversifies the types of traps for enhanced maneuverability [40], such as twin, vortex, and bottle traps [10], beyond the conventional focal beam created by IB and TRM. To date, HAEF's iterative phase computation is quite time-consuming and faces challenges in mathematically converging [42]. Additionally, the tissues inevitably induce phase distortion to the beams and weaken the accuracy of phase computation. To seek precision, Cao et al. positioned a hydrophone at the target for phase calibration and produced a twin trap after waves passed through a porcine rib [40], but this device is not suitable for non-invasive treatments. Worse yet, HAEF does not compute the pressure amplitude, which means it cannot correct the non-uniform wavefront. Thereby, the weakened acoustic intensity profiles in tissues cannot be improved by amplitude modulation (AM) [44]. To sum up, these methods still present substantial limitations for medical potential.

To advance the medical potential of acoustic traps for increasing MRI-guided MB accumulation, we propose a future-proof technique that leverages machine learning to deliver trap generation in heterogeneous media. As shown in Figure 1c, the acoustic trap located at the predefined microvessel will accumulate the MBs and counteract the blood pulsatile, thereby potentially prolonging the retention time of MBs. The 2D segmented MR slices contain various tissue materials, such as skin, fat, muscle, liver, and bone, acting as heterogeneous media (Figure 1d). Considering the computation inefficiency of both TRM and HAEF, machine learning is expected to rapidly predict key parameters of beam propagation. Recently, a learning-based model has validated its efficacy and efficiency (~200 ms) for holographical phase modulation (PM) in a single medium [42]. To date, learning-based PM in heterogeneous media has not been studied before. Therefore, the proposed learning-based model is designed to predict beams' ToF and amplitude in 2D heterogeneous media, accommodating multiple trap types (e.g., focal beam and twin trap). The predicted amplitude will be applied to modulate elements' emission pressure, thereby correcting the non-uniform wavefront. To this end, two artificial neural networks (ANN) are used to create the inverse mapping from the given trap position to elements' actuation signals (i.e., ToF and amplitude, respectively). These two ANNs can achieve high computation efficiency (<10 ms per target) much faster than the existing methods (e.g., TRM). Presently, as a pioneering work on learning-empowered trap generation, we collect the training dataset via 2D finite element (FE) modeling. The key work contributions are differentiated as follows:

(1) Development of the first machine learning-empowered model to facilitate the generation of acoustic traps in 2D heterogeneous media. This approach delivers accurate generation of multiple types of traps (e.g., focal beam and twin trap).
(2) High computation efficiency for enabling rapid phase-amplitude (PA) modulation on transducer array. The model can predict time-of-flight (ToF) and pressure amplitude within 10 ms to modulate all elements, which is significantly faster (four orders of magnitude) than the existing method (e.g., TRM).

(3) FE-based validations on the prediction performance using MR images, based on three factors, i.e., sample density, element diameters, and penetration depths, supporting our model's potential for medical applications. FE modeling further demonstrates the capacity of twin traps for trapping microbubbles (negative acoustic contrast factor).

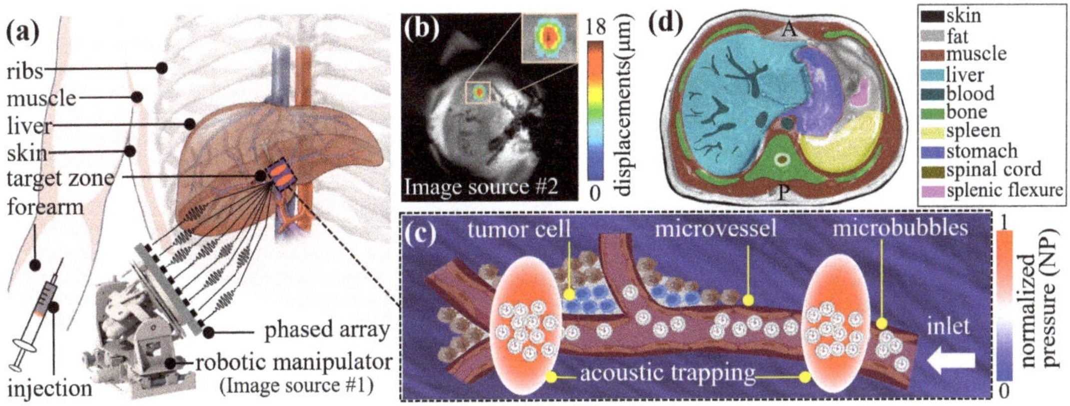

Figure 1. Overview of trapping drug carriers using acoustic trapping. (**a**) Acoustic trap generation within human liver using phased array. MR-compatible robotic manipulator (Image source #1) [27] positions the array towards the liver. The emitted beams pass through tissues (e.g., skin, fat, muscle, and ribs) to generate a trap at the given target zone. MBs are injected via radial artery of the forearm [45]. (**b**) Example of focal spot visualization in *ex vivo* porcine kidney via MR-ARFI (Image source #2 [29]). (**c**) Close-up illustration of MB accumulation in a microvessel due to acoustic trapping. Two finger-like high-pressure (warmer color) regions locate MBs around the tumor cells. (**d**) Segmentation of MR T2 image for FE modeling in both data acquisition and performance validation process. Letters "A" and "P" in black color indicate the anterior and posterior abdominal walls, respectively.

2. Materials and Methods

This section details the learning-based method dedicated to acoustic trap generation. We provide comprehensive information for model architecture, dataset collection, model training, and validation. To facilitate the reproducibility of our work, all designed files are accessible on the GitHub repository (https://github.com/mengjwu/acoustictrap, accessed on 6 September 2024), such as FE modeling files, MR anatomy models, and Python codes.

2.1. Learning-Based Trap Generation Model

An acoustic trap is expected to be formed at a predefined target position; therefore, elements' actuation signals, i.e., phase and amplitude, will have to be modulated rapidly and accurately. By TRM, the phase can be directly converted from ToF (Equations (3) and (4)). Thus, the proposed learning-based model aims to create the inverse mapping from the prescribed trap position to ToF and amplitude. As shown in Figure 2, the trap position and the elements' position are the input variables, denoted as $\boldsymbol{P} \in \mathbb{R}^2$ and $\boldsymbol{P} \in \mathbb{R}^{2n}$, respectively, where n is the element number. Two ANNs with n output nodes were designed to predict ToF ($\boldsymbol{T} \in \mathbb{R}^n$) and amplitude ($\boldsymbol{A} \in \mathbb{R}^n$), respectively. Note that they feature different node layouts in hidden layers. Five hidden layers' node numbers are denoted as $N_1, N_2, \ldots, N_5$.

Upon finalizing the prediction, phase–amplitude modulation to the phased array was performed as illustrated in step 2. $\boldsymbol{T}$ was used for computing the focal beam's phase patterns $\boldsymbol{H} \in \mathbb{R}^n$ (Equation (3)), and $\boldsymbol{A}$ was for computing the amplitude pattern $\boldsymbol{W} \in \mathbb{R}^n$ (Equation (5)). Specifically, a time reversal process was conducted in PM, which converts ToF to phase for generating the phase patterns ($\boldsymbol{H}$) of focal beam [11,43]. Following HAEF [10], $\boldsymbol{H}$ acts as an acoustic lens that focuses beams at target, functioning as the "single-beam trap" to ensnare objects [6,14]. Then, by adding a fixed holographic signature, the focal beam can be transformed into diverse acoustic structures, e.g., vortex, bottle, and twin [10]. For twin trap, the signature (i.e., π-phase difference) can be interpreted as a phase shift of $+\pi/2$ and $-\pi/2$, and they are added to the left and right half of array, respectively, yielding phase patterns of the twin trap (Equation (4)), denoted as $\boldsymbol{M} \in \mathbb{R}^n$. Due to the different attenuation levels across all transmission pathways, the wavefront arriving at the target cannot remain uniform. As a result, the acoustic intensity profiles become weakened, such as the lower focusing quality in single-beam trap. Therefore, we introduced the amplitude patterns $\boldsymbol{W}$, consisting of n coefficients, to modulate element's emission pressure, aiming to recreate a uniform wavefront and optimize the trap's intensity profiles.

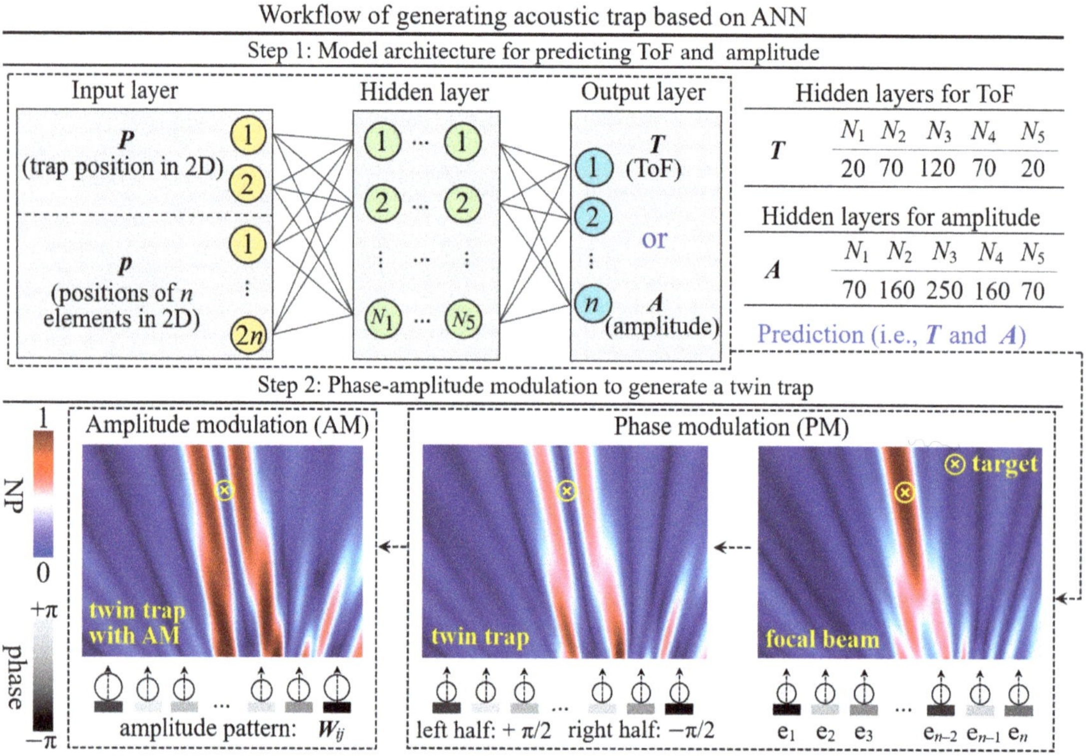

Figure 2. ANN-based workflow for acoustic trap generation. Step 1 illustrates the architecture of a learning-based model for predicting ToF or amplitude. The variable n is the element number. The proposed two ANNs have $2n + 2$ input variables, five hidden layers and n outputs but diverge in hidden layers' node layout (i.e., $N_1, N_2, \ldots, N_5$) for $\boldsymbol{T}$ and $\boldsymbol{A}$ prediction. Step 2 depicts the phase–amplitude modulation process applied to n elements ($e_1, e_2, \ldots, e_n$), including phase modulation (PA) and amplitude modulation (AM). Circles' radii are proportional to the element's emission pressure, and the grey scale of elements represents the phase pattern. A blue–red color bar is used to characterize the normalized pressure (NP) across all acoustic fields.

2.2. FE-Based Dataset Collection

To train the proposed trapping modulation model, dataset will have to be collected, consisting of the target positions, elements' positions, ToF, and amplitude. ToF and amplitude were computed using Pressure Acoustics, Transient module from COMSOL Multiphysics. FE-based experiments have demonstrated considerable accuracy for acoustic trapping, affirming its reliability as a ground truth [46]. To ensure our simulation accuracy, the maximum mesh size was set at 1/8 wavelength [47]. The data collection was conducted based on the time reversal principle, as it can correct the phase and amplitude distortion [44]. The detailed steps were as follows: (1) A point (Øλ/20) was positioned at the target to emit waves, implying that each sample point represents a sound source, as shown in Figure 3a; (2) Elements acted as receivers to capture ToF and amplitudes from their received signals. Given the reciprocity of the beam transmission pathway, elements retransmit waveforms in the time-reversed order of ToF, allowing these waveforms to be refocused at the position of initial sound source [43].

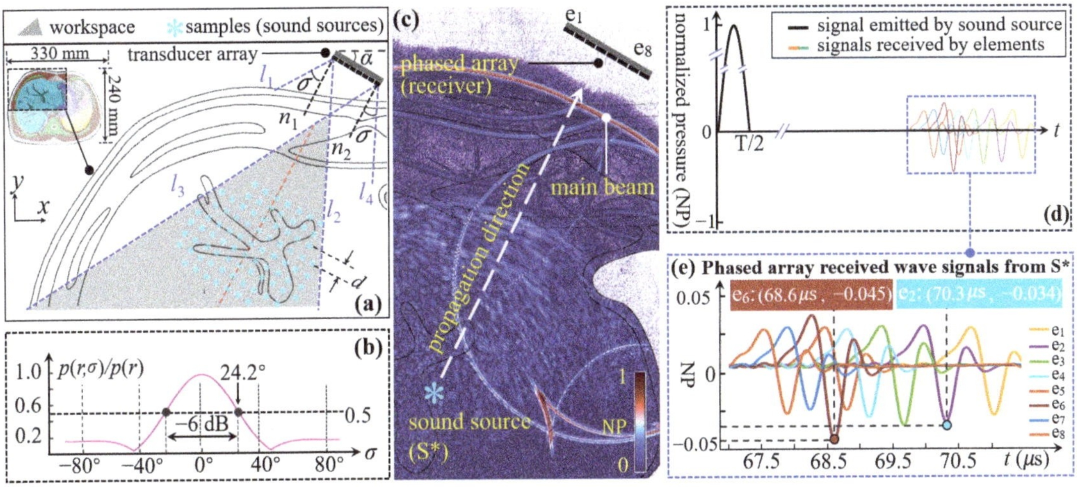

Figure 3. Overview of FE-based data collection process. (**a**) Planning of array position and workspace for data collection. An MR T2 image (330 mm × 240 mm) was segmented to build a 2D geometry model for FE modeling. Boundaries l_2 and l_3 outlined the workspace, and these samples were distributed around the vessel with a spacing of d. (**b**) Pressure attenuation in relation to spread angle (2σ). At the critical angle of σ = 24.2°, the pressure attenuates by half (−6 dB) over the same propagating distance r. (**c**) Simulated acoustic field in COMSOL. Eight elements acted as receivers to capture wave signals from one representative sound source S* (**d**) Signal emitted by sound source S* and received signals by the array. (**e**) Close-up illustration of the received wave signals. Eight peak amplitudes and their timestamps denote, respectively, $\boldsymbol{A}$ and $\boldsymbol{T}$.

We aimed to produce acoustic traps within the liver anatomy; thus, an open-access accessible abdominal MR image [48] was manually segmented to act as the geometry of FE modeling. The array was placed near the subxiphoid to reduce ribs' disturbance (e.g., 10–20 times higher absorption than soft tissues [18]) [49]. In Figure 3a, the predefined targets are situated around a vessel so that the formed trap would be located at its branching microvessels. Each column of samples is parallel to normal line (e.g., n_1). Moreover, referring to parameters outlined in Table 1, our elements have a diameter of 3.7 mm and operate at a frequency of 1.0 MHz. To increase the efficiency of energy delivery to the target region, the array plane should be parallel to the vessel by adjusting angle α (α = 30° in our study). Some MR-compatible robots, embedded with MR-based positional sensing markers, can be used to facilitate the precise adjustment of array pose [27,50,51]. Upon finalizing the array's position and orientation, we outlined the effective workspace using two boundaries l_2 and l_3 for dataset collection. The boundaries were determined by the beam spread angle (2σ), which defines the −6 dB pressure attenuation over an identical propagating distance r, comparing the pressure $p(r, \sigma)$ at the boundary to pressure $p(r)$ along the normal line [52]:

$$p(r, \sigma) = p(r)\frac{2J_1(ke\sin(\sigma))}{ke\sin(\sigma)} \tag{1}$$

where $J_1(ke\sin(\sigma))$ is the Bessel function of first order; k (=$2\pi/\lambda$) is the wave number; e is the element radius. As shown in Figure 3b, the critical attenuation of −6 dB corresponds to the angle of σ= 24.2°. After outlining the workspace, we introduced the sample density (D), defined as the reciprocal of the spacing (d) between two adjacent samples, and conducted the data collection at $D = 1/\lambda$. Figure 3d shows a half cycle of sinusoidal wave emitted by the sound source (S*) and the signals received by eight elements. Figure 3e displays the close-up illustration of the received wave signals, in which eight peak amplitudes and their timestamps represent amplitude and ToF, respectively. All FE modeling was conducted on a processing computer (Ryzen Threadripper 3990X, 32 GB RAM, AMD, USA).

Table 1. Major parameters of single-side phased array in previous works.

Literature	Frequency [MHz]	Wavelength * [mm]	Element Diameter [mm]	Element Gap [mm]	Traversed Media
Kang et al. (2010) [53]	1.00	1.50	5.08 × 5.08	0.51	Water
Ghanem et al. (2020) [34]	1.50	1.00	7.0	0.50	Pig bladder, water
Hu et al. (2021) [11]	1.04	1.44	2.6 × 2.6	0.20	Water
Yang et al. (2022) [43]	1.04	1.44	2.6 × 2.6	0.20	Macaque skull, water

* denotes the acoustic speed in water.

2.3. Model Training

After the data collection, the model needs to be trained to predict the elements' actuation signals so as to generate an acoustic trap at the target. All training and prediction were conducted on a notebook with an Intel i5-7360U CPU and 8 GB RAM. A total of m pairs of samples were acquired, and each pair included $\boldsymbol{P}$, $\boldsymbol{p}$, $\boldsymbol{T}$, and $\boldsymbol{A}$. A total of 74.5%, 13%, and 12.5% of samples were used for model training, validation, and hold-out testing, respectively. We implemented the model using Scikit-learn with L2 loss (L)

$$L = \frac{1}{m}\sum_{i=1}^{m}||\boldsymbol{x}_i - \boldsymbol{x}_i^*||_2 \tag{2}$$

where $\boldsymbol{x}_i$ represents the predicted values and $\boldsymbol{x}_i^*$ represents the labeled values; $i = 1, 2, \ldots,$ m is the index of samples. ReLU was selected as the activation function. The trained model has 18 input variables. Five hidden layers feature a total of 300 nodes for $\boldsymbol{T}$ prediction and 720 nodes for $\boldsymbol{A}$ prediction (Figure 2). Then, the prediction results were applied to phase–amplitude modulation on array.

2.4. FE-Based Trap Visualization

FE modeling is a typical method to validate whether the modulated acoustic traps/HIFU align with the target [15,54]. We used COMSOL Multiphysics to visualize the acoustic field pattern based on the predicted results. Eight elements were activated via phase–amplitude modulation following the four-step process outlined below: (1) computing the maximum ToF ($\boldsymbol{T}_{\max}$) and the maximum amplitude ($\boldsymbol{A}_{\max}$) of the eight beams originating from the same sound source; (2) converting ToF to focal beam's phase patterns based on time-reversed order. The phase patterns are expressed as:

$$\boldsymbol{H}_{ij} = \frac{2\pi(\boldsymbol{T}_{\max} - \boldsymbol{T}_{ij})}{\mathrm{T}} \tag{3}$$

where T (=1 μs in this study) is the period of ultrasound beams; (3) and then adding the holographic signature to form the twin trap's phase patterns. The twin trap's phase patterns are expressed as:

$$\boldsymbol{M}_{ij} = \boldsymbol{H}_{ij} + \frac{\pi}{2} - \pi \cdot \text{Heaviside}(j - \frac{n+1}{2}) \tag{4}$$

(4) The amplitude modulation patterns were calculated ($\boldsymbol{W}$) as follows:

$$\boldsymbol{W}_{ij} \cdot \boldsymbol{A}_{ij} = \boldsymbol{A}_{\max} \tag{5}$$

These coefficients modulate the elements' emission pressure based on their respective attenuation levels along the transmission pathways, allowing all beams to contribute equal pressure to the focal beam. Through the phase–amplitude modulation, the element's actuation signals $\boldsymbol{\Psi}_{ij}$ for focal beam generation are expressed as:

$$\boldsymbol{\Psi}_{ij} = \vartheta \boldsymbol{W}_{ij} \cdot \exp(\mathrm{j} \cdot \boldsymbol{H}_{ij}) \tag{6}$$

where j is the imaginary number and ϑ is a constant. This constant is set at 1000, indicating that the default emission pressure of elements is 1000 pascals. This setting permits modification, enabling linear scaling of the entire field pressure. Then, signals $\boldsymbol{\Psi}_{ij}$ were imported to the Pressure Acoustics, Frequency Domain module from COMSOL to actuate transducer elements.

2.5. FE-Based Microbubbles Trapping

We used numerical simulations to display the MB trapping process in a twin trap. A simple 2D model was set up in COMSOL, which included the trap's ARF, gravity, and drag force of fluid. ARF was produced by eight modulated transducers, and the maximum pressure of the acoustic field was about 340 KPa [55]. Drag force was exerted by a pulsatile flow with a period of one second. The pulsatile' inflow velocity was expressed as signal *Pul(t)*:

$$\begin{aligned} Pul(t) = {} & 13.83 \times vmin + 1.383 \times (vmax - vmin) \times (3\sin(2\pi t)^2 + \mathrm{abs}(\sin(2\pi t)) \times \sin(2\pi t) + \sin(\pi t)^2 + \\ & 42 \times (\sin(2\pi t) + \mathrm{abs}(\sin(2\pi t))) \times \exp(-20 \times (t - \mathrm{round}(t - 0.5))) \times 1.35/(1 + \exp(130t + 10))) \end{aligned} \tag{7}$$

where *vmin* (=0.5 mm/s in our work) is the minimum inflow velocity and *vmax* (=1.54 mm/s in our work) is the maximum inflow velocity. Due to this pulsatile, the maximum flow speed in the microvessel (Ø40 μm [56]) rose up to 2.13 cm/s. The MB is typically filled with gas (e.g., perfluoropropane gas) in the biocompatible shell (e.g., protein) [57]. Thus, the surface tension of MB was set as 0.1 N/m. In addition, Wrede et al. have demonstrated that the ratio (τ) between shell thickness and MB radius affects the ARF direction [6]. When $\tau < 0.10$, the acoustic contrast factor (ACF) of MB is negative, and MB would move towards the high-pressure region (i.e., pressure anti-node). Conversely, MB would move towards the low-pressure regions (i.e., pressure node). In our simulation, MB radius was set to 10 μm, and the shell thickness was 50 nm [57]. Further, the mean density (ρ_{mb}) of hollow MB can be calculated as [6]:

$$\rho_{mb} = \rho_{shell} \times (1 - (1 - \tau)^3) \tag{8}$$

where ρ_{shell} is 1050 kg/m^3 [58].

3. Results and Discussion

Our work presents a machine learning-based model to modulate phase and amplitude patterns of elements capable of rapidly generating an acoustic trap at the predefined target. The model was trained separately using datasets collected at four sample densities, assessing the impact of sample size on model prediction. The resulting performance was analyzed, with details described in the following section. A threshold for sample density

(D) was identified, which trades off the prediction accuracy against the computational expense. At this density, the accurate ToF prediction ensures trap generation at given targets with different penetration depths. Then, the predicted amplitudes were used to modulate the elements' emission pressure, updating the trap's acoustic intensity profiles.

3.1. ToF Prediction

ToF is the most significant parameter, so its prediction accuracy directly determines the trap generation. Considering that the accuracy of the learning-based model is generally affected by the sample size [59], a large number of samples are expected for our model training, but this concurrently demands excessive computing resources to collect samples. Specifically, each FE simulation run, which employs massive geometric meshes to discretize the 2D MR anatomy, takes several hours to collect one sample (about 75 min in our study). Note that the sampling method itself is not the core focus of our work. Both FE modeling and the aforementioned MR-ARFI can serve as sampling techniques. Our primary concern on dataset collection is to identify an optimal sample density to balance the prediction accuracy and sample size. To date, the sample density for learning-empowered trap generation has yet to be delved into. Inspired by the sampling strategy in our previous learning-based FUS work [54], we initiated the dataset collection at a sample density of $D = 1/\lambda$. A total of 920 samples were collected, with 115 samples for hold-out testing. Out of the remaining samples, three subsets consisting of 244, 122, and 65 training samples were selected, corresponding to sparser sample densities of $1/2\lambda$, $1/3\lambda$, and $1/4\lambda$, respectively. These four datasets were used for model training separately. In Figure 4a, at densities of $1/\lambda$, except for a few outliers (only 11 outliers out of 920 predictions), the errors remain within −0.45 μs to 0.67 μs, demonstrating high prediction accuracy. When reducing densities, the maximum errors increased notably by approximately tenfold. Moreover, the presence of numerous outliers indicates large fluctuations in accuracy. Thus, it is evident that a larger sample size indeed raises our model's prediction accuracy. The prediction efficacy based on these four groups will be validated via FE modeling in later sections (see Sections 3.2 and 3.3 for details).

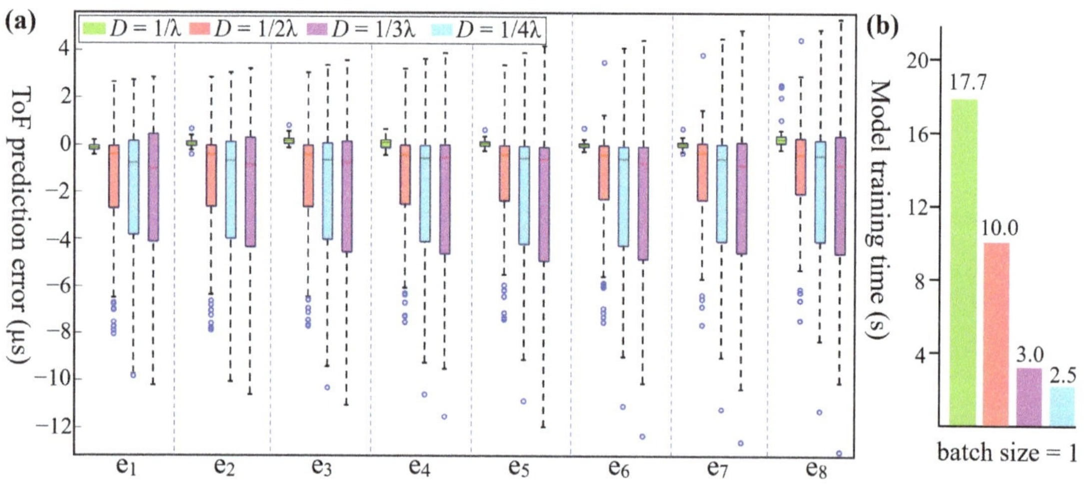

Figure 4. ToF prediction performance of the model trained separately using datasets with four sample sizes. (**a**) Prediction errors of eight elements (i.e., e_1, e_2, ..., e_8). The higher sample density (D) corresponds to the larger sample size. When $D = 1/\lambda$, the errors remain range from −0.45 μs to 0.67 μs, with few outliers. Below this density, the errors surge, and many outliers occur. (**b**) Average model training time over 10 runs using datasets with four sample sizes.

In addition to accuracy, model training time also needs to be evaluated. In future clinical practice, the training process must be completed quickly once the data collection is finished using MR-ARFI. This is because patients cannot remain still for long periods to avoid position shifts between the array and body. In the training process, a batch size of 1 was used for the above four training datasets. Figure 4b shows the average model training times over 10 runs, which were 17.7 s, 10.0 s, 3.0 s, and 2.5 s, respectively. This indicates that the proposed model for ToF prediction can deliver rapid model training, offering significant potential for clinical practice.

3.2. Phase-Only Modulation for Focal Beam

Since correct PM can ensure trap generation [10,43], we used the predicted ToF to compute phase patterns and input them into FE modeling to visualize the acoustic field. This approach can examine if the ToF accuracy is adequate for trap generation at the target. Given the widespread application of single-beam traps to concentrate MBs [6], we initially validated the generation of a focal beam via PM. This modulation used eight elements, each with a representative diameter of 3.7 mm, referring to the custom-made phased arrays listed in Table 1. This kind of small element is increasingly preferred for recent lab-based research, attributed to their flexible sonication delivery, improved trapping stiffness, and high trapping resolution [18,41]. In contrast, commercial transducers usually have larger diameters due to manufacturing processes, production costs, etc. To investigate our model's adaptability for the commercial transducer, PM was repeatedly performed using eight elements with a diameter of 7 mm, paralleling the specifications of transducer PK4GA7P1 (Thorlabs Inc., Newton, NJ, USA).

Figure 5a shows four acoustic fields formed on the same 2D abdominal anatomy. The warmer colors indicate the higher pressures. At sample densities of $1/\lambda$, a finger-like cylindrical region was successfully generated along the propagation direction and covered the target position, signifying a focal beam. At densities of $1/2\lambda$, the focal beam was still generated but drifted from the target position, indicating that ToF prediction accuracy becomes unreliable. Notably, the unwanted side lobes occur in both focal beams, which potentially induce unnecessary tissue injury [54]. Unlike the heat deposition within the focal spot of FUS, the acoustic energy of a focal beam distributes throughout a finger-like region, effectively moderating the temperature rise. Thus, the side effects of lobes in trapping applications are not as severe as FUS. An *in vivo* study on acoustic traps used low-duty cycle (e.g., 1–10%) actuation signals to regulate the energy delivery, showing commendable safety [14]. As the density reduces to $1/3\lambda$, two off-target focal beams emerge, and at $1/4\lambda$, the focal beam fully disappears. Figure 5b shows four acoustic fields using eight larger elements (Ø7 mm). Another representative point, with deeper penetration depth inside the liver, was predefined as the new target. Consistent with the trends observed in Figure 5a, PM again enabled an accurate generation of the focal beam, as sample density was not below $1/\lambda$. Moreover, the side lobes were also observed. Given the above comparison based on different element sizes and penetration depths, the threshold ($1/\lambda$) can be a precedent for dataset collection in subsequent studies, reaching a balance between the prediction accuracy and sample size. In addition, the results demonstrate that our model can adapt to both custom-made and commercial elements, indicating its practical applicability.

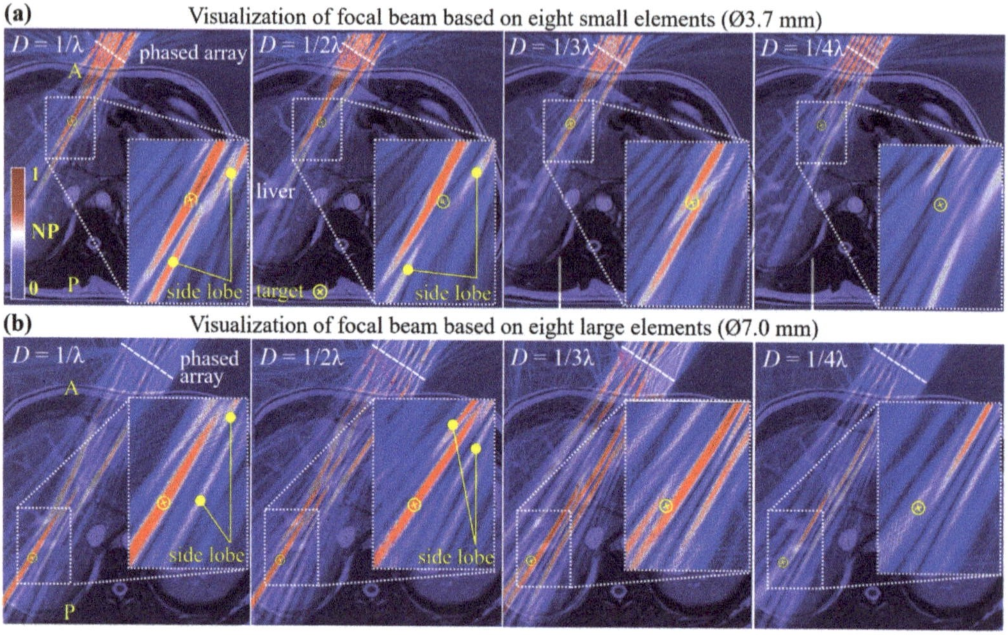

Figure 5. Visualization of focal beam based on two phased arrays with different apertures. The size of the (**a**) small elements is 3.7 mm, and the (**b**) large element is 7.0 mm. When D was not less than $1/\lambda$, both kinds of phased arrays formed a focal beam at the given position. Letters "A" and "P" in yellow color indicate the anterior and posterior abdominal walls, respectively.

3.3. Phase-Only Modulation for Twin Trap

Apart from the focal beam acting as a trap, vortex-, bottle-, and twin-shaped acoustic structures also function effectively as acoustic traps. Based on HAEF, these traps can be transformed from a focal beam [10]. Among them, the vortex trap has been successfully applied to concentrate MBs within a mouse's microvessels [14]. Considering the stronger pulsatility and larger tumor volume in humans, our work focuses on a twin trap, which offers superior lateral force and a bigger working volume compared to a vortex trap [10,15]. A twin trap is interpreted as the combination of two focal beams generated at two designated control points, typically spaced 2λ apart (Figure 6a) [11]. This acoustic structure suggests that MBs with negative ACF will be aggregated in the two focal beams (see Section 3.6 for details). Specifically, MBs will be confined to form two clusters within two focal beams. Once the clusters attain a certain size, the Stokes force they experience exceeds ARF, and they escape the focal beams and then disintegrate into smaller clusters or individual MB [16]. We believe that two microbubble seeding sites created by the twin trap around the tumor potentially increase both the absorption rate and speed compared to a single seeding site facilitated by a single-beam trap. Therefore, in this section, we engage in exploring the generation of twin traps via PM. All acoustic fields were visualized via FE modeling, taking into account three factors: sample sizes, element sizes, and penetration depth.

In Figure 6a, at sample densities of $1/\lambda$, two finger-like cylindrical regions signify high-pressure zones, which shape a twin trap with the low-pressure region in between. Notably, at $1/\lambda$, despite that the twin-shaped acoustic structure is complete, the pronounced amplitude difference between the two focal beams is evident. This phenomenon has not been presented in a single medium [11,15]. It suggests that tissues' heterogeneity leads to a non-uniform wavefront, thereby making elements contribute unequal pressure to the dual focal beams. As the density decreased, two finger-like regions around the target contracted substantially and eventually vanished. The fields depicted in Figure 6b reaffirm our model's efficacy for large-sized transducers and deeper penetration. The sample density of 1λ remained a reliable threshold for twin trap generation. At 1/2λ and 1/3λ, while two focal beams still persisted, they increasingly deviated from the target position, and the acoustic intensity of side lobes turned stronger than the twin trap. Note that these two elongated finger-like focal beams, which almost traversed the liver, will not block red blood cells, thus not causing potential hemolytic lysis. The previous experiment has proven that the ARF exerted on red blood cells is far less than the forces exerted on MBs [16].

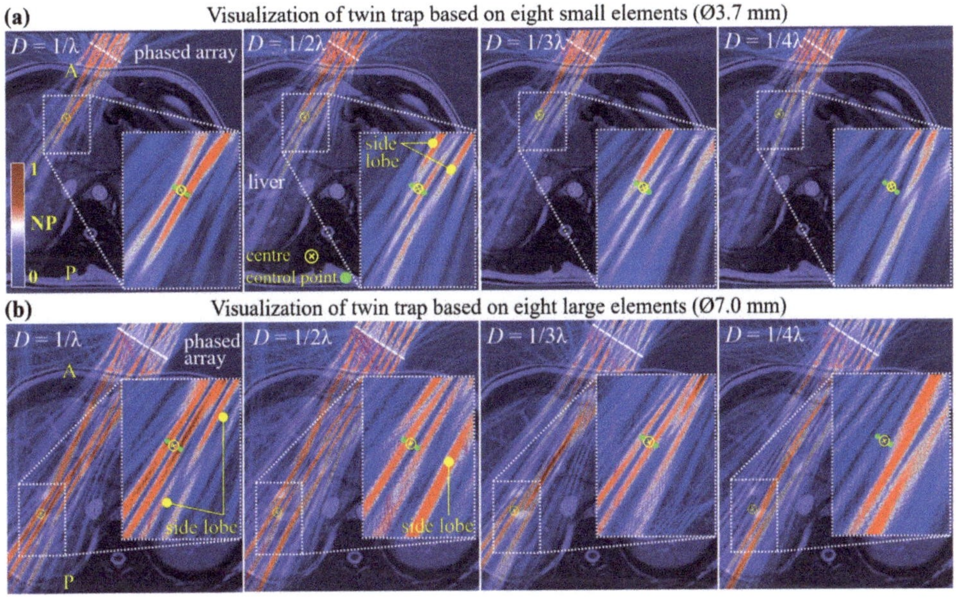

Figure 6. Visualization of twin traps based on two-phased arrays with different apertures. The size of the (**a**) small elements was 3.7 mm, and the (**b**) large elements was 7.0 mm. When D was not less than $1/\lambda$, both arrays formed twin traps at the given positions. A pair of solid green circles represented twin trap's two control points.

3.4. Phase–Amplitude Modulation

Previous works predominantly focused on PA [9–11], with limited effort dedicated to AM. Preliminary attempts have been made to alter the trap's acoustic intensity profile by equally scaling elements' emission pressure in a single-medium environment [40]. This method is easy, as it does not require prior knowledge of wave attenuation in beam transmission channels. However, in heterogeneous media, such simplistic AM becomes inapplicable due to different attenuation across all channels. The works in FUS successfully realized an amplitude compensation for attenuating effects and recreated a uniform acoustic wavefront [43,44]. This AM is acknowledged for reducing sidelobes and refining the quality of the focus [44]. Drawing inspiration from these works, we applied this AM strategy in our learning-empowered traps.

Figure 7a shows that the average errors of 115 runs of amplitude prediction stay below ±2% at $1/\lambda$. This indicates the training dataset, collected at or above threshold density (i.e., $D \geq 1/\lambda$), ensures high precision for amplitude prediction. Note that the prediction has a few outliers, which potentially lead to inaccurate amplitude modulation on certain element(s). The cases for three sparser sample densities (i.e., $1/2\lambda$, $1/3\lambda$, and $1/4\lambda$) were not further discussed. Figure 7b shows the average training time over 10 runs. Along with the above model training for phase prediction, our model can train two ANNs within 2 min. Figure 7c illustrates the focal beams formed without and with AM at the same representative target, where the phase-only modulation served as a baseline. Two focal beams exhibited notable differences, wherein the latter showed warmer colors. We plotted their beam profiles in a lateral direction across a 3.5λ span. The stacked figures reflected changes in normalized pressure when three randomly selected representative elements (i.e., e_1, e_5, and e_6) were activated sequentially alongside the constant activation of e_2, e_3, e_4, e_7, and e_8. This method can verify if the AM can recreate a uniform wavefront. In the baseline, the increases in normalized pressure were different: 0.13 by e_1, 0.11 by e_5, and 0.09 by e_6. In comparison, AM made the increases uniform, all increases (i.e., 0.13, 0.12, 0.11) closely approximated the ideal value of 0.125 (=1/8). These values proved that AM could create a uniform wavefront in heterogeneous media. As a result, their full-width-at-half-maximum of the acoustic intensity profile at the target decreased from 4.59 mm to 4.04 mm. This improvement indicates a higher focusing quality, which is consistent with the results observed in FUS [44]. However, we did not find a reduction in sidelobes. The possible reason may be attributed to the different arrangement of elements, where the FUS uses elements arranged on a spherical surface [44,54]. Regarding the twin trap, each element contributed different pressure to two control points rather than to a single focal spot. Therefore, while the current AM can still be used to change the twin trap's acoustic intensity profile (as shown in Figure 2), it is unable to recreate two uniform wavefronts at both control points.

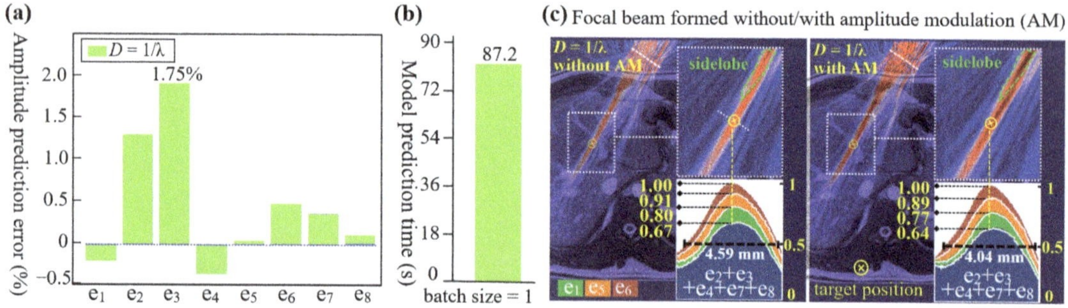

Figure 7. Performance evaluation on focal beam formed without and with AM. (**a**) Average prediction errors of eight elements at the validated sample densities (i.e., 1/λ). Their maximum errors were less than 1.75%. (**b**) Average model training time over 10 runs. (**c**) Comparison of two focal beams formed without and with AM. The left one formed without AM served as a baseline. Two stacked figures showed the normalized pressure changes at target's lateral direction across a 3.5λ span, as elements were activated in sequential order. The bottom blue layer represents the beam profile formed when five elements (i.e., e_2, e_3, e_4, e_7, and e_8) were simultaneously activated. The turquoise, orange, and red layers represent the beam profiles of e_1, e_5, and e_6, respectively. Their full-width-at-half-maximum of the acoustic intensity profile at the target was 4.59 mm and 4.04 mm, respectively.

3.5. Computation Efficiency

Since the medical potential of acoustic traps necessitates prompt phase–amplitude modulation, our model's computational efficiency needs to be evaluated. Among the existing approaches, IB currently cannot solve complicated acoustic fields in thick tissues [14,42]. In addition, HAEF has low accuracy in heterogeneous media and needs a hydrophone to calibrate the phase, making it unsuitable for non-invasive scenarios [40]. Therefore, we compared the computation efficiency between the proposed learning-based model and TRM. The time for 10 runs was averaged for evaluation, and each run included 115 testing samples. Note that we only compared the model trained using the dataset collected at a sampling density of $D = 1/\lambda$. The TRM computation was conducted in COMSOL, and its maximum mesh size was set at 1/8 wavelength. The scripts of the TRM-based model were run on a workstation (AMD 3990X, 32 GB RAM) without using GPU acceleration.

As shown in Table 2, the machine learning-based model is much more efficient in computation, differing by four orders of magnitude compared to TRM. Our model can update the phase–amplitude modulation in real-time (<10 ms), indicating the potential of acoustic traps in future clinical practice.

Table 2. Computation efficiency comparison.

Approach	Average Time for ToF [ms]	Average Time for Amplitude [ms]
Learning-based model	3.5	5.7
TRM	1.5 × e5	

3.6. Microbubble Trapping Capacity

Focal beam has been validated to be capable of trapping hollow MBs with negative ACF at the pressure anti-nodes [6]. Regarding twin trap, the current works only demonstrated it can trap solid particles (ACF > 0) between two focal beams [10,11]. However, the effect on hollow MB with negative ACF has not been investigated. In this section, FE modeling was built to display the process of MB trapping using a twin trap. The trajectory of each MB was shown dynamically. As shown in Figure 8a, it shows the curve of inflow velocity. Its period is 1 s, which was applied to the inlet of microvessel. In the straight microvessel (Ø40 μm), the flow speed can rise to 2.13 cm/s (Video S1, see Supplementary Materials), matching with the mean velocity of blood flow in capillary [16]. In Figure 8b, it shows the part of acoustic field pattern around the microvessel. Its maximum pressure is about 340 KPa, which is a typical acoustic pressure level in *in vivo* study [55]. After the twin trap was generated, the MBs experienced ARF. As shown in Figure 8c, at four different time points (i.e., 0 s, 0.08 s, 0.12 s, and 0.18 s), the MBs gradually moved towards adjacent pressure anti-nodes. Note that the MB diameter in Figure 8c was enlarged by a scale factor of 12 after completing the computation process only for clear demonstration. This result proves that twin trap can accumulate MBs (ACF < 0) in the encounter flow condition (Video S2, see Supplementary Materials). Compared with the single accumulation spot produced by focal beam, the twin trap can provide two spots for MB accumulation.

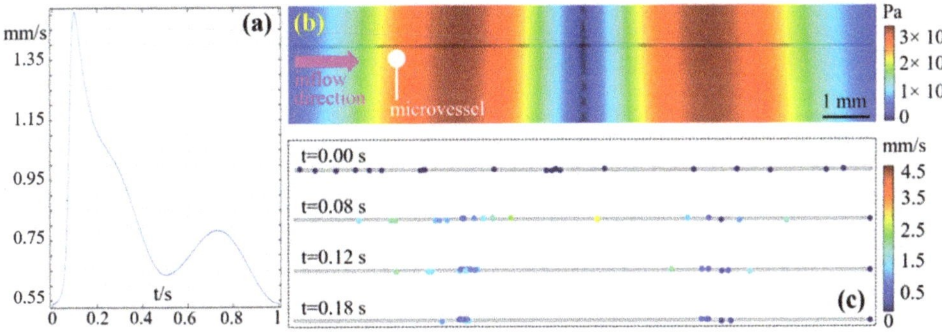

Figure 8. Numerical simulation on the microbubble trapping in pulsatile flow. (**a**) Curve of inflow velocity. Its period is 1 s, and the maximum velocity is up to 1.54 mm/s. It was used to drive the fluid inside microvessel. (**b**) Acoustic field pattern of the twin traps around the predefined microvessel. The maximum pressure is about 340 KPa. The microvessel diameter is 40 μm, and the fluid enters from the left side and flows to the right. (**c**) MB transient distribution at t = 0 s, 0.08 s, 0.12 s, and 0.18 s in COMSOL simulation. At t = 0 s, all MBs do not experience ARF and move due to fluid dynamics. After triggering the twin trap, the MBs (ACF < 0) gradually accumulated in two highest pressure spots.

4. Conclusions

In this study, we propose a pre-concept that uses machine learning to rapidly generate acoustic traps in 2D abdominal anatomy. It aims to advance the trap's medical potential of concentrating drug carriers (e.g., microbubbles) in pulsatile flows. Accredited to the advances in MRI for sensing acoustic interference, the acoustic trap would be visualized using MRI. In our work, we currently use FE modeling to visualize the trap's acoustic field pattern. By mapping from trap position to array actuation signals, two ANNs can be trained rapidly (<2 min). This would reduce the patient waiting time and prevent relative movement between body and array in future medical scenarios. After model training, the results suggest that a sampling density of $D = 1/\lambda$ can act as a reliable reference to balance the sampling time and prediction accuracy. The achieved high ToF prediction accuracy, with error margins ranging from −0.45 µs to 0.67 µs (with only a few isolated outliers), will ensure that the trap position aligns with the predefined microvessel. Meanwhile, the real-time MRI-guided MB tracking will provide visual feedback to surgeons in future medical scenarios, allowing surgeons to select a more suitable microvessel on-site. Our model's high efficiency in prediction satisfies the rapid adjustment to the trap position. It delivers fast updates to phase modulation in real-time. Such computation efficiency far surpasses that of HAEF and TRM. The predicted amplitude can be used to modulate the transducers' emission pressure, thereby optimizing the acoustic intensity profiles of traps. The AM results demonstrated an improvement in the focusing quality of the single-beam trap but did not show a reduction in the sidelobes. Furthermore, the trap generation has been successfully replicated when changing the transducer sizes and target penetration depth, which indicates our model's applicability for future clinical practice.

Our study demonstrates that machine learning can deliver accurate and rapid prediction for ToF and amplitude, enabling the modulation of acoustic trap patterns. However, this pioneering work is only realized in 2D anatomy, and it currently falls short in real clinical applicability. Future efforts in acoustic traps will focus on 3D media, including the dataset collection and model training in 3D, to enhance its practical operability. In the data collection process, we currently use MR anatomy to simulate heterogeneous media, which is insufficient to reflect the complexity of human tissues. Such datasets will weaken the model's accuracy and reliability. MR-ARFI is likely to be a promising method for data collection. It would visualize the trap's acoustic field and measure its amplitude in phantoms or the human body so that the mapping between traps and array elements (i.e., phase and amplitude) can be created without requiring prior knowledge of tissue structures. Moreover, MR-ARFI can also consider human physiological activities (e.g., respiration), thereby enriching the dataset's diversity to improve the trained model's robustness.

Supplementary Materials: The following are available online at https://www.mdpi.com/article/10.3390/s24196342/s1, Video S1: Simulated fluid flow velocity in a microvessel (Ø40 µm) using COMSOL. Video S2: Microbubbles (with negative ACF) trapping process in a microvessel (Ø40 µm) via a twin trap.

Author Contributions: Conceptualization, M.W.; methodology, M.W. and W.L.; software, M.W. and W.L.; validation, M.W.; formal analysis, M.W.; investigation, M.W.; resources, M.W.; data curation, M.W.; writing—original draft preparation, M.W.; writing—review and editing, M.W.; visualization, M.W.; supervision, M.W.; project administration, M.W. All authors have read and agreed to the published version of the manuscript.

Funding: This research received no external funding.

Informed Consent Statement: Not applicable.

Data Availability Statement: Our designed files are accessible on the GitHub repository (https://github.com/mengjwu/acoustictrap, accessed on 6 September 2024), such as FE modeling files, anatomy models, and Python codes.

Conflicts of Interest: The authors declare no conflicts of interest.

References

1. Graham, S.J.; Bronskill, M.J.; Henkelman, R.M. Time and temperature dependence of MR parameters during thermal coagulation of ex vivo rabbit muscle. *Magn. Reson. Med.* **1998**, *39*, 198–203. [CrossRef] [PubMed]
2. Jagannathan, J.; Sanghvi, N.K.; Crum, L.A.; Yen, C.-P.; Medel, R.; Dumont, A.S.; Sheehan, J.P.; Steiner, L.; Jolesz, F.; Kassell, N.F. High intensity focused ultrasound surgery (HIFU) of the brain: A historical perspective: With modern applications. *Neurosurgery* **2009**, *64*, 201–210. [CrossRef] [PubMed]
3. Wijlemans, J.; Bartels, L.; Deckers, R.; Ries, M.; Mali, W.; Moonen, C.; van den Bosch, M.A. Magnetic resonance-guided high-intensity focused ultrasound (MR-HIFU) ablation of liver tumours. *Cancer Imaging* **2012**, *12*, 387–394. [CrossRef]
4. Liu, L.; Wang, T.; Lei, B. High-intensity focused ultrasound (HIFU) ablation versus surgical interventions for the treatment of symptomatic uterine fibroids: A meta-analysis. *Eur. Radiol.* **2022**, *32*, 1195–1204. [CrossRef]
5. Mehta, R.I.; Carpenter, J.S.; Mehta, R.I.; Haut, M.W.; Wang, P.; Ranjan, M.; Najib, U.; D'haese, P.-F.; Rezai, A.R. Ultrasound-mediated blood–brain barrier opening uncovers an intracerebral perivenous fluid network in persons with Alzheimer's disease. *Fluids Barriers CNS* **2023**, *20*, 46. [CrossRef]
6. Wrede, P.; Aghakhani, A.; Bozuyuk, U.; Yildiz, E.; Sitti, M. Acoustic Trapping and Manipulation of Hollow Microparticles under Fluid Flow Using a Single-Lens Focused Ultrasound Transducer. *ACS Appl. Mater. Interfaces* **2023**, *15*, 52224–52236. [CrossRef] [PubMed]
7. Wu, J. Acoustical tweezers. *J. Acoust. Soc. Am.* **1991**, *89*, 2140–2143. [CrossRef]
8. Yuan, Z.; Lu, C.; Liu, C.; Bai, X.; Zhao, L.; Feng, S.; Liu, Y. Ultrasonic tweezer for multifunctional droplet manipulation. *Sci. Adv.* **2023**, *9*, eadg2352. [CrossRef]
9. Ozcelik, A.; Rufo, J.; Guo, F.; Gu, Y.; Li, P.; Lata, J.; Huang, T.J. Acoustic tweezers for the life sciences. *Nat. Methods* **2018**, *15*, 1021–1028. [CrossRef]
10. Marzo, A.; Seah, S.A.; Drinkwater, B.W.; Sahoo, D.R.; Long, B.; Subramanian, S. Holographic acoustic elements for manipulation of levitated objects. *Nat. Commun.* **2015**, *6*, 8661. [CrossRef]
11. Hu, Q.; Ma, T.; Zhang, Q.; Wang, J.; Yang, Y.; Cai, F.; Zheng, H. 3-D acoustic tweezers using a 2-D matrix array with time-multiplexed traps. *IEEE Trans. Ultrason. Ferroelectr. Freq. Control* **2021**, *68*, 3646–3653. [CrossRef] [PubMed]
12. Collins, D.J.; Morahan, B.; Garcia-Bustos, J.; Doerig, C.; Plebanski, M.; Neild, A. Two-dimensional single-cell patterning with one cell per well driven by surface acoustic waves. *Nat. Commun.* **2015**, *6*, 8686. [CrossRef] [PubMed]
13. Jung, H.; Shung, K.K.; Lim, H.G. Ultrasonic High-Resolution Imaging and Acoustic Tweezers Using Ultrahigh Frequency Transducer: Integrative Single-Cell Analysis. *Sensors* **2023**, *23*, 1916. [CrossRef] [PubMed]
14. Yang, Y.; Yang, Y.; Liu, D.; Wang, Y.; Lu, M.; Zhang, Q.; Huang, J.; Li, Y.; Ma, T.; Yan, F.; et al. In-vivo programmable acoustic manipulation of genetically engineered bacteria. *Nat. Commun.* **2023**, *14*, 3297. [CrossRef]
15. Yang, Y.; Ma, T.; Zhang, Q.; Huang, J.; Hu, Q.; Li, Y.; Wang, C.; Zheng, H. 3D Acoustic Manipulation of Living Cells and Organisms Based on 2D Array. *IEEE Trans. Biomed. Eng.* **2022**, *69*, 2342–2352. [CrossRef] [PubMed]
16. Lo, W.-C.; Fan, C.-H.; Ho, Y.-J.; Lin, C.-W.; Yeh, C.-K. Tornado-inspired acoustic vortex tweezer for trapping and manipulating microbubbles. *Proc. Natl. Acad. Sci. USA* **2021**, *118*, e2023188118. [CrossRef]
17. Chowdhury, S.M.; Abou-Elkacem, L.; Lee, T.; Dahl, J.; Lutz, A.M. Ultrasound and microbubble mediated therapeutic delivery: Underlying mechanisms and future outlook. *J. Control. Release* **2020**, *326*, 75–90. [CrossRef]
18. Thanou, M.; Gedroyc, W. MRI-Guided Focused Ultrasound as a New Method of Drug Delivery. *J. Drug Deliv.* **2013**, *2013*, 616197. [CrossRef]
19. Conteduca, D.; Brunetti, G.; Barth, I.; Quinn, S.D.; Ciminelli, C.; Krauss, T.F. Multiplexed Near-Field Optical Trapping Exploiting Anapole States. *ACS Nano* **2023**, *17*, 16695–16702. [CrossRef]
20. Brunetti, G.; Sasanelli, N.; Armenise, M.N.; Ciminelli, C. Nanoscale Optical Trapping by Means of Dielectric Bowtie. *Photonics* **2022**, *9*, 425. [CrossRef]
21. Zeng, B.F.; Deng, R.; Zou, Y.L.; Huo, C.A.; Wang, J.Y.; Yang, W.M.; Liang, Q.M.; Qiu, S.J.; Feng, A.; Shi, J.; et al. Optical Trapping of a Single Molecule of Length Sub-1 nm in Solution. *CCS Chem.* **2023**, *5*, 830–840. [CrossRef]
22. Hölzel, R.; Calander, N.; Chiragwandi, Z.; Willander, M.; Bier, F.F. Trapping Single Molecules by Dielectrophoresis. *Phys. Rev. Lett.* **2005**, *95*, 128102. [CrossRef] [PubMed]
23. Bespalova, M.I.; Mahanta, S.; Krishnan, M. Single-molecule trapping and measurement in solution. *Curr. Opin. Chem. Biol.* **2019**, *51*, 113–121. [CrossRef] [PubMed]
24. Wang, X.; Li, Y.; Wu, M.; Hao, Y.; Tian, L.; He, Z.; Au, K.W.S.; Taylor, R.H.; Iordachita, I.; Chan, J.Y.K.; et al. Intra-operative image-guided interventional robotics—Where are we now and where are we going? In *Machine Learning, Medical AI and Robotics: Translating Theory into the Clinic*; Vardhanabhuti, V., Kwok, K.-W., Chan, J.Y.K., Dou, Q., Eds.; IOP Publishing: Bristol, UK, 2023.
25. Rizzitelli, S.; Giustetto, P.; Cutrin, J.; Castelli, D.D.; Boffa, C.; Ruzza, M.; Menchise, V.; Molinari, F.; Aime, S.; Terreno, E. Sonosensitive theranostic liposomes for preclinical in vivo MRI-guided visualization of doxorubicin release stimulated by pulsed low intensity non-focused ultrasound. *J. Control. Release* **2015**, *202*, 21–30. [CrossRef]
26. Baek, H.; Lockwood, D.; Mason, E.J.; Obusez, E.; Poturalski, M.; Rammo, R.; Nagel, S.J.; Jones, S.E. Clinical intervention using focused ultrasound (FUS) stimulation of the brain in diverse neurological disorders. *Front. Neurol.* **2022**, *13*, 880814. [CrossRef]
27. Dai, J.; He, Z.; Fang, G.; Wang, X.; Li, Y.; Cheung, C.-L.; Liang, L.; Iordachita, I.I.; Chang, H.-C.; Kwok, K.-W. A Robotic Platform to Navigate MRI-guided Focused Ultrasound System. *IEEE Robot. Autom. Lett.* **2021**, *6*, 5137–5144. [CrossRef]

28. Kyriakou, A.; Neufeld, E.; Werner, B.; Paulides, M.M.; Szekely, G.; Kuster, N. A review of numerical and experimental compensation techniques for skull-induced phase aberrations in transcranial focused ultrasound. *Int. J. Hyperth.* **2014**, *30*, 36–46. [CrossRef]
29. Holman, R.; Lorton, O.; Guillemin, P.C.; Desgranges, S.; Santini, F.; Preso, D.B.; Farhat, M.; Contino-Pépin, C.; Salomir, R. Perfluorocarbon emulsion enhances MR-ARFI displacement and temperature in vitro: Evaluating the response with MRI, NMR, and hydrophone. *Front. Oncol.* **2023**, *12*, 1025481. [CrossRef] [PubMed]
30. Rieke, V.; Butts Pauly, K. MR thermometry. *J. Magn. Reson. Imaging* **2008**, *27*, 376–390. [CrossRef]
31. Gu, J.; Jing, Y. mSOUND: An Open Source Toolbox for Modeling Acoustic Wave Propagation in Heterogeneous Media. *IEEE Trans. Ultrason. Ferroelectr. Freq. Control* **2021**, *68*, 1476–1486. [CrossRef]
32. Schneider, M. Characteristics of SonoVuetrade mark. *Echocardiography* **1999**, *16*, 743–746. [CrossRef]
33. Melde, K.; Mark, A.G.; Qiu, T.; Fischer, P. Holograms for acoustics. *Nature* **2016**, *537*, 518–522. [CrossRef] [PubMed]
34. Ghanem, M.A.; Maxwell, A.D.; Wang, Y.-N.; Cunitz, B.W.; Khokhlova, V.A.; Sapozhnikov, O.A.; Bailey, M.R. Noninvasive acoustic manipulation of objects in a living body. *Proc. Natl. Acad. Sci. USA* **2020**, *117*, 16848–16855. [CrossRef]
35. Jooss, V.M.; Bolten, J.S.; Huwyler, J.; Ahmed, D. In vivo acoustic manipulation of microparticles in zebrafish embryos. *Sci. Adv.* **2022**, *8*, eabm2785. [CrossRef] [PubMed]
36. Watanabe, A.; Hasegawa, K.; Abe, Y. Contactless Fluid Manipulation in Air: Droplet Coalescence and Active Mixing by Acoustic Levitation. *Sci. Rep.* **2018**, *8*, 10221. [CrossRef]
37. Tian, Z.; Wang, Z.; Zhang, P.; Naquin, T.D.; Mai, J.; Wu, Y.; Yang, S.; Gu, Y.; Bachman, H.; Liang, Y.; et al. Generating multifunctional acoustic tweezers in Petri dishes for contactless, precise manipulation of bioparticles. *Sci. Adv.* **2020**, *6*, eabb0494. [CrossRef] [PubMed]
38. Ochiai, Y.; Hoshi, T.; Rekimoto, J. Three-dimensional mid-air acoustic manipulation by ultrasonic phased arrays. *PLoS ONE* **2014**, *9*, e97590. [CrossRef]
39. Zhao, L.; Zhou, S. Compact Acoustic Rainbow Trapping in a Bioinspired Spiral Array of Graded Locally Resonant Metamaterials. *Sensors* **2019**, *19*, 788. [CrossRef]
40. Cao, H.X.; Jung, D.; Lee, H.-S.; Du Nguyen, V.; Choi, E.; Kang, B.; Park, J.-O.; Kim, C.-S. Holographic Acoustic Tweezers for 5-DoF Manipulation of Nanocarrier Clusters toward Targeted Drug Delivery. *Pharmaceutics* **2022**, *14*, 1490. [CrossRef]
41. Marzo, A.; Drinkwater, B.W. Holographic acoustic tweezers. *Proc. Natl. Acad. Sci. USA* **2019**, *116*, 84–89.
42. Zhong, C.; Jia, Y.; Jeong, D.C.; Guo, Y.; Liu, S. AcousNet: A Deep Learning Based Approach to Dynamic 3D Holographic Acoustic Field Generation from Phased Transducer Array. *IEEE Robot. Autom. Lett.* **2022**, *7*, 666–673. [CrossRef]
43. Yang, Y.; Ma, T.; Li, S.; Zhang, Q.; Huang, J.; Liu, Y.; Zhuang, J.; Li, Y.; Du, X.; Niu, L.; et al. Self-Navigated 3D Acoustic Tweezers in Complex Media Based on Time Reversal. *Research* **2021**, *2021*, 9781394. [CrossRef]
44. White, J.; Clement, G.T.; Hynynen, K. Transcranial ultrasound focus reconstruction with phase and amplitude correction. *IEEE Trans. Ultrason. Ferroelectr. Freq. Control* **2005**, *52*, 1518–1522. [CrossRef]
45. Pedersoli, F.; Fang, J.; Boas, E.; Park, J.J. Operator radiation exposure during radioembolisation of the liver: Transfemoral versus transradial access using real-time dose monitoring. *Radiat. Prot. Dosim.* **2023**, *19*, 2344–2348. [CrossRef] [PubMed]
46. Gralinski, I.; Alan, T.; Neild, A. Non-contact acoustic trapping in circular cross-section glass capillaries: A numerical study. *J. Acoust. Soc. Am.* **2012**, *132*, 2978–2987. [CrossRef]
47. Bancel, T.; Houdouin, A.; Annic, P.; Rachmilevitch, I.; Shapira, Y.; Tanter, M.; Aubry, J.-F. Comparison Between Ray-Tracing and Full-Wave Simulation for Transcranial Ultrasound Focusing on a Clinical System Using the Transfer Matrix Formalism. *IEEE Trans. Ultrason. Ferroelectr. Freq. Control* **2021**, *68*, 2554–2565. [CrossRef] [PubMed]
48. Lukovic, J.; Henke, L.; Gani, C.; Kim, T.K.; Stanescu, T.; Hosni, A.; Lindsay, P.; Erickson, B.; Khor, R.; Eccles, C.; et al. MRI-Based Upper Abdominal Organs-at-Risk Atlas for Radiation Oncology. *Int. J. Radiat. Oncol. Biol. Phys.* **2020**, *106*, 743–753. [CrossRef]
49. Akshaya, R.; Tej Prakash, S. POCUS in Abdominal Trauma: Old Gadget, New Insights. In *Abdominal Trauma—New Solutions to Old Problems*; Garbuzenko, V.D., Ed.; IntechOpen: Rijeka, Croatia, 2022.
50. Cheung, C.-L.; Ho, J.D.-L.; Vardhanabhuti, V.; Chang, H.-C.; Kwok, K.-W. Design and Fabrication of Wireless Multilayer Tracking Marker for Intraoperative MRI-Guided Interventions. *IEEE/ASME Trans. Mechatron.* **2020**, *25*, 1016–1025. [CrossRef]
51. Cheung, C.-L.; Wu, M.; Fang, G.; Ho, J.D.L.; Liang, L.; Tan, K.V.; Lin, F.-H.; Chang, H.-C.; Kwok, K.-W. Omnidirectional Monolithic Marker for Intra-Operative MR-Based Positional Sensing in Closed MRI. *IEEE Trans. Med. Imaging* **2024**, *43*, 439–448. [CrossRef]
52. Prado, T.d.A.; Moura, H.L.; Passarin, T.A.; Guarneri, G.A.; Pires, G.P.; Pipa, D.R. A straightforward method to evaluate the directivity function of ultrasound imaging systems. *NDT E Int.* **2021**, *119*, 102402. [CrossRef]
53. Kang, S.-T.; Yeh, C.-K. Potential-well model in acoustic tweezers. *IEEE Trans. Ultrason. Ferroelectr. Freq. Control* **2010**, *57*, 1451–1459. [CrossRef] [PubMed]
54. Dai, J.; Zhu, B.; Wang, X.; Jiang, Z.; Wu, M.; Liang, L.; Xie, X.; Lam, J.; Chang, H.-C.; Kwok, K.-W. Learning-Based Efficient Phase-Amplitude Modulation and Hybrid Control for MRI-Guided Focused Ultrasound Treatment. *IEEE Robot. Autom. Lett.* **2024**, *9*, 995–1002. [CrossRef]
55. Hosseinkhah, N.; Goertz, D.E.; Hynynen, K. Microbubbles and Blood–Brain Barrier Opening: A Numerical Study on Acoustic Emissions and Wall Stress Predictions. *IEEE Trans. Biomed. Eng.* **2015**, *62*, 1293–1304. [CrossRef] [PubMed]

56. Schot, M.; Becker, M.; Paggi, C.A.; Gomes, F.; Koch, T.; Gensheimer, T.; Johnbosco, C.; Nogueira, L.P.; van der Meer, A.; Carlson, A.; et al. Photoannealing of Microtissues Creates High-Density Capillary Network Containing Living Matter in a Volumetric-Independent Manner. *Adv. Mater.* **2024**, *36*, 2308949. [CrossRef]
57. Navarro-Becerra, J.A.; Borden, M.A. Targeted Microbubbles for Drug, Gene, and Cell Delivery in Therapy and Immunotherapy. *Pharmaceutics* **2023**, *15*, 1625. [CrossRef] [PubMed]
58. Fossey, D.J.; Smith, C.H.; Wischmann, K.B. A New Potting Material—Expandable Polystyrene Bead Foam. *J. Cell. Plast.* **1977**, *13*, 347–353. [CrossRef]
59. Figueroa, R.L.; Zeng-Treitler, Q.; Kandula, S.; Ngo, L.H. Predicting sample size required for classification performance. *BMC Med. Inform. Decis. Mak.* **2012**, *12*, 8. [CrossRef]

sensors

MDPI

Article

Phonocardiogram (PCG) Murmur Detection Based on the Mean Teacher Method

Yi Luo [1,†], Zuoming Fu [2,†], Yantian Ding [3], Xiaojian Chen [1] and Kai Ding [4,*]

[1] Department of Biomedical Engineering, Johns Hopkins University, Baltimore, MD 21287, USA; yluo62@jh.edu (Y.L.); xchen279@jhu.edu (X.C.)
[2] Department of Electrical and Computer Engineering, Cornell University, Ithaca, NY 14853, USA; zf242@cornell.edu
[3] Department of Applied Mathematics and Statistics, Johns Hopkins University, Baltimore, MD 21287, USA; yding72@jh.edu
[4] Department of Radiation Oncology and Molecular Radiation Sciences, Johns Hopkins University, Baltimore, MD 21287, USA
* Correspondence: kai@jhu.edu
† These authors contributed equally to this work.

Abstract: Cardiovascular diseases (CVDs) are among the primary causes of mortality globally, highlighting the critical need for early detection to mitigate their impact. Phonocardiograms (PCGs), which record heart sounds, are essential for the non-invasive assessment of cardiac function, enabling the early identification of abnormalities such as murmurs. Particularly in underprivileged regions with high birth rates, the absence of early diagnosis poses a significant public health challenge. In pediatric populations, the analysis of PCG signals is invaluable for detecting abnormal sound waves indicative of congenital and acquired heart diseases, such as septal defects and defective cardiac valves. In the PhysioNet 2022 challenge, the murmur score is a weighted accuracy metric that reflects detection accuracy based on clinical significance. In our research, we proposed a mean teacher method tailored for murmur detection, making full use of the Phyionet2022 and Phyionet2016 PCG datasets, achieving the SOTA (State of Art) performance with a murmur score of 0.82 and an AUC score of 0.90, providing an accessible and high accuracy non-invasive early stage CVD assessment tool, especially for low and middle-income countries (LMICs).

Keywords: cardiovascular disease (CVD); phonocardiogram (PCG); murmur detection; semi-supervised learning; mean teacher; non-invasive assessment; low and middle-income countries (LMICs)

1. Introduction

The leading cause of mortality worldwide are CVDs [1]. Cardiac auscultation is the most accessible method for early-stage, non-invasive assessment of CVDs, particularly in LMICs. This technique involves the analysis of heart sounds, crucial indicators of heart function, and potential abnormalities. In the realm of digital health technology, PCGs elevate this basic assessment by recording these heart sounds electronically. PCGs capture the acoustic data of the heart's beats, which can be analyzed to detect irregular patterns such as murmurs, gallops, and rubs.

In PCG recordings, the first heart sound (S1) and the second heart sound (S2) are fundamental components. S1 is produced by the closing of the atrioventricular valves, signaling the beginning of systole. In contrast, S2 is generated by the closing of the semilunar valves, marking the end of systole and the start of diastole. Beyond these, heart murmurs detected in PCG signals provide essential clinical insights into heart malfunctions due to congenital and acquired diseases in pediatric populations [2,3]. Consequently, developing machine learning algorithms that can precisely detect murmurs from PCG signals presents a significant challenge.

One major challenge in applying deep learning techniques to cardiovascular signals is the lack of annotated data. Creating reliable diagnoses requires considerable time and the expertise of highly skilled professionals, making the process demanding and resource-intensive.

Although researchers have conducted lots of work to make full use of the labeled PCG data [4–6], a considerable portion of unlabeled data remain underutilized. Semi-supervised learning paradigms have demonstrated notable efficacy in leveraging unlabeled data alongside a limited subset of labeled data [7,8].

The PhysioNet 2022 [9] dataset comprises heart sound recordings collected from 1568 patients, most of whom were pediatric patients from rural Brazil. These recordings were gathered using an electronic stethoscope at four key auscultation points (aortic valve, pulmonic valve, tricuspid valve, and mitral valve). A total of 5272 recordings were made, with detailed annotations provided by cardiologists. The dataset includes demographic and clinical information such as age, sex, height, and weight, as well as clinical outcomes like heart murmurs and abnormal cardiac function. The recordings are labeled into three categories—Present, Absent, and Unknown for murmurs—providing a rich source of supervised learning data. However, only 60% of this dataset (2982 recordings) was released as the public portion for model training, which we utilized in our study.

In contrast, the PhysioNet 2016 dataset [10] was primarily collected for the classification of normal versus abnormal heart sounds and includes 4430 recordings from 1072 subjects. This dataset features recordings from a diverse population of both children and adults, gathered across various environments using different recording devices. While the primary objective was to differentiate between normal and abnormal heart function, the dataset lacks the detailed murmur annotations present in the PhysioNet 2022 dataset, making it well-suited as unlabeled data for our semi-supervised learning framework. The challenge's training set comprises 3153 heart sound recordings from 764 patients. Notably, the high-quality 2016-e subset, which has even been used for generative tasks [11], was included in our experiments and consists of 2054 PCG recordings.

In this study, we utilized the mean teacher model that leverages the labeled PCG data from Physionet 2022 in conjunction with the unlabeled PCG data in Physionet 2016, our method exhibits a comparable performance to fully supervised models. Additionally, by incorporating unlabeled data sourced from an external dataset alongside the labeled dataset, we achieve an enhanced model performance relative to utilizing labeled data alone. Our findings underscore the utility of integrating unlabeled data through a semi-supervised framework, thereby augmenting model performance and facilitating clinical decision-making by harnessing the wealth of readily available unlabeled data.

2. Related Work

Computer-aided classification in PCGs has captivated researchers for over 50 years, yet it continues to present challenges. Gerbarg et al. were pioneers in this field, using a threshold-based method to tackle the problem [12]. Early approaches in heart sound murmur detection involved enhancing acoustic signals through digital stethoscopes and basic electronic filtering techniques. Subsequently, techniques like the Fourier transform were employed to break down heart sounds into individual frequency components, facilitating the identification of abnormal frequencies [13,14]. Ahlström's work comprehensively summarized the physics, analysis, and processing of heart sounds [15]. More sophisticated approaches later integrated complex signal processing methodologies, such as spectral analysis and wavelet transformations [16–18]. Despite these advancements, traditional methods were limited by their dependency on high-quality acoustic signals and struggled in noisy clinical environments or with low-amplitude murmur sounds.

The rise of machine learning, especially deep learning, has transformed the landscape of murmur detection. The George B. Moody PhysioNet Challenge in 2016 [10] attracted many researchers to the field, leading to significant contributions in CVD assessment. Cristhian Potes et al. [19] achieved the highest score in the challenge, which combined the results from an AdaBoost algorithm and a convolutional neural network (CNN). The AdaBoost algorithm was trained on 124 time–frequency features, while the CNN was trained on four frequency bands extracted from the PCG signals. Subsequent research utilized methods such as ResNet [20], Mel-frequency features [21], feature fusion methods [22], CNN-LSTM networks [23,24], and transfer learning [25] have achieved even higher performance.

The George B. Moody PhysioNet Challenge in 2022 [9] provided more detailed classifications for typical diagnoses, including heart murmurs. Hui Lu et al. achieved SOTA performance in murmur detection using a lightweight CNN with wide features and Mel-spectrogram [4], while LCNN, dual Bayesian ResNet, hidden semi-Markov model, and hierarchical multi-scale convolutional neural networks were also deployed to achieve competitive performance in the Physionet 2022 challenge [5,6,26,27]. Derivative works include the use of a new wav2vec model [28], multi-scale CNN with LSTM [29], and complex-valued networks [30]. Beyond the pursuit of higher classification performance, efforts have also been made to enhance model interpretability. For instance, M. Rohr et al. employed a multiple-instance learning framework to facilitate model interpretability [31], while G. Zhou et al. explored different data augmentation methods for PCG classification [32].

In the realm of semi-supervised methods, Bachman et al. [33] introduced the concept of "Pseudo-Ensembles" in 2014, which created an ensemble of models by perturbing the data and model parameters during training. This approach aims to enhance diversity in the learned representations. Further advancements were made by M Sajjadi et al. [34] in 2016, who introduced randomness into the training process to promote smoothness in the model's predictions and increase robustness to variations in the input data [35]. Building upon this, Laine and Valpola proposed "Temporal Ensembling" in 2016, where the model's predictions on unlabeled data were averaged over time to provide consistent pseudo-labels for regularization. These developments culminated in the introduction of the mean teacher concept by Tarvainen and Valpola in 2017 [8]. The mean teacher approach enhanced consistency regularization by encouraging the model to maintain stable predictions even when input data were slightly altered. This approach leverages both labeled and unlabeled data by comparing the student model's current predictions with those of the teacher model, enabling simultaneous learning from both data types. This strategy greatly boosts the model's ability to generalize and perform well on unseen data.

Despite the success of semi-supervised learning in various domains, its application in PCG murmur detection remains largely unexplored. The complexities of heart sound and the scarcity of labeled data pose significant challenges for traditional semi-supervised methods. However, integrating semi-supervised learning into PCG murmur detection could potentially alleviate the need for large labeled datasets and further improve classification performance in real-world clinical settings.

3. Method

Our method comprises two main components: the mean teacher model and a lightweight CNN architecture that serves as the base model.

The mean teacher method is a semi-supervised learning strategy that harnesses unlabeled data to significantly improve model generalization and robustness [8]. Compared to traditional semi-supervised learning approaches, the mean teacher approach employs a teacher model to enhance supervision, thereby reducing the generalization error and improving the model performance; the architecture of the mean teacher model is shown in Figure 1.

In the training process of the mean teacher method, student and teacher models both utilize the same architectural framework. During each iteration, a batch of labeled data is fed into the student model, generating predictions $p_{s;1}$. Unlabeled data are also input into the student model and subsequently into the teacher model, producing predictions $p_{s;2}$ and $p_{t;2}$, respectively. The classification loss between $p_{s;1}$ and the true labels, as well as the consistency loss between $p_{s;2}$ and $p_{t;2}$, are then calculated. These losses are weighted and combined to compute the total loss, which is used to optimize the student model, as indicated by the dotted line in Figure 1. Finally, the teacher model is updated through the exponential moving average according to the parameters of the student model, completing one iteration. This process is repeated multiple times during a training cycle. As the iterations progress, the model's performance stabilizes, marking the completion of training. At the end of this process, the student model is employed for testing.

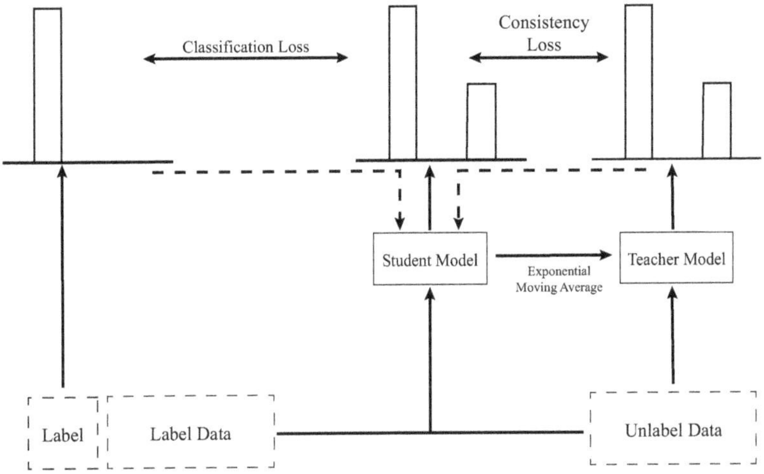

Figure 1. Architecture of Mean teacher model.

In the optimization process, the relative weighting of the classification and consistency losses dictates how supervised and unsupervised learning contribute to the model's performance. The mathematical expressions for these classification and consistency losses are outlined as follows.

$$Classification\ Loss = w_1 \sum_{i=1}^{L} y_i log(p_{s;i}) \tag{1}$$

$$Consistency\ Loss = w_2 \sum_{i=1}^{L} (p_{t;i}\breve{}p_{s;i})^2 \tag{2}$$

In this project, weights w_1 and w_2 are assigned to the classification and consistency losses, respectively. The classification task involves three classes $(L = 3)$. The label for class i is denoted as y_i, while $p_{s;i}$ and $p_{t;i}$ are the probabilities predicted by the student and teacher models for class i. The weight for the consistency loss is calculated using the function $e^{-r(1-x)^2} * (w_e - w_i) + w_i$, where $x = \frac{ep}{eps}$, with r representing the coefficient for the consistency loss weight. The classification loss weight is determined by a function given by $1 - \frac{ep-1}{eps}$, where ep represents the current epoch and eps is the total number of epochs. This function gradually decreases from 1 to 0 throughout the training period, allowing the model to initially converge quickly on labeled data and then progressively learn from unlabeled data.

The formula for Exponential Moving Average (EMA) is as follows:

$$\alpha_t = k\alpha_t + (1-k)\alpha_s \tag{3}$$

In this context, α_t and α_s represent the trainable parameters of the teacher and student models, respectively. The smoothing coefficient k is set to 0.98 throughout the training process.

The following represents the foundational architecture of the teacher and student models, a lightweight CNN [4]; the architecture is shown in Figure 2. The lightweight CNN is the SOTA model in the Physionet 2022 challenge, outperforming other models due to its simple architecture, strong robustness, and generalization ability across different PCG signal data and populations. This makes it highly practical for use in PCG murmur detection.

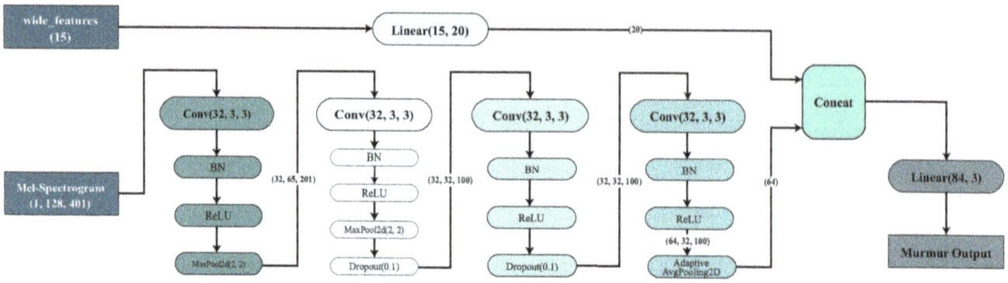

Figure 2. Architecture of the lightweight CNN model employed in mean teacher.

This lightweight convolutional neural network architecture is specifically designed for processing audio data, particularly for identifying cardiac murmurs. The primary input to the model is a Mel-spectrogram with dimensions of (1, 128, 401), where 128 represents the frequency dimension and 401 represents the time dimension. Additionally, the model receives an auxiliary input called "wide features", which is a 15-dimensional feature vector. This vector is processed through a linear layer to transform it into a 20-dimensional feature.

The model architecture consists of four convolutional blocks followed by a linear layer. Each convolutional block includes batch normalization and a ReLU activation function to enhance audio feature extraction. The first and second convolutional blocks have max-pooling layers with a (2, 2) window, which reduces the spatial dimensions of the feature maps and lessens the computational burden. Additionally, the second convolutional block features a dropout layer with a 0.1 dropout rate to mitigate overfitting. Following the convolutional blocks, an adaptive average pooling layer is used to further condense the feature space before passing it to the linear layer.

Finally, the model combines 20-dimensional "wide features" processed through a linear layer with the 64-dimensional features derived from the Mel-spectrogram via the adaptive average pooling layer, resulting in an 84-dimensional feature vector. This vector is then passed through another linear layer, which maps it to one of three output categories: "Present", "Absent", or "Unknown". The design of the entire model aims to optimize audio classification accuracy by learning the intrinsic features of the Mel-spectrogram and the directly inputted "wide features". Additionally, the simplified network structure maintains computational efficiency.

4. Experiment

The development environment for our project was built using Python 3.9.0, Nvidia A100 GPUs, Pytorch 2.1.0, and CUDA 12.1.0. For the training process, we implemented the ADAM optimizer, setting the learning rate to 1e-3 and using a batch size of 16. The detailed explanation is shown in Table 1.

Table 1. Key parameter configuration.

Parameter Name	Parameter Value	Parameter Description
Optimizer	Adam	Neural Network optimizer
Learning Rate	1×10^{-3}	learning rate for optimizer
Batch Size	16	Number of samples per iteration.
EMA k	0.98	Smoothing coefficient in Exponential Moving Average.
Weight con init	0.7	Initial value for consistency loss in mean teacher.
Weight con end	1	Final value for consistency loss in mean teacher.

4.1. Dataset and Preprocessing

In our research, we primarily utilized data from the George B. Moody PhysioNet Challenge for the years 2016 and 2022. The 2022 dataset comprises heart sound recordings from 874 patients, totaling 2982 entries collected from various auscultation points. Additionally, this dataset includes annotated patient information such as age, sex, height, weight, and other relevant variables. The recordings are categorized into three main groups based on the presence of murmurs in the PCG signals: Present, Absent, and Unknown. These categories serve as the labeled data for the model training. The 2016 dataset, designed for normal/abnormal classification rather than murmur detection, includes the 2016-e datasets renowned for their high-quality 2054 PCG recordings, which we used as unlabeled data in our analysis.

In our project, we standardized the duration of audio recordings, which originally ranged from 4.8 s to 120 s, by setting a fixed duration of 10 s for each recording. During preprocessing, recordings longer than 10 s were randomly cropped, while shorter recordings were padded with zeros to achieve a consistent length. This preprocessing method was uniformly applied during both the training and validation phases.

For each recording, we extracted a Mel-spectrogram with 128 Mel bands, using a window size of 50 ms and a frame shift of 25 ms, covering a frequency range from 25 to 2000 Hz, as illustrated in Figure 3. In addition to spectrogram extraction, we computed wide features, which were then categorized and encoded into one-hot vectors. To further enrich the dataset, we extracted statistical features from the audio signals, including the zero-crossing rate, spectral centroids, and spectral bandwidth.

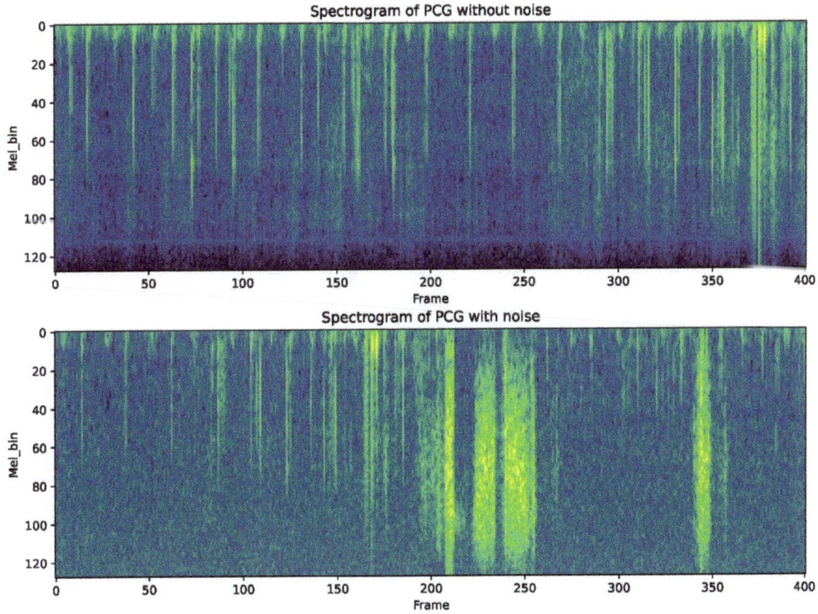

Figure 3. (**Top**): Mel-spectrogram of original PCG, (**Bottom**): Mel-spectrogram of PCG with Noise.

To enhance the robustness of our model against overfitting and to simulate real-world noisy environments, we implemented two primary data augmentation techniques on the Mel-spectrograms. Initially, Gaussian noise with a zero mean and a standard deviation of 15 dB was added to each recording during training, as shown in Figure 4, with a probability of 0.5 for each signal. Additionally, we applied masking in both frequency and time domains to the spectrograms, as shown in Figure 3. This involved selectively blocking out multiple blocks of frequency channels and time steps. These augmentation strategies were designed to introduce variability and improve the generalizability of our model across diverse scenarios.

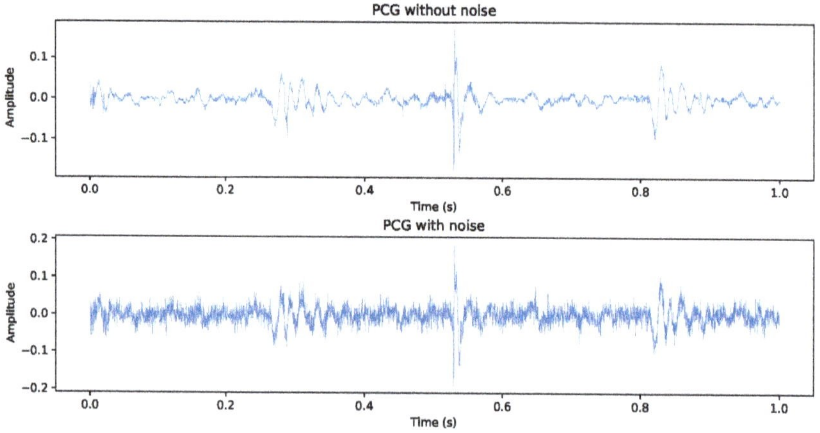

Figure 4. Top: Original PCG recording; **Bottom**: PCG recording after adding Gaussian noise.

4.2. Results

Due to significant class imbalance in the dataset, we used the murmur score from the PhysioNet Challenge 2022 [9] and AUC indicators to provide an unbiased evaluation of our model's performance. The murmur score is a weighted accuracy and can be defined by the confusion matrix shown in Table 2 and Formulation (4).

$$s_{\mathrm{m}} = \frac{5m_{PP}+3m_{UU}+m_{AA}}{5(m_{PP}+m_{UP}+m_{AP})+3(m_{PU}+m_{UU}+m_{AU})+(m_{PA}+m_{UA}+m_{AA})} \quad (4)$$

To demonstrate the robustness and practicality of our methodology, we refer to Table 3, which presents both the average values and their corresponding standard deviations. Initially, the 2022 dataset was segmented into five folds, ensuring that all recordings from the same subject were confined to a single fold. This subject-wise cross-validation approach guaranteed that no recordings from the same patient appeared simultaneously in both the training and validation/test sets, thereby mitigating the risk of data leakage and providing a more reliable evaluation of the model performance. Using one fold for labeled training and another for validation, this approach achieved a five-fold cross-validation murmur score of 0.7020 and an AUC of 0.7951. Subsequently, applying our mean teacher model, we designated one fold labeled as training data while gradually increasing the unlabeled data from one fold to three folds, with the remaining fold reserved for validation. This semi-supervised strategy improved the model's performance, achieving a murmur score of 0.7540 and an AUC of 0.8562. As expected, we observed that as the amount of unlabeled data increased, the model's performance improved accordingly. The *t*-test show a *t*-value of -18.88 leading to a *p*-value of 1.36×10^{-4} on Murmur Score, and a *t*-value of -99.55 leading to a *p*-value of 7.79×10^{-8} on AUC. These results indicate that incorporating an additional 60% of unlabeled data significantly improved the model's performance over the baseline.

Table 2. Confusion matrix for murmur detection.

		Murmur Expert		
		Present	Unknown	Absent
Murmur Classifier	Present	m_{PP}	m_{PU}	m_{PA}
	Unknown	m_{UP}	m_{UU}	m_{UA}
	Absent	m_{AP}	m_{AU}	m_{AA}

Table 3. Comparative analysis of various training modes across different data splits.

Model	Data Split			Result Measurements	
	Labeled Data	Unlabeled Data	Validation Data	Murmur Score	AUC
Full Supervise	20% v2022	/	20% v2022	0.7020 ± 0.0026	0.7951 ± 0.0008
Semi Supervise	20% v2022	20% v2022	20% v2022	0.7298 ± 0.0035	0.8217 ± 0.0009
Semi Supervise	20% v2022	40% v2022	20% v2022	0.7418 ± 0.0021	0.8398 ± 0.0013
Semi Supervise	20% v2022	60% v2022	20% v2022	0.7540 ± 0.0040	0.8562 ± 0.0007
Full Supervise **(Current SOTA)**	80% v2022	/	20% v2022	0.8074 ± 0.0010	0.8936 ± 0.0011
Semi Supervise **(Ours)**	80% v2022	100% v2016-e	20% v2022	**0.8180 ± 0.0006**	**0.9004 ± 0.0019**

Furthermore, we conducted an experiment using a five-fold cross-validation approach, where four folds were used as labeled data and the fifth fold served as the validation set. This setup resulted in a murmur score of 0.8074 and an AUC of 0.8936, highlighting the effectiveness and robustness of the method, and establishing it as the current SOTA approach in this domain.

To further enhance our semi-supervised learning strategy, we extended our approach by incorporating both the Physionet 2016 and 2022 challenge datasets. Specifically, we utilized 80% of the 2022 dataset as labeled data while using the entire 2016-e dataset as unlabeled data. The remaining 20% of the 2022 dataset was reserved for validation. This comprehensive configuration led to an improved murmur score of 0.8180 and an AUC of 0.9004. As shown in Figure 5, our model performs well in detecting murmurs in both the "absent" and "present" classes. However, it struggles significantly with the "unknown" class, which is the minority class in our dataset. This poor performance can also be attributed to the inherent clinical difficulty of the "unknown" class, even for clinical experts. Compared to the current SOTA model, our model significantly outperformed in both the present and absent class classifications.

These outcomes clearly demonstrate that our mean teacher method not only delivers more robust performance but also surpasses the results obtained by solely utilizing the 2022 challenge data. This underscores the effectiveness of incorporating unlabeled data to achieve a superior overall performance, highlighting the potential of our approach in advancing murmur detection and enhancing clinical decision-making in cardiovascular health. Consequently, our method establishes a new SOTA model for PCG murmur detection.

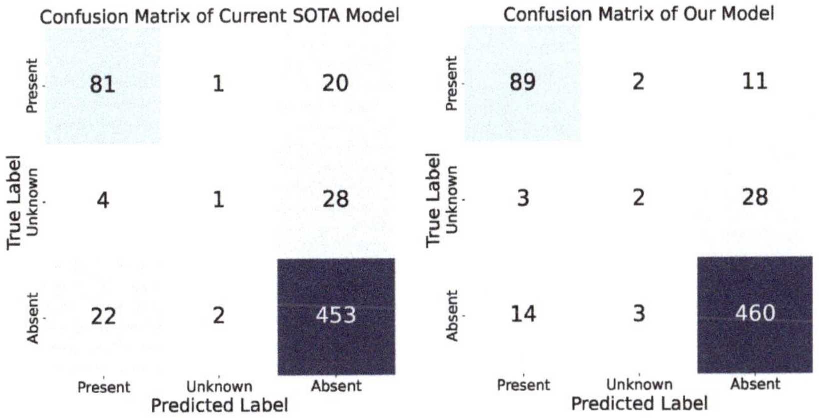

Figure 5. Comparison of the confusion matrix between the current SOTA model and our model.

5. Discussion

While our research demonstrates significant advancements in PCG murmur detection using a semi-supervised mean teacher method, certain limitations persist. Typically, semi-supervised learning frameworks are most effective when the volume of unlabeled data substantially exceeds that of the labeled data. In our study, the volume of unlabeled data was comparable to that of the labeled data, potentially capping the performance improvements that could be achieved. This suggests that acquiring a larger pool of unlabeled data could further enhance the model performance by better leveraging the strengths of semi-supervised learning. In our preprocessing strategy, recordings longer than 10 s were randomly cropped. Incorporating all available 10-second segments and applying additional data augmentation techniques could help to expand the dataset and improve performance. Future efforts could also focus on expanding the dataset, particularly with high-quality unlabeled PCG recordings, to fully exploit the potential of our semi-supervised approach and push the boundaries of the current SOTA method.

Additionally, the Physionet Challenge 2022 and 2016 datasets may have certain flaws that need to be addressed. Although expert labels are always considered the ground truths in our training and validation processes, emerging research has indicated that relabeling can significantly improve model performance [36]. This insight suggests that expert labels, while highly reliable, may still contain some inaccuracies that can impact the model's efficacy. A potential solution to this issue is to pinpoint ambiguous recordings within the dataset using the model's initial outputs. Once these ambiguous cases are identified, they can be relabeled with the assistance of expert knowledge. By double-checking and refining the labels, this approach can help to eliminate any discrepancies and improve the overall quality of the dataset. Besides, previous studies on the Physionet Challenge 2022 primarily concentrated on the performance of the murmur score [4–6,26]. While we have incorporated AUC performance in our experimental evaluation, other metrics such as sensitivity and specificity could also be considered for a more comprehensive assessment. Moreover, using additional high-quality PCG datasets, even simultaneous PCG and ECG (Electrocardiogram) dataset [37] for multi-signal source training may further elaborate our model. Finally, traditional murmur detection systems focus primarily on identifying the presence or absence of murmurs without differentiating their underlying cause or severity. However, heart murmurs, while indicative of abnormal heart activity, can stem from a variety of causes. Some of them are benign and others associated with serious valvular heart diseases such as aortic stenosis or mitral regurgitation. It is crucial clinically to distinguish between murmurs that are significant and those that are less concerning. As such, enhancing murmur detection algorithms to automatically classify murmurs based on their clinical relevance would also mark a significant step forward.

Also, the model's application to pediatric heart murmur detection and in LMICs presents certain limitations. As noted by [38], detecting heart murmurs in neonates and infants poses unique challenges due to the lower amplitude and higher variability of their heart sounds. Even though the PhysioNet2022 data focus on patients under 21 years old [9,39], for younger patients and children, the change in PCG signals would post challenge on our model. In addition, noisy environments are common in low-resource settings, where medical equipment may not meet optimal standards, can further degrade the model performance. The significant GPU computational requirements of the current model may also be difficult to meet in such environments. To address these challenges, future improvements should focus on creating pediatric-specific heart murmur datasets, enhancing the model's robustness to noise, and reducing its hardware and computational demands.

6. Conclusions

Our study showcases the mean teacher model in PCG signal murmur detection, achieving SOTA performance. By integrating both labeled and unlabeled data from the Physionet 2022 and 2016 challenges, we significantly enhanced detection accuracy, culminating in an impressive murmur score of 0.8180 and an AUC of 0.9004. These results not only affirm our model's superiority over existing methods, but also demonstrate its effectiveness in managing the sparse datasets typical in medical applications. The study underscores the critical role of extending the dataset with additional unlabeled data, which is pivotal in advancing semi-supervised learning to achieve better performance and robustness in clinical diagnostics. Furthermore, our method provides an accessible, highly accurate, and cost-effective approach for early-stage CVD diagnosis, especially in LMICs, aiming to improve clinical equity worldwide.

Author Contributions: Conceptualization, Y.L., Z.F. and K.D.; methodology, Y.L. and Z.F.; software, Y.L., Z.F. and Y.D.; investigation, Y.L., Z.F., Y.D. and X.C.; writing—original draft, Y.L. and Z.F.; writing—review and editing, Y.L. and Z.F.; visualization, Y.L., Z.F., Y.D. and X.C.; supervision, K.D.; project administration, K.D.; funding acquisition, K.D. All authors have read and agreed to the published version of the manuscript.

Funding: Research reported in this publication was supported by the National Institutes of Health (award numbers R37CA229417).

Institutional Review Board Statement: The content is solely the responsibility of the authors and does not necessarily represent the official views of the National Institutes of Health.

Informed Consent Statement: Informed consent for participation was not required, as the dataset used in this research is publicly available.

Data Availability Statement: The original data used in this study are openly available at https://physionet.org/content/challenge-2016/1.0.0/ and https://physionet.org/content/challenge-2022/1.0.0/. The models presented in this study are available upon request from the corresponding author.

Acknowledgments: We acknowledge the invaluable guidance from Vishal Patel throughout the research.

Conflicts of Interest: The authors declare no conflict of interest.

References

1. World Health Organization. *Cardiovascular Diseases (CVDs)*; World Health Organization: Geneva, Switzerland, 2017.
2. Dornbush, S.; Turnquest, A.E. Physiology, heart sounds 2023. In *StatPearls [Internet]*; StatPearls Publishing: Treasure Island, FL, USA, January 2024. [PubMed]
3. Hoeting, N.M.; McCracken, C.E.; McConnell, M.; Sallee, D.; Iannucci, G.J.; Oster, M.E. Systolic ejection click versus split first heart sound: Are our ears deceiving us? *Congenit. Heart Dis.* **2017**, *12*, 417–420. [CrossRef] [PubMed]
4. Lu, H.; Yip, J.B.; Steigleder, T.; Grießhammer, S.; Heckel, M.; Jami, N.V.S.J.; Eskofier, B.; Ostgathe, C.; Koelpin, A. A Lightweight Robust Approach for Automatic Heart Murmurs and Clinical Outcomes Classification from Phonocardiogram Recordings. In Proceedings of the 2022 Computing in Cardiology (CinC), Tampere, Finland, 4–7 September 2022; Volume 498, pp. 1–4. [CrossRef]
5. Xu, Y.; Bao, X.; Lam, H.K.; Kamavuako, E.N. Hierarchical Multi-Scale Convolutional Network for Murmurs Detection on PCG Signals. In Proceedings of the 2022 Computing in Cardiology (CinC), Tampere, Finland, 4–7 September 2022; Volume 498, pp. 1–4. [CrossRef]
6. McDonald, A.; Gales, M.J.; Agarwal, A. Detection of Heart Murmurs in Phonocardiograms with Parallel Hidden Semi-Markov Models. In Proceedings of the 2022 Computing in Cardiology (CinC), Tampere, Finland, 4–7 September 2022; Volume 498, pp. 1–4. [CrossRef]
7. Cai, Q.; Pan, Y.; Ngo, C.W.; Tian, X.; Duan, L.; Yao, T. Exploring object relation in mean teacher for cross-domain detection. In Proceedings of the IEEE/CVF Conference on Computer Vision and Pattern Recognition, Long Beach, CA, USA, 15–20 June 2019; pp. 11457–11466.
8. Tarvainen, A.; Valpola, H. Mean teachers are better role models: Weight-averaged consistency targets improve semi-supervised deep learning results. *Adv. Neural Inf. Process. Syst.* **2017**, *30*, 1195–1204.
9. Reyna, M.A.; Kiarashi, Y.; Elola, A.; Oliveira, J.; Renna, F.; Gu, A.; Perez Alday, E.A.; Sadr, N.; Sharma, A.; Kpodonu, J.; et al. Heart murmur detection from phonocardiogram recordings: The george b. moody physionet challenge 2022. *PLoS Digit. Health* **2023**, *2*, e0000324. [CrossRef] [PubMed]
10. Clifford, G.D.; Liu, C.; Moody, B.; Springer, D.; Silva, I.; Li, Q.; Mark, R.G. Classification of normal/abnormal heart sound recordings: The PhysioNet/Computing in Cardiology Challenge 2016. In Proceedings of the 2016 Computing in cardiology conference (CinC), Vancouver, BC, Canada, 11–14 September 2016; pp. 609–612.
11. Dissanayake, T.; Fernando, T.; Denman, S.; Sridharan, S.; Fookes, C. Generalized generative deep learning models for biosignal synthesis and modality transfer. *IEEE J. Biomed. Health Inform.* **2022**, *27*, 968–979. [CrossRef] [PubMed]

12. Gerbarg, D.S.; Taranta, A.; Spagnuolo, M.; Hofler, J.J. Computer analysis of phonocardiograms. *Prog. Cardiovasc. Dis.* **1963**, *5*, 393–405. [CrossRef]
13. El-Segaier, M.; Lilja, O.; Lukkarinen, S.; Sörnmo, L.; Sepponen, R.; Pesonen, E. Computer-based detection and analysis of heart sound and murmur. *Ann. Biomed. Eng.* **2005**, *33*, 937–942. [CrossRef]
14. Delgado-Trejos, E.; Quiceno-Manrique, A.; Godino-Llorente, J.; Blanco-Velasco, M.; Castellanos-Dominguez, G. Digital auscultation analysis for heart murmur detection. *Ann. Biomed. Eng.* **2009**, *37*, 337–353. [CrossRef]
15. Ahlström, C. Processing of the Phonocardiographic Signal: Methods for the Intelligent Stethoscope. Ph.D. Thesis, Institutionen för Medicinsk Teknik, Linköping, Sweden, 2006.
16. Kumar, D.; Carvalho, P.; Antunes, M.; Henriques, J.; Melo, A.S.E.; Schmidt, R.; Habetha, J. Third Heart Sound Detection Using Wavelet Transform-Simplicity Filter. In Proceedings of the 2007 29th Annual International Conference of the IEEE Engineering in Medicine and Biology Society, Lyon, France, 22–26 August 2007; pp. 1277–1281. [CrossRef]
17. Eslamizadeh, G.; Barati, R. Heart murmur detection based on wavelet transformation and a synergy between artificial neural network and modified neighbor annealing methods. *Artif. Intell. Med.* **2017**, *78*, 23–40. [CrossRef]
18. Varghees, V.N.; Ramachandran, K. Effective heart sound segmentation and murmur classification using empirical wavelet transform and instantaneous phase for electronic stethoscope. *IEEE Sens. J.* **2017**, *17*, 3861–3872. [CrossRef]
19. Potes, C.; Parvaneh, S.; Rahman, A.; Conroy, B. Ensemble of feature-based and deep learning-based classifiers for detection of abnormal heart sounds. In Proceedings of the 2016 Computing in Cardiology Conference (CinC), Vancouver, BC, Canada, 11–14 September 2016; pp. 621–624.
20. Chorba, J.S.; Shapiro, A.M.; Le, L.; Maidens, J.; Prince, J.; Pham, S.; Kanzawa, M.M.; Barbosa, D.N.; Currie, C.; Brooks, C.; et al. Deep learning algorithm for automated cardiac murmur detection via a digital stethoscope platform. *J. Am. Heart Assoc.* **2021**, *10*, e019905. [CrossRef]
21. Abduh, Z.; Nehary, E.A.; Wahed, M.A.; Kadah, Y.M. Classification of heart sounds using fractional fourier transform based mel-frequency spectral coefficients and traditional classifiers. *Biomed. Signal Process. Control* **2020**, *57*, 101788. [CrossRef]
22. Li, H.; Wang, X.; Liu, C.; Zeng, Q.; Zheng, Y.; Chu, X.; Yao, L.; Wang, J.; Jiao, Y.; Karmakar, C. A fusion framework based on multi-domain features and deep learning features of phonocardiogram for coronary artery disease detection. *Comput. Biol. Med.* **2020**, *120*, 103733. [CrossRef] [PubMed]
23. Megalmani, D.R.; Shailesh, B.; Rao, A.; Jeevannavar, S.S.; Ghosh, P.K. Unsegmented heart sound classification using hybrid CNN-LSTM neural networks. In Proceedings of the 2021 43rd Annual International Conference of the IEEE Engineering in Medicine & Biology Society (EMBC), Virtual, 31Oct–4 November 2021; pp. 713–717.
24. Bao, X.; Xu, Y.; Kamavuako, E.N. The effect of signal duration on the classification of heart sounds: A deep learning approach. *Sensors* **2022**, *22*, 2261. [CrossRef] [PubMed]
25. Almanifi, O.R.A.; Ab Nasir, A.F.; Razman, M.A.M.; Musa, R.M.; Majeed, A.P.A. Heartbeat murmurs detection in phonocardiogram recordings via transfer learning. *Alex. Eng. J.* **2022**, *61*, 10995–11002. [CrossRef]
26. Lee, J.; Kang, T.; Kim, N.; Han, S.; Won, H.; Gong, W.; Kwak, I.Y. Deep learning based heart murmur detection using frequency-time domain features of heartbeat sounds. In Proceedings of the 2022 Computing in Cardiology (CinC), Tampere, Finland, 4–7 September 2022; Volume 498, pp. 1–4.
27. Walker, B.; Krones, F.; Kiskin, I.; Parsons, G.; Lyons, T.; Mahdi, A. Dual Bayesian ResNet: A deep learning approach to heart murmur detection. In Proceedings of the 2022 Computing in Cardiology (CinC), Tampere, Finland, 4–7 September 2022; Volume 498, pp. 1–4.
28. Panah, D.S.; Hines, A.; McKeever, S. Exploring wav2vec 2.0 model for heart murmur detection. In Proceedings of the 2023 31st European Signal Processing Conference (EUSIPCO), Helsinki, Finland, 4–8 September 2023; pp. 1010–1014.
29. Kamson, A.P.; Lewis, M.C.; Sunil, B.V.; Jeevannavar, S.S.; Sawant, A.; Ghosh, P.K. E2E Multi-Scale CNN with LSTM for Murmur Detection in PCG or Noise Identification. In Proceedings of the 2023 International Conference on Electrical, Communication and Computer Engineering (ICECCE), Dubai, United Arab Emirates, 30–31 December 2023; pp. 1–6.
30. Bondareva, E.; Rizos, G.; Han, J.; Mascolo, C. Embracing the Imaginary: Deep Complex-valued Networks for Heart Murmur Detection. In Proceedings of the 2023 Computing in Cardiology (CinC), Atlanta, GA, USA, 1–4 October 2023; Volume 50, pp. 1–4.
31. Rohr, M.; Müller, B.; Dill, S.; Güney, G.; Hoog Antink, C. Multiple instance learning framework can facilitate explainability in murmur detection. *PLoS Digit. Health* **2024**, *3*, e0000461. [CrossRef] [PubMed]
32. Zhou, G.; Chen, Y.; Chien, C. On the analysis of data augmentation methods for spectral imaged based heart sound classification using convolutional neural networks. *BMC Med Inform. Decis. Mak.* **2022**, *22*, 226. [CrossRef]
33. Bachman, P.; Alsharif, O.; Precup, D. Learning with pseudo-ensembles. *Adv. Neural Inf. Process. Syst.* **2014**, *27*, 3365–3373.
34. Laine, S.; Aila, T. Temporal ensembling for semi-supervised learning. *arXiv* **2016**, arXiv:1610.02242.
35. Sajjadi, M.; Javanmardi, M.; Tasdizen, T. Regularization with stochastic transformations and perturbations for deep semi-supervised learning. *Adv. Neural Inf. Process. Syst.* **2016**, *29*.
36. Patwa, A.; Rahman, M.M.U.; Al-Naffouri, T.Y. Heart murmur and abnormal pcg detection via wavelet scattering transform & a 1d-cnn. *arXiv* **2023**, arXiv:2303.11423.
37. Kazemnejad, A.; Gordany, P.; Sameni, R. EPHNOGRAM: A simultaneous electrocardiogram and phonocardiogram database. *PhysioNet* **2021**. [CrossRef]

38. Noponen, A.L.; Lukkarinen, S.; Angerla, A.; Sepponen, R. Phono-spectrographic analysis of heart murmur in children. *BMC Pediatr.* **2007**, *7*, 1–10. [CrossRef] [PubMed]
39. Oliveira, J.; Renna, F.; Costa, P.D.; Nogueira, M.; Oliveira, C.; Ferreira, C.; Jorge, A.; Mattos, S.; Hatem, T.; Tavares, T.; et al. The CirCor DigiScope dataset: From murmur detection to murmur classification. *IEEE J. Biomed. Health Inform.* **2021**, *26*, 2524–2535. [CrossRef] [PubMed]

sensors

MDPI

Article

Design of a Cost-Effective Ultrasound Force Sensor and Force Control System for Robotic Extra-Body Ultrasound Imaging

Yixuan Zheng [1,*], Hongyuan Ning [1], Eason Rangarajan [1], Aban Merali [1], Adam Geale [1], Lukas Lindenroth [1], Zhouyang Xu [1], Weizhao Wang [1], Philipp Kruse [1], Steven Morris [1], Liang Ye [2], Xinyi Fu [2], Kawal Rhode [1,†] and Richard James Housden [1,†]

1 School of Biomedical Engineering and Imaging Sciences, King's College London, London SE1 7EH, UK; hongyuan.ning@kcl.ac.uk (H.N.); eason.rangarajan@kcl.ac.uk (E.R.); aban.merali@kcl.ac.uk (A.M.); adam.t.geale@kcl.ac.uk (A.G.); lukas.lindenroth@kcl.ac.uk (L.L.); zhouyang.xu@kcl.ac.uk (Z.X.); weizhao.wang@kcl.ac.uk (W.W.); philipp.kruse@kcl.ac.uk (P.K.); steven.c.morris@kcl.ac.uk (S.M.); kawal.rhode@kcl.ac.uk (K.R.); richard.housden@kcl.ac.uk (R.J.H.)

2 School of Medicine, Southern University of Science and Technology, Shenzhen 518055, China; 12112644@mail.sustech.edu.cn (L.Y.); 12112743@mail.sustech.edu.cn (X.F.)

* Correspondence: yixuan.zheng@kcl.ac.uk

† These authors contributed equally to this work.

Abstract: Ultrasound imaging is widely valued for its safety, non-invasiveness, and real-time capabilities but is often limited by operator variability, affecting image quality and reproducibility. Robot-assisted ultrasound may provide a solution by delivering more consistent, precise, and faster scans, potentially reducing human error and healthcare costs. Effective force control is crucial in robotic ultrasound scanning to ensure consistent image quality and patient safety. However, existing robotic ultrasound systems rely heavily on expensive commercial force sensors or the integrated sensors of commercial robotic arms, limiting their accessibility. To address these challenges, we developed a cost-effective, lightweight, 3D-printed force sensor and a hybrid position–force control strategy tailored for robotic ultrasound scanning. The system integrates patient-to-robot registration, automated scanning path planning, and multi-sensor data fusion, allowing the robot to autonomously locate the patient, target the region of interest, and maintain optimal contact force during scanning. Validation was conducted using an ultrasound-compatible abdominal aortic aneurysm (AAA) phantom created from patient CT data and healthy volunteer testing. For the volunteer testing, during a 1-min scan, 65% of the forces were within the good image range. Both volunteers reported no discomfort or pain during the whole procedure. These results demonstrate the potential of the system to provide safe, precise, and autonomous robotic ultrasound imaging in real-world conditions.

Keywords: robot-assisted ultrasound imaging; force control for robotic systems; force sensor design; medical robotics; multi-sensor fusion; ultrasound-compatible phantom design

1. Introduction

Ultrasound imaging, first introduced in the 1940s, has become a widely utilized diagnostic modality due to its radiation-free nature, portability, real-time imaging capabilities, and greater accessibility compared to computed tomography (CT) scans and magnetic resonance imaging (MRI) [1]. However, traditional ultrasonography depends heavily on skilled operators, resulting in variability in image quality and reproducibility [2]. Additionally, it poses risks to operators, such as musculoskeletal disorders and increased infection risk when scanning patients with infectious diseases [3]. Robot-assisted ultrasound imaging may addresses these challenges by enabling consistent, precise, and faster scans, reducing human error and lowering healthcare costs [4] while simultaneously mitigating the risk of operator injury and infection. A critical factor in advancing robotic ultrasound systems is the implementation of effective force control, which ensures optimal contact between the probe and the patient's body. Excessive contact force can lead to patient injury, whereas insufficient force compromises image quality [5]. Researchers adopt diverse approaches to force control in ultrasound robots, involving varying choices of sensors, control strategies, and implementation methodologies.

Robotic ultrasound systems employ various types of force sensors to ensure precise and safe interactions between the ultrasound probe and the patient's body. Among these, external commercial force/torque sensors are widely used due to their accuracy and reliability, making them a preferred choice for precise force measurements. For instance, many systems utilize six-axis force/torque sensors, typically positioned between the robot arm and the probe, as demonstrated in studies by Merouche et al., Jiang et al., Kim et al., and Chen et al. [6–9]. Alternatively, some systems adopt different configurations, such as the approach described by Huang et al., which employs two one-axis sensors mounted on the front face of the ultrasound probe to measure the contact force between the probe and the patient [10]. Beyond commercial sensors, custom-built sensors are also explored. Noh et al. developed a six-axis force/torque sensor using 3D-printing technology, demonstrating the feasibility of designing sensors tailored to specific tool shapes and functional requirements; however, this sensor has not yet been validated in clinical settings [11]. Additionally, certain robotic ultrasound systems utilize built-in sensors integrated into the robot's structure, offering a compact and streamlined solution for force feedback. Examples of such systems are detailed in the works of Goel et al., Suligoj et al., Virga et al., and Hennersperger et al. [12–15]. Each type of sensor—whether commercial, custom-built, or built-in—plays a pivotal role in enhancing the performance and accuracy of robotic ultrasound systems, with the choice of sensor often guided by the specific application requirements and system constraints.

The force control strategies for maintaining appropriate contact force during robotic ultrasound procedures vary across systems, each offering distinct advantages. A common strategy is applying a constant force, as implemented in [6,9,13]. This approach provides simplicity and reliability, ensuring uniform probe–patient interaction that supports consistent imaging outcomes and basic patient comfort. Another approach involves maintaining contact force within a specific range, as seen in [10,15], where the force is dynamically adjusted to predefined limits. This strategy allows for a balance between safety and effectiveness by ensuring sufficient tissue contact to capture clear images while preventing excessive pressure that could lead to discomfort or injury. A more advanced method is applying a patient-specific contact force, utilized in [7,8,14]. These systems tailor the force based on real-time feedback, such as ultrasound image quality or confidence maps, optimizing diagnostic accuracy by adapting to individual patient anatomy and imaging needs. Each of these strategies reflects a trade-off between simplicity, adaptability, and personalization in robotic ultrasound applications.

Various force control methods are employed to regulate the interaction force between the ultrasound probe and the patient, each offering specific advantages. Position control, as implemented in [6], relies on precise robotic arm movements to maintain the desired force, providing a straightforward approach for force regulation. Impedance control, highlighted in [9,13], adjusts the robot's compliance based on measured forces, enabling safer and more consistent contact by dynamically responding to force variations. Hybrid position–force control, used in [8,10,12], combines the strengths of both position and force control to dynamically modulate the probe's position and applied force. This approach ensures optimal contact with the patient and enhances image quality by balancing movement precision with force regulation. Compliance control, described in [15], maintains a constant desired force within a specified range, allowing the system to adapt to varying tissue characteristics while ensuring both patient safety and imaging accuracy.

In addition to these force control strategies, some studies incorporate additional mechanical safety features to limit the maximum force, providing an extra layer of protection. Housden et al. [16] developed a spring-ball-based mechanical clutch to mechanically restrict the force to a safe limit. Similarly, Tsumura et al. [17] employed a constant-force spring mechanism that, combined with a counterweight system, ensures a steady and limited contact force by offsetting the probe's weight and directly controlling the applied force. These mechanical safety features ensure that the interaction force remains within safe limits, enhancing patient safety during robotic ultrasound procedures.

In conclusion, current robotic ultrasound systems exhibit a diverse range of force sensors and control strategies. However, most of these systems rely on either standardized and costly commercial force sensors or the built-in sensors of commercial robot arms. There is a notable lack of solutions specifically designed for ultrasound scanning that utilize lightweight and cost-effective sensing systems for custom-made robots, which often lack sophisticated internal force sensors.

In this paper, we present a solution for automated extra-body ultrasound scanning, with a comprehensive robot control framework as the core contribution. Supporting this framework are innovations in sensor design, algorithm development, and validation through realistic testing, collectively advancing the feasibility of automated ultrasound in clinical applications. The key contributions of our work are as follows:

(1) Robot Control Framework: The primary contribution of this work is the development of a comprehensive robot control framework for automated ultrasound scanning. This framework integrates patient-specific surface-based path planning, operator inputs, and force-control algorithms, ensuring safe and consistent acquisition of high-quality ultrasound images. This framework represents a complete and robust solution for automated scanning in real-world scenarios.
(2) Force Sensor Design: To support the control framework, we designed a compact, single-axis force sensor using load cells that can be seamlessly integrated into the ultrasound probe. This sensor is optimized for 3D printing, reducing fabrication costs while providing reliable and accurate force measurements necessary for implementing the control algorithms.
(3) Phantom Design: We developed a two-layer abdomen phantom, featuring a stiff inner layer with an abdominal aortic aneurysm (AAA) and inferior vena cava (IVC) and a soft adipose-mimicking outer layer. This phantom closely resembles human tissue, allowing realistic testing of force-control algorithms and ultrasound imaging in a simulated clinical setting.
(4) System Validation: We validated the system's performance through both phantom and healthy volunteer experiments, demonstrating its effectiveness in real-world scenarios.

This paper is organized into two main parts: sensor design and characterization, followed by robot control design and system performance testing. In the methodology section, Sections 2.2 and 2.5.1 focus on the design and characterization of the custom force sensor, detailing its mechanical structure, signal processing, and calibration experiments. Sections 2.3, 2.4, 2.5.2 and 2.5.3 describe the robot control algorithms, phantom design, and experimental setups for validating the system's performance. In the results section, Section 3.1 presents the outcomes of the sensor characterization experiments, including calibration, accuracy, and performance under real-world conditions. Sections 3.2 and 3.3

evaluate the robot control algorithms and system performance, showcasing the integration of force control, path planning, and automated ultrasound image acquisition through both phantom and volunteer experiments. Finally, Sections 4 and 5 discuss and conclude this research, respectively.

2. Materials and Methods

The aim of this research is to develop a system for automated ultrasound scanning of a target region, ensuring consistent force application within a predefined safe range. To achieve this, we began by implementing system registration and calibration to ensure precise alignment and tracking between the patient and the robotic system. We then designed and fabricated a cost-effective, custom-built force sensor tailored for ultrasound applications. Building on this, we developed a force control strategy to maintain optimal probe contact during scanning, ensuring patient safety and high-quality imaging. Finally, we constructed a detailed, ultrasound-compatible phantom as a realistic testing platform to validate the system's functionality before advancing to patient trials.

2.1. Robotic System Setup and Workflow

This research advances a dual-arm robotic system, originally developed by our research team at King's College London, for enhanced prenatal ultrasound screenings as detailed in [18]. Featuring 17 degrees of freedom, the robot's end effector is equipped with an ultrasound holder that mimics the dexterity of a human sonographer's wrist. It is controlled through custom software developed in C++ using Qt Creator 8.0.2 (based on Qt 6.3.2, GCC 10.3.1, The Qt Company Ltd., Espoo, Finland). The software not only allows real-time visualization of the robot's pose and the patient's surface mesh but also supports various control methodologies, including motor and endpoint control. This makes it highly suitable for both research and clinical applications. Additionally, the system is ethically approved for abdominal scans, including use with pregnant women. The system incorporates patient-to-robot registration and real-time movement detection using two 3 cm × 3 cm AprilTag markers—one placed on the patient and the other on the robot—along with an Intel RealSense RGB-D camera, as shown in Figure 1. This setup enables the robot to continuously track the patient's position and dynamically adjust for any patient movement. Additional details can be found in our previous work [19].

The general workflow of our current system for autonomous robotic ultrasound acquisitions is illustrated in Figure 2. The system operates in two main stages: Before Acquisition and During Acquisition. In the Before Acquisition stage, an external RGB-D camera captures the patient's surface information, which enables real-time patient-to-robot registration using AprilTag markers. This process ensures that the scanned patient surface aligns correctly within the robot control's simulation environment, allowing the sonographer to precisely select the desired scanning region in the software. The robot then moves to the corresponding location on the patient's body to perform the scan. During the image acquisition stage, the robot autonomously follows a scanning path defined by the sonographer while actively monitoring and adjusting contact force via force sensors. It dynamically modifies the path in response to patient movement and force feedback, ensuring safe and accurate image acquisition. Upon completion, the acquired ultrasound images can be transferred to healthcare professionals or AI-based algorithms for further assessment, facilitating diagnostic support or advanced analysis.

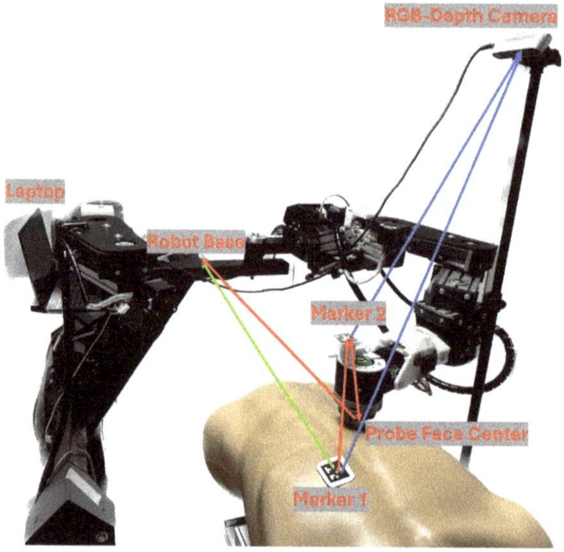

Figure 1. Patient-to-robot base registration process, along with three transformations utilized in this study shown with red arrows, relative marker transformation calculation with the help of the camera with blue arrows, and final patient-to-robot base transformation with the green arrow [19].

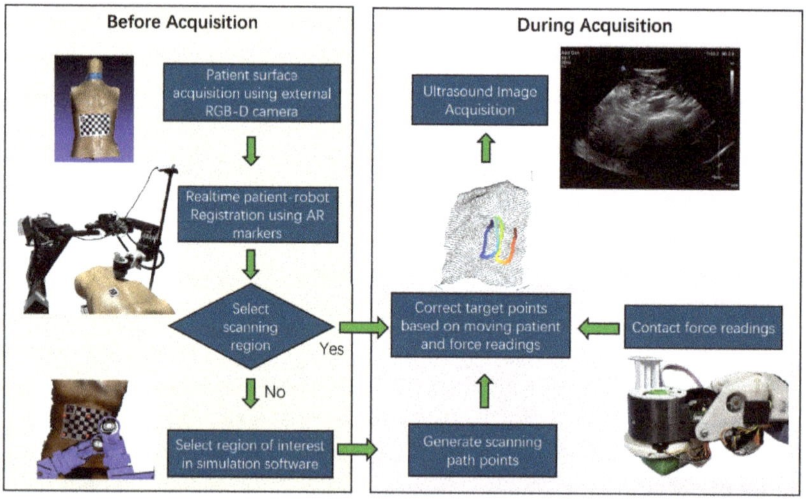

Figure 2. Workflow for automatic ultrasound image acquisition of the current system. The colorful path in the 'Correct Target Points' section represents the corrected path plan, with blue indicating the start and red indicating the stop point. For detailed explanation, see Figure 20.

2.2. Force Sensor Design for Ultrasound Probe Compatibility

2.2.1. Mechanical Structure Design

In most robotic ultrasound systems, the force sensor is typically positioned either at the probe face to measure the direct force between the patient and the probe [10] or at the top of the probe [6,7,9], between the robot end and the probe, for ease of mounting on the robot. However, placing the sensor on the probe face limits the sensor size and can cause shadowing on the ultrasound image. On the other hand, mounting the sensor at the top of the probe leads to indirect and less accurate force measurements due to the complexity of force transmission.

To overcome these challenges, our goal is to design a round-shaped force sensor that holds the probe similarly to a human sonographer, inspired by and building upon our previous model [11]. We aim to create a sensor that is not only affordable and easy to manufacture but also widely accessible to researchers and industries around the world. By focusing on a straightforward structure and force-sensing principle, we can ensure that the sensor is easy to repair if needed. Additionally, since this sensor is intended for clinical use, it is designed to be robust and capable of long-term operation without failure. Lastly, the sensor is designed to be adaptable to a wide range of commonly used ultrasound probe types and robotic systems, ensuring broad applicability for various research and clinical applications.

To meet these requirements, we have selected the FX29 Compact Compression Load Cell from TE Connectivity as the basic element. These load cells are cost-effective, readily available, and can be easily replaced if damaged, avoiding the need to rebuild the entire sensor. The proposed sensor consists of four parts, as is shown in Figure 3. The left and right parts are separate to fit the probe, while the upper and lower parts are separate to facilitate easy replacement of the load cells. The lower part is designed to hold the probe firmly, with its inner shape derived from the probe's CT scan. The upper part is designed to fit the sensor into the robot's end effector based on the robot's CAD model. The left-right and upper-lower parts of the sensor can be easily modified and reprinted to accommodate different probe shapes and robotic configurations, making the sensor highly adaptable.

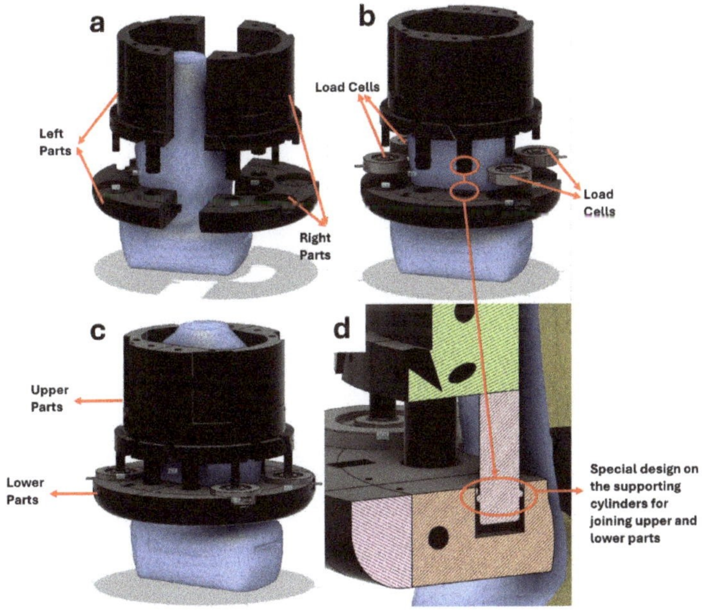

Figure 3. Custom-made force sensor mechanical structure design. (**a**) Left and right parts are separate to fit the ultrasound probe; (**b**) upper and lower parts are separate for easy replacement of the load cells; (**c**) complete sensor structure with probe and load cells; (**d**) special design on the supporting cylinders.

The load cells are evenly distributed on the lower plate, providing a stable structure. Eight cylinders extend from the top half; four are designed to push the load cells at their centres, with lengths matching the upper and lower gap to prevent undue pressure on the load cells when unloaded. The remaining four cylinders are supporting cylinders with tiny bulges at their ends (Figure 3d). By applying pressure, these supporting cylinders fit securely into corresponding holes in the bottom plate, ensuring that upper and lower parts fit together during scans. The holes for the supporting cylinders are designed to be 2 mm deeper than the cylinders themselves. This 2 mm gap exceeds the maximum deformation expected in the load cells, allowing the supporting cylinders to move freely downward under pressure without transferring force. The entire sensor structure is fabricated using 3D-printed PLA material, optimizing for both function and manufacturing efficiency.

The sensor is calibrated as a single-axis force sensor, meaning it only measures forces perpendicular to the plane of the four load cells. The total force (F_z) is computed by summing the readings from the four load cells. This calibration ensures accurate measurement of vertical forces, which is used in subsequent force-control algorithms.

2.2.2. Force Sensor Signal Processing

In this research, we implemented a load cell system using four load cells with I2C communication, each with four wires: power, ground, clock, and data. The load cells were assigned unique I2C addresses and connected to a shared I2C bus using ribbon cables for improved organization and reliability. The system features a custom-made printed circuit board (PCB) equipped with pull-up resistors and a zero (tare) button, which is crucial for calibrating the load cells before operation (Figure 4b). The I2C signals are processed by a Teensy 4.1 microcontroller (PJRC, Sherwood, Oregon, USA) housed in a custom 3D-printed enclosure mounted beneath the robot arm (Figure 4c). The Teensy 4.1 reads the load cell data from the PCB and transmits them to the robot's main microcontroller for force control. This setup not only ensures seamless integration of load cell data with the robot's control system but also allows for easy replacement of damaged load cells and direct signal access for debugging purposes (Figure 4d).

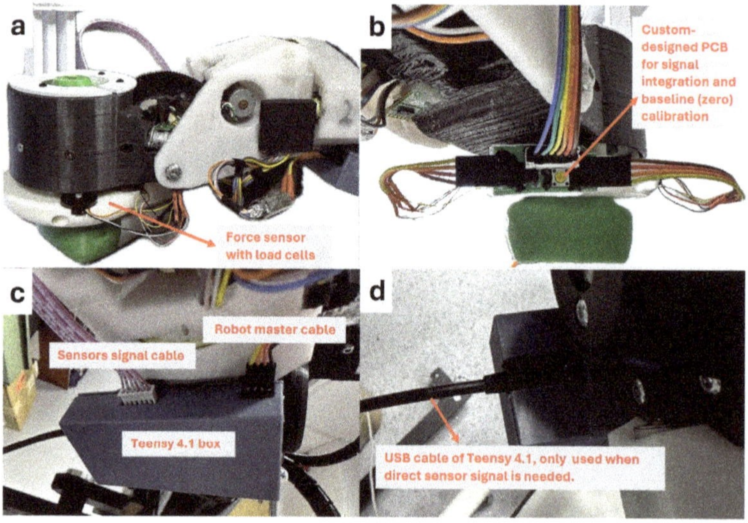

Figure 4. Force sensor signal processing setup. (**a**) Front view of the force sensor with integrated load cells. (**b**) Back view of the force sensor showing the custom-designed printed circuit board (PCB) for signal integration and zero calibration, along with two 4-pin input cables (each carrying signals from two load cells) and a 6-pin output cable connected to the Teensy 4.1. (**c**) Bottom view of the 3D-printed box for the Teensy 4.1 board, with sensor signal and robot master cables. (**d**) Top view of the Teensy 4.1 box, highlighting the USB cable used for direct sensor signal access when needed for debugging.

2.3. *Robot Control Algorithm Design*

2.3.1. Force Control Strategy Design

To ensure safe and effective scanning, we implemented a force control strategy that dynamically adjusts the probe's position to maintain the contact force within a patient-specific, predefined range. This range is designed to balance safety and image quality during extra-body ultrasound procedures. The "force" referred to here is (F_z), the total vertical force calculated as the sum of the four load cells, as described in Section 2.2.1.

Our system employs a proportional (P) control strategy, a simple yet effective approach for maintaining consistent force. The system continuously monitors the contact force and calculates the deviation from the desired range. If the current force is within the range, no adjustment is needed. However, if the force exceeds or falls below the desired range, an adjustment is applied to the probe's position to correct the error.

The adjustment distance, d_{adjusted}, is calculated as:

$$d_{\text{adjusted}} = d_{\text{current}} + k_p \cdot F_{\text{error}}$$

where F_{error} represents the deviation from the target force range, and k_p is the proportional control coefficient tuned based on the softness of the tissue. Softer tissues require higher k_p values for greater sensitivity to force changes, while firmer tissues require lower k_p values to prevent overcorrections.

To enhance the system's precision, the robot calculates the normal direction of the surface at each contact point. Adjustments to the probe's position are made along this normal direction, ensuring that the ultrasound probe maintains optimal alignment with the scanning surface. The new target position is determined as:

$$\text{TargetPosition} = \text{SurfacePointPosition} + \text{TargetSurfaceNormal} \cdot d_{\text{adjusted}}$$

2.3.2. Path Planning

To achieve a higher level of automation, we equipped our robot with autonomous scanning path planning, implementing two distinct methods: direct path planning and area coverage path planning.

We first designed a user interaction interface to enable users to select path points and save them in sequence. The interface, shown in Figure 5, includes a 3D model of the patient's abdomen and a simulated robotic arm, facilitating intuitive interaction. Using the Visualization Toolkit 7.1 (VTK) in the C++ library, we created an interactive window that visualizes the path point selection process.

In direct path planning, users manually select specific points on the 3D model by right-clicking on the desired locations. For each selected point, the system retrieves its 3D pose as a 4 × 4 transformation matrix, which encodes both the position and orientation information. These poses are stored sequentially to define a predefined scanning path. A red sphere visually marks the selected points on the 3D model, providing immediate feedback to the user, as shown in Figure 5a. This method is particularly suitable for small scanning areas with clear anatomical landmarks, such as the belly button or nipples, allowing for precise path customization.

In area coverage path planning, users define a rectangular scanning region by clicking and dragging the mouse over the desired area on the model. This action prompts VTK to capture the start and end points of the drag, which are then used to compute the four corner coordinates of the rectangle (Figure 5b). Based on these coordinates and the size of the ultrasound probe, a serpentine (zig-zag) path is generated to ensure complete coverage of the selected region. These path points are automatically calculated and stored in sequence for the robot to follow during scanning. This method is particularly advantageous for larger regions without clear anatomical landmarks, offering a more automated solution.

By integrating path planning with force control, the robot dynamically maintains safe and accurate alignment with the surface during scanning, ensuring optimal image quality and patient safety.

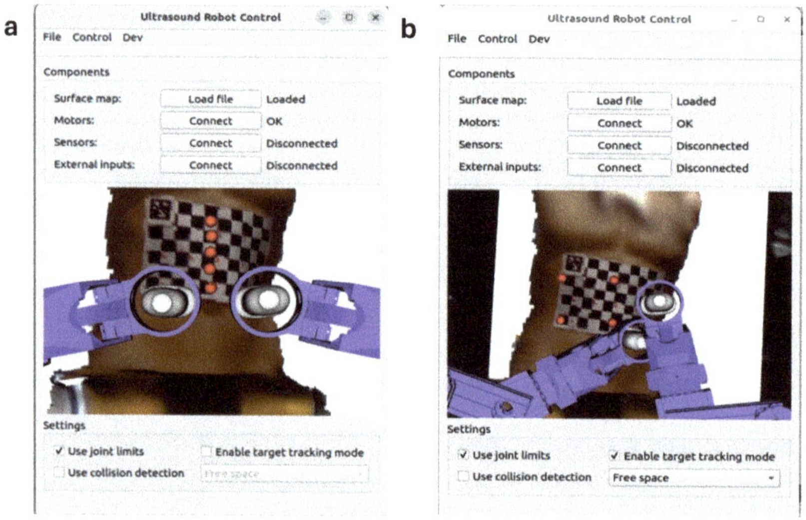

Figure 5. User interface for two path planning methods: (**a**) direct path planning method; (**b**) area coverage path planning method.

2.4. Tissue-Mimicking Phantom

To validate the force control mechanism of the robot, an abdominal phantom was developed. This model is critical for assessing abdominal aortic aneurysm (AAA), combining an idealized AAA model and an inferior vena cava (IVC) based on published vessel measurements from computed tomography and ultrasound scans. The abdomen phantom includes a stiffer layer containing the AAA and IVC, along with an outer adipose tissue layer to interact with the robotically controlled ultrasound (US) probe (Figure 6a).

Three-dimensional models of the lumens of the AAA and IVC were created based on aggregate dimensions from CT scan measurements. Models were 3D-printed using a desktop fused filamented fabrication printer. The AAA and its branches were printed in polyvinyl alcohol (PVA), a water-soluble synthetic resin, while the box and the unbranched IVC were printed in polylactic acid (PLA). Both pieces were coated with epoxy resin to create a smooth surface finish.

To create the stiffer inner layer of the phantom, a polyurethane rubber, Vytaflex 20 (Smooth-On Inc., Macungie, PA, USA), was chosen due to its Shore hardness of 20A and elastic modulus of $E = 337$ kPa. The urethane rubber was degassed in a vacuum chamber to eliminate air bubbles—a process essential to minimizing artefacts in imaging. The mixture was then poured into the box until the AAA and IVC lumen models were just covered and left to cure.

Following curing, the PVA model was dissolved in a heated and stirred water bath. The PLA model was removed by applying force to one end of the model.

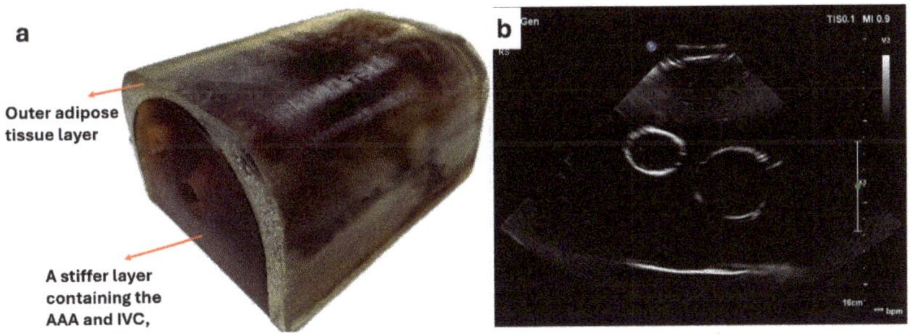

Figure 6. Ultrasound-compatible abdominal aortic aneurysm phantom: (**a**) external appearance and (**b**) ultrasound visualization of internal structures. The vertical dotted line in (**b**) represent a depth scale bar, with each segment indicating 1 cm.

To make the phantom more realistic to human abdomen tissue, the outer adipose layer simulating subcutaneous fat was added. A mixture of BJB F116 and BJB SC22 polyurethane rubber (BJB Enterprises, Tustin, CA, USA) was used, resulting in a flexible, realistic layer with a shore hardness of 00-15.

An ultrasound image of the phantom is shown in Figure 6b. For optimal imaging, ultrasound gel was applied between the two layers, and the phantom was submerged in water to a level above the vessels and below the probe. Scanning was performed using an EPIQ 7G Philips ultrasound machine (Bothell, WA, USA) and the Philips X6-1 ultrasound probe, optimized for abdominal imaging with a frequency range of 6–1 MHz. Minimal force was applied to the probe, which was sufficient to clearly delineate the boundaries of the two embedded vessels, mimicking realistic visualization conditions.

The ultrasound image reveals distinct vessel boundaries with minimal signal artefacts, indicating an effective simulation of abdominal structures. The contrast and clarity in the image suggest that the phantom's materials and layer composition are appropriate for testing the force control and imaging capabilities of the robot, especially for applications like AAA assessment. This setup provides a reliable platform for evaluating the interaction between the robotically controlled ultrasound probe and simulated soft tissue, enhancing the realism and utility of the testing environment.

2.5. Experimental Design and Setup

This section outlines the design and setup of experiments conducted to validate the system's performance, focusing on sensor accuracy and robotic control. The force sensor testing experiments ensured the reliability of the custom-designed sensor when integrated with the robotic system. The force-control algorithm testing experiments were designed to validate the robot's ability to maintain target forces during sweeping motions, a key factor for ensuring patient safety. Finally, the automated ultrasound image acquisition experiments on phantom and volunteer subjects were designed to validate the system's ability to acquire clear, clinically usable images under real-world conditions, such as patient breathing and varying tissue properties.

2.5.1. Experimental Setting for Force Sensor Calibration and Performance Evaluation

As mentioned in Section 2.2.1, four load cells are evenly spaced on the same level and positioned on the lower part of the force sensor. As the four load cells share the total Z-axis force, the final sensor reading is obtained by summing the individual readings from each load cell. The force sensor calibration was conducted by applying known weights ranging from 0 g to 3500 g in increments of 500 g (Figure 7a). For each weight, the corresponding sensor readings were recorded. The calibration function was determined by calculating the linear fit of the sensor's raw data against the known weights. After deriving the calibration function, the sensor's accuracy, stability, and hysteresis were evaluated through further testing. To assess stability, known weights were applied to the sensor and held for 5 s, during which all sensor readings were recorded. The standard deviation of these readings was calculated to quantify the sensor's stability. Accuracy was evaluated by averaging the sensor's readings over the same period and comparing the result with the applied weight. Finally, hysteresis testing involved sequentially applying and removing weights one by one over five cycles. The sensor readings during the loading and unloading phases were compared to identify any discrepancies or lag in the sensor's response.

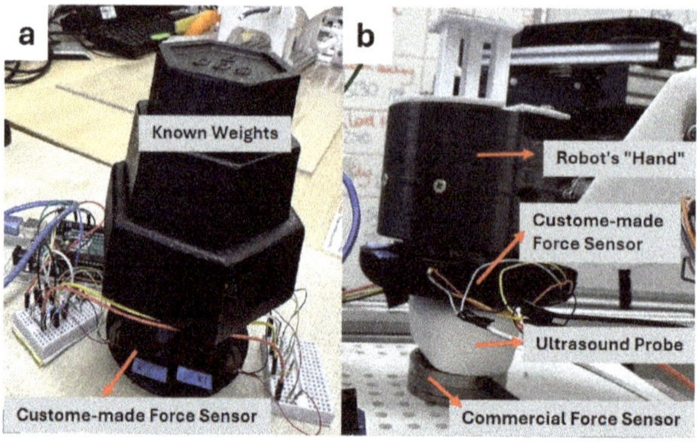

Figure 7. Sensor calibration and performance assessment: (**a**) on its own with known weights; (**b**) on the robot with an ultrasound probe and commercial force sensor.

We designed another experiment to test the sensor performance under the real ultrasound scanning scenario. In this experiment, the sensor was mounted onto the robot's end effector to assess its performance under real ultrasound scanning conditions. In this experiment, the force applied by the robot arm was unknown, so an ATI Mini40 force sensor (ATI Industrial Automation, Apex, NC, USA) was used as a reference. As shown in Figure 7b, the ATI sensor was securely mounted on a custom-designed support platform to ensure stability and prevent slippage. The robot arm was positioned at various angles along different axes, including both in-plane and out-of-plane rotations, to push the ATI sensor and mimic real ultrasound scanning scenarios. A total of 32 poses were tested to evaluate the performance of our custom-designed force sensor. These poses included straight pushes along the Z-axis, lateral angles ranging from −25° to +25° in increments of 5°, elevational angles from −25° to +25° in 5° increments, and axial rotations from −60° to +60° in increments of 20°. Because the ATI sensor measures force along the global Z-axis, while our custom sensor measures force relative to the ultrasound probe's axis, we calculated the projected force component on the global Z-axis for comparison with the ATI readings. It is important to note that the ATI sensor's readings include the weight of the probe and the sensor itself, in addition to the force applied by the robot arm. These weights were premeasured and subtracted from the ATI's readings.

2.5.2. Experimental Setup for Force-Control Algorithm

In order to verify the usability and reliability of the surface force-following mode in the Z direction, we designed a platform for conducting the standalone force control experiment, as shown in Figure 8. The whole experiment was divided into the following steps: (1). One AprilTag marker and an ATI sensor were securely mounted on a 3D-printed platform, with the sensor fixed using screws. Another AprilTag marker was attached to the robot arm. (2). A depth camera was used to collect the pose data of the AprilTag markers on the platform and on the wrist of the robot in real time. (3). The platform was moved up and down by hand and we observed whether the robot follows the movement of the platform. In this process, we set the desired force range to be 3–6 N. (4). We observed the sensor data, i.e., the change in contact force, during the control of the platform moving up and down, and tried to keep the force values at less than 3 N, 3 N–6 N, and more than 6N, each for a certain period of time.

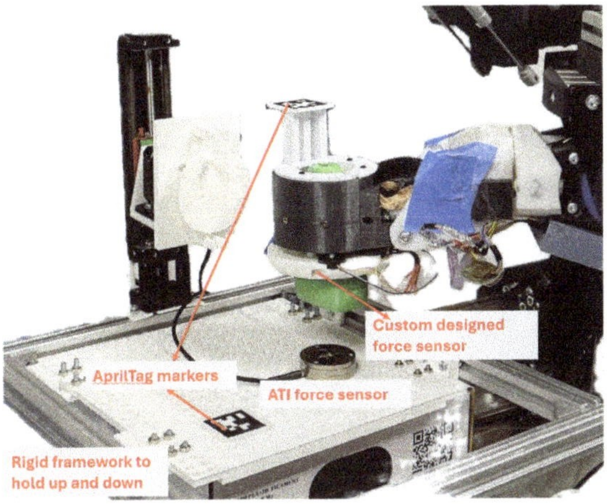

Figure 8. Experimental setup for validating the surface force-following algorithm in the Z-direction using an ATI force sensor, a 3D-printed framework, and two AprilTag markers used for position registration.

To assess the surface force-following function with the path-planning algorithm, we performed an automated scan on our AAA phantom (without the adipose layer). Initially, the 3D surface of the phantom was scanned and imported into the software. Users can employ the two-point rectangle method to manually select the upper-left and lower-right corners of the rectangle, after which the software automatically calculates the remaining two corners to define the full scanning area, as shown in Figure 9. The robot was programmed to execute a full-coverage scan within this designated area, aiming to maintain a consistent contact force ranging between 3 N and 6 N throughout the process. This force range was manually set to ensure optimal contact between the probe and the phantom surface. However, since the ideal force range varies among individuals, this range will be adjusted accordingly for volunteer scans.

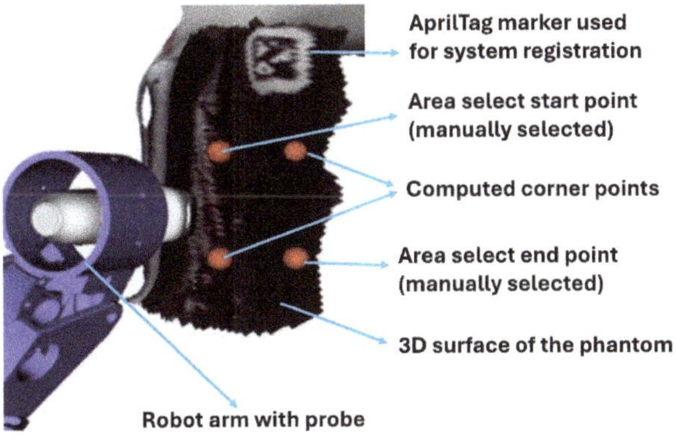

Figure 9. Selected area on abdominal phantom in simulation window.

2.5.3. Experimental Setup for Automated Ultrasound Image Acquisition

Following the successful validation of the integrated functionality of the path-planning-based force-control algorithm, the system was deemed ready for automated ultrasound image acquisition using a real ultrasound probe. The setup for this acquisition process is illustrated in Figure 10. To mimic clinical imaging conditions and minimize surface friction, the AAA phantom was submerged in water, which eliminated air bubbles that could interfere with ultrasound signal transmission. The water level was carefully maintained below the ultrasound probe to ensure smooth probe movement without submerging the probe. Ultrasound gel was also applied between the probe and the phantom to ensure optimal contact. The custom-designed force sensor securely housed the ultrasound probe and was mounted on the robot's end effector. Two AprilTag markers and one external camera were used to register the phantom with the robot. A standby operator was standing near the robot and phantom to monitor the entire process, ready to press the emergency stop button if necessary. For added safety, this emergency stop button will be given to volunteers or patients in human trials.

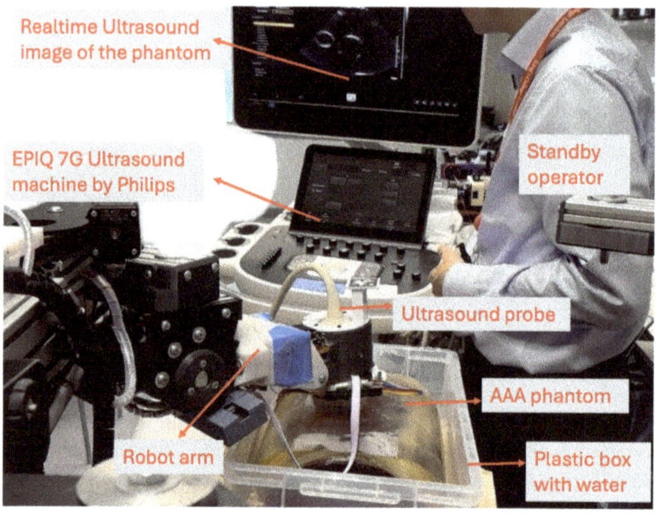

Figure 10. Automated ultrasound image acquisition on phantom experiment setup.

The phantom test procedure follows the same methodology as in Section 2.5.2, where the robot autonomously plans its path to cover the user-defined region while maintaining applied force within a specified range. In this experiment, however, our primary focus shifted to evaluating the system's ability to acquire ultrasound images and analyzing the image quality obtained under varying force levels.

Lastly, we conducted volunteer tests to evaluate the system's performance under real-world conditions, including patient breathing and variations in tissue stiffness. We obtained ethical approval from the King's College London local ethics committee to test our robotic ultrasound system on volunteers for general abdominal scans (study title: Investigating Robotic Abdominal Ultrasound Imaging, study reference: HR-22/23-5412). Although our long-term goal is to develop a universal automatic ultrasound scanning protocol, this study first focused on the imaging of the abdominal aorta and common iliac arteries as an initial application due to the relatively straightforward motion planning requirements.

Two volunteers with different body shapes participated in the study: Volunteer 1 with a body mass index (BMI) of 21 and Volunteer 2 with a BMI of 33. The experimental setup, illustrated in Figure 11, was similar to the phantom test. In this configuration, the volunteer lay on a bed, with the RGB-D camera and robot positioned on their left side. The robot held an ultrasound probe connected to the ultrasound machine, which was located on the volunteer's right side. The ultrasound system used was the Philips EPIQ 7G paired with the X6-1 probe (Philips Healthcare, Amsterdam, Netherlands), consistent with the phantom test setup. For safety, each volunteer was provided with an emergency stop button to hold during the procedure. If pressed due to discomfort, the robot's power was immediately cut off, halting all operations to ensure the volunteer's safety and comfort.

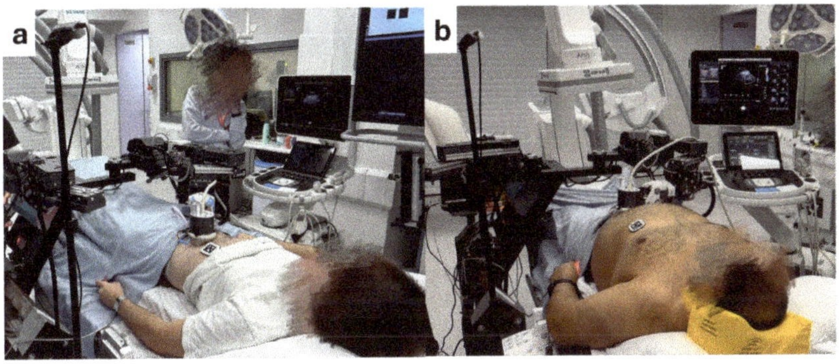

Figure 11. Setup for automated ultrasound image acquisition in healthy volunteer tests: (**a**) Volunteer 1 with BMI = 21, (**b**) Volunteer 2 with BMI = 33.

The volunteer test consisted of three distinct stages: (1) a manual scan performed by the operator, (2) a robotic scan conducted at the same position with varying levels of applied force, and (3) an automated robotic sweep designed to cover the entire region of interest.

Unlike scanning a phantom, which contains only two vessels and is composed of materials specifically designed for ultrasound imaging, the anatomy of the human abdomen is far more complex and varies between individuals. Therefore, the initial step is to locate the abdominal aorta or the common iliac arteries for each volunteer. To achieve this, we performed a freehand ultrasound scan. The scan is conducted exclusively in the transverse plane (where the aorta appears as a circular structure in the ultrasound image), as the diameter of the aorta is a critical criterion for detecting aneurysms. We initially positioned the probe on the volunteer's abdomen, several centimetres to the left of the umbilicus. In the resulting ultrasound image, the top corresponds to the volunteer's skin surface, while the bottom represents the spine and posterior structures. There are a number of key features used to identify the aorta. It appears as a circular structure with bright walls and a dark lumen. This is distinguished from the inferior vena cava (IVC), which has thinner walls and therefore appears less bright. The IVC is also less circular than the aorta due to the thinner walls and reduced intraluminal pressure (Figure 12a). This distinction can be further confirmed by the pulsing motion observed in B mode in the aorta, which is particularly clear in slimmer individuals. The pulsatile nature of the aorta can also be demonstrated with colour and spectral Doppler (Figure 12b), which contrasts to the respiratory modulated phasic flow demonstrated in the IVC. Anatomically, the aorta is situated just superficially to the spine, and to the left of the IVC. The spine appears as a curved structure at the bottom of the image, with a bright boundary and dark interior. Throughout this process, the sonographer must also adjust the ultrasound machine settings, including gain, depth, and focus, to optimize the image quality. These parameters are patient-specific and are carefully fine-tuned to ensure clear visualization of the aorta and surrounding structures.

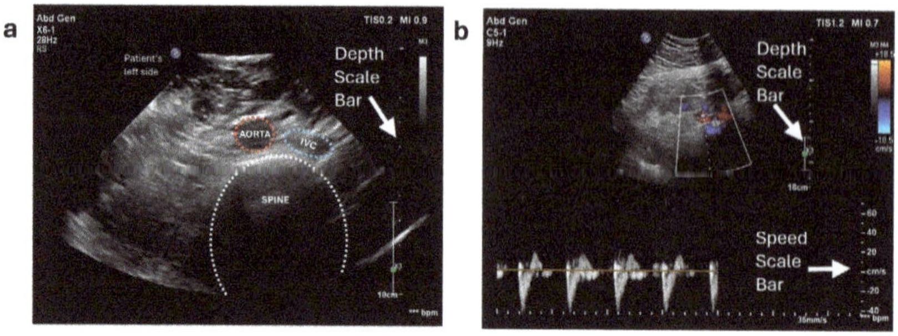

Figure 12. Abdominal aorta and iliac artery localization in volunteer testing: (**a**) B−mode image of the abdominal aorta annotated by a clinical vascular scientist (Volunteer 1) and (**b**) pulsed wave Doppler of the iliac artery (Volunteer 2).

After identifying the location of the aorta or iliac artery, the next step was to determine the optimal force range required to acquire high-quality ultrasound images for each volunteer. This force range was later used during the automatic scanning process. In this step, the ultrasound probe was mounted on the robot, and the system was operated in "kinematics control" mode to vary the probe's depth in the Z-direction. Force values and their corresponding ultrasound images were recorded across a range of forces, from 2 N (representing minimal full contact) to 10 N. The results are presented in Figure 13. For Volunteer 1 with BMI = 21, optimal image quality was observed within the force range of 3–6 N. For Volunteer 2 with BMI = 33, optimal images appeared at 4–10 N. However, to balance image quality with patient comfort, the desired force range for automatic scanning was set to 4–7 N. As the applied force increases, depth to the centre of the aorta/iliac artery decreases. This occurs because the increased force causes the abdominal tissue to deform downward, compressing and displacing any gas pockets, which brings the probe closer to the aorta/iliac artery.

The third step mirrored the process used in the phantom test: selecting a region of interest within the robot control software and allowing the robot to automatically scan that area while maintaining the applied force within the predetermined range. For both volunteers, the region around the umbilicus was selected, with efforts to position the manual scan point at the centre of the targeted area.

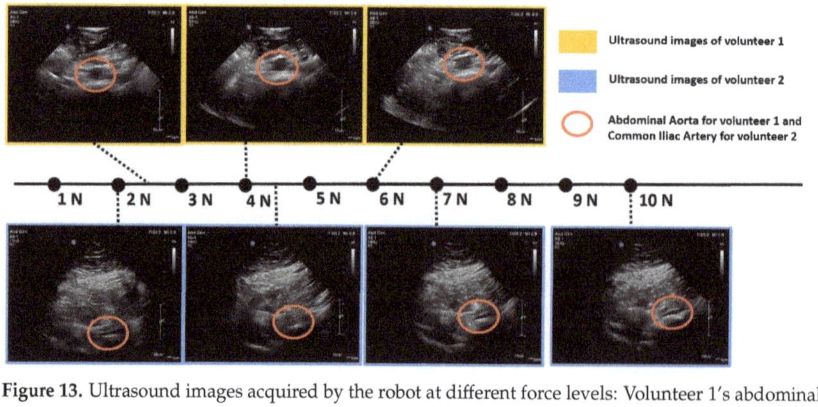

Figure 13. Ultrasound images acquired by the robot at different force levels: Volunteer 1's abdominal aorta (yellow border) and Volunteer 2's iliac artery (blue border). The vertical dotted line in all the ultrasound images represent a depth scale bar.

3. Experiments and Results

This section mirrors the structure outlined in Section 2.5, presenting the corresponding results of the experiments described earlier. It systematically reports the outcomes of force sensor accuracy testing, the performance of the force-control and path-planning algorithms, and the system's capability to acquire ultrasound images on both a phantom and a healthy volunteer.

3.1. *Force Sensor Testing Experiments and Results*

3.1.1. Sensor Calibration and Sensor Performance Using Known Weights

Table 1 provides the results of the force sensor's calibration and performance evaluation. The Measured Weight column lists the force values recorded by the sensor when known weights were applied, showing close agreement with the ground truth across all tested weights. The Avg. Error (N) and Avg. Error (g) columns show the average deviation between the measured weight and the ground truth, which ranges from +0.015 N (1.5 g) at 4.907 N to +0.18 N (18.3 g) at 0 N, with no noticeable trend as the applied weight increases. Accordingly, the Percentage Error column expresses the average error as a percentage of the ground truth, ranging from 0.1% at 34.349 N to 0.69% at 19.628 N, showing relatively consistent accuracy across the weights. The STD (N) column represents the standard deviation of sensor readings over 5 s, which increases with the applied weight, from 0.006 N at 0 N to 0.17 N at 19.628 N, reflecting a slight decrease in stability with higher forces. Overall, the results demonstrate the sensor's reliable accuracy and stability across the tested weight range, with only minor variations.

Table 1. Accuracy and stability test results using known weight.

Ground Truth	Measured Weight	Avg. Error (N)	Avg. Error (g)	Percentage Error	STD (N)
0 N	0.18 N	+0.18 N	18.3 g	0.31%	0.006 N
4.907 N	4.922 N	+0.015 N	1.5 g	0.31%	0.031 N
9.814 N	9.840 N	+0.026 N	2.7 g	0.26%	0.049 N
14.721 N	14.815 N	+0.094 N	9.6 g	0.64%	0.155 N
19.628 N	19.764 N	+0.136 N	13.7 g	0.69%	0.17 N
24.535 N	24.653 N	+0.118 N	12.0 g	0.48%	0.065 N
29.442 N	29.581 N	+0.139 N	14.2 g	0.47%	0.124 N
34.349 N	34.313 N	−0.036 N	3.7 g	0.1%	0.14 N

As shown in Figure 14a, the sensor exhibited excellent linearity, achieving an R^2 value of 0.9994. Additionally, the hysteresis behaviour, illustrated in Figure 14b, was minimal, with the hysteresis loops nearly overlapping across all five testing cycles. Figure 15 provides separate hysteresis maps for the five testing cycles, offering a detailed view of the sensor's performance in each cycle. The maximum observed deviation was 0.19 N during the third cycle, which corresponds to only 0.52% of the full-scale output. This small deviation demonstrates the sensor's high precision, repeatability, and consistency across multiple testing cycles.

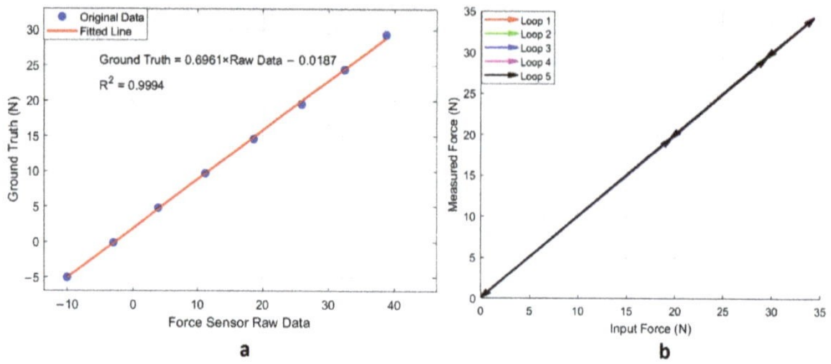

Figure 14. Performance testing of the force sensor using known weights: (**a**) linearity and calibration equation relating sensor raw data to ground truth and (**b**) hysteresis evaluation over five loading–unloading cycles, comparing input and measured weights. The five colored lines overlap, resulting in a single visible black line. Detailed zoomed-in graphs are provided in Figure 15.

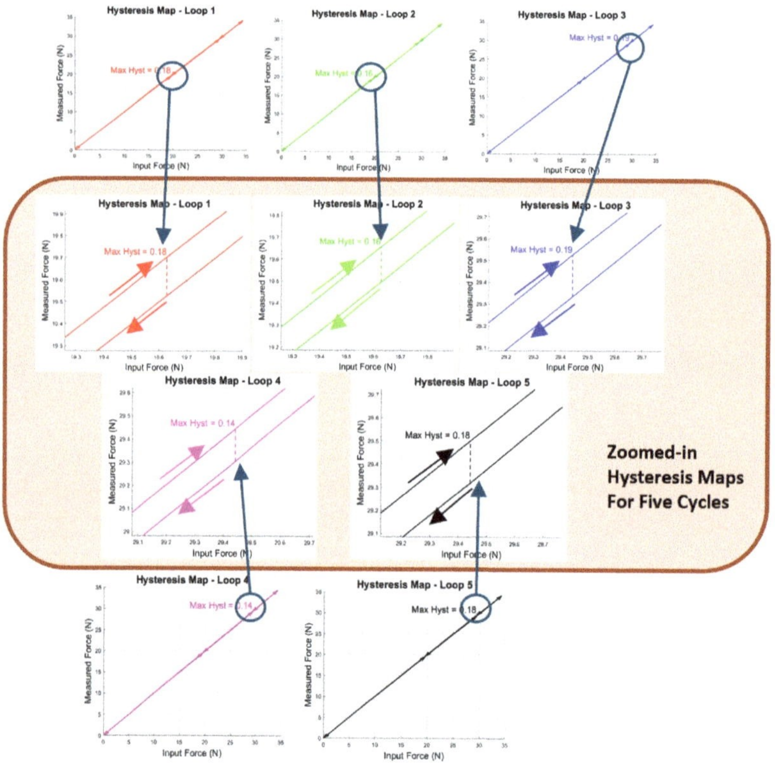

Figure 15. Separate hysteresis maps for five cycles with zoomed-in largest hysteresis points highlighted in the orange rectangle.

3.1.2. Sensor Performance on Robot Under Real-World Scenario

The root mean square error (RMSE) between our sensor readings and those from the ATI sensor was 0.71 N, with a standard deviation of 0.70 N. Some error may be attributed to slight deformation of the robotic arm under higher force levels, which can alter the effective pushing angle from the preset values. Despite this potential source of variation, a strong linear correlation was observed between the two sensors, as shown in Figure 16, with an R^2 value of 0.97. This high correlation indicates that our sensor can reliably measure the force exerted by the ultrasound probe on the patient across a range of pressures and poses. The error observed remains within the tolerance required for safe and effective abdominal scanning.

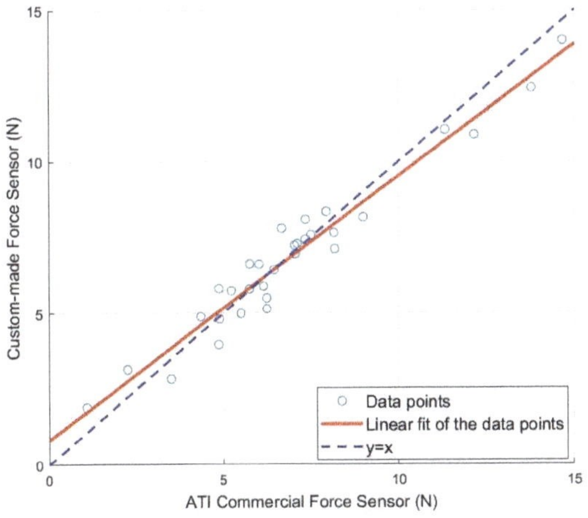

Figure 16. Correlation between the ATI commercial force sensor readings (*x*-axis) and our custom-made ultrasound force sensor readings (*y*-axis). Each blue point represents a measurement instance where the same force was applied to both sensors. The red solid line indicates the linear fit of these data points, while the blue dashed line represents the ideal 1:1 correspondence (y = x) between these two sensors.

3.2. *Force-Control Testing and Experiments*

3.2.1. Standalone Experiment on Force Control

To evaluate the sensor performance in real force-control scenarios, we cross-validated the ATI sensor readings on the platform with those of the force sensor integrated into the ultrasound probe, as shown in Figure 17. The comparison reveals a good agreement between the two sensors, particularly when contact forces are low; both sensors produce nearly identical readings under these conditions. As the contact force exceeds 8 N, a slight discrepancy emerges, with the ATI sensor registering slightly higher values than the custom force sensor. This minor difference is likely due to the custom sensor measuring force along the Z-axis of the ultrasound probe. At higher forces, any tilt in the platform may introduce a misalignment between the probe's axis and the true force direction, resulting in a slight underestimation by the custom sensor. This alignment under complex conditions suggests the robustness of our force sensor in measuring contact forces accurately.

Next, we plotted the Z-coordinate trajectories of both the robot arm and the platform in the camera coordinate system, as shown in Figure 18. Throughout the process, the gap between the two positions consistently remained at 80 mm (±10 mm), which corresponds to the vertical distance between the two markers when the robot arm and the ATI sensor are in proper contact. This consistency demonstrates the robot arm's ability to move upward under high force and downward under low force, effectively tracking the platform's vertical movement. However, around the 65-second mark, a rapid decrease in the platform's Z-coordinate led to a significant delay in the robot's response, with the gap exceeding 90 mm. This discrepancy is attributed to the fixed proportional gain in the hybrid force–position control, which limits the robot's ability to adjust its response rate to rapid changes in force. Consequently, when the platform descends quickly, the robot is unable to immediately follow. Additionally, occasional zero-force readings were observed due to electromagnetic interference from nearby motors. However, the control software is programmed to ignore these sporadic readings, ensuring they do not affect the robot's performance.

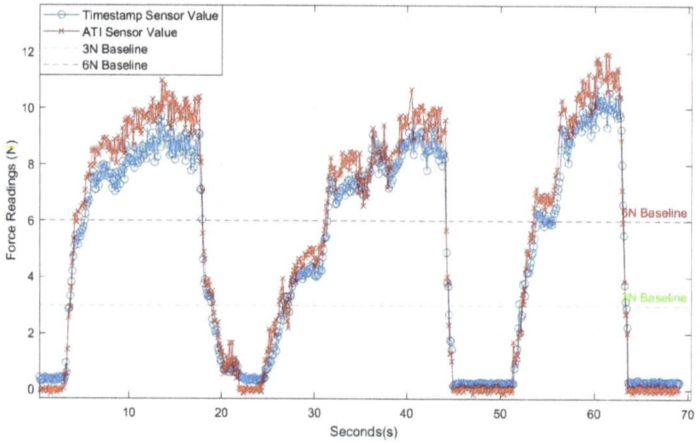

Figure 17. Force comparison between custom-made force sensor and ATI force sensor.

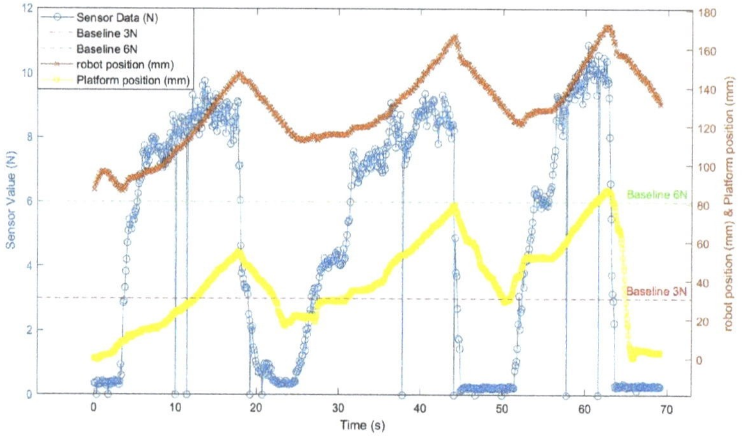

Figure 18. Force sensor data, platform position, and robot position over time.

3.2.2. Combined Path-Planning and Force-Control Experiment on Phantom

For the combined path-planning and force-control experiment, the recorded force sensor data, as illustrated in the upper graph of Figure 19, reveal the variations in the probe–surface contact force during the area coverage scan. During the scan, the force-control system dynamically adjusts the probe's contact force in real-time as it follows the designated path, operating continuously alongside path planning. As the robot approaches each new point on the path, the system compensates for variations in distance, trying to maintain the force within the target range. Throughout the process, the force fluctuates between 0 and 12 N, generally increasing when below 3 N and decreasing when above 6 N. The standard deviation of the force values is 1.9 N.

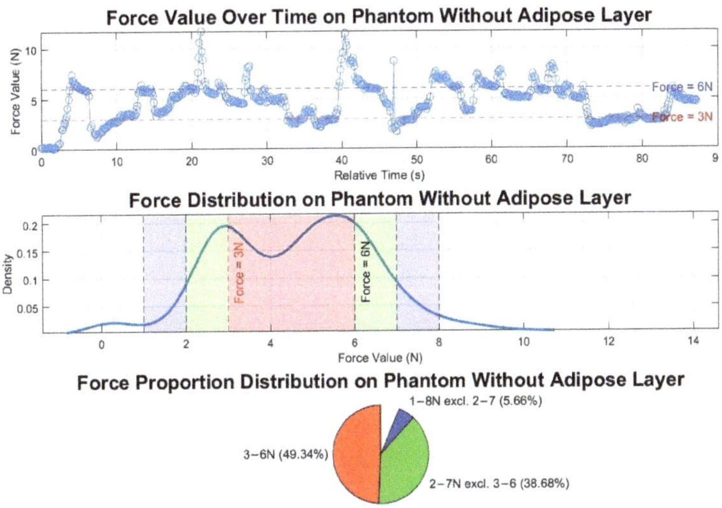

Figure 19. Force values and distributions in the combined path-planning and force-control experiment on phantom without adipose layer. The top subfigure shows force variation over time, with 3 N and 6 N thresholds marked; the middle subfigure depicts the force value distribution, highlighting ranges below 3 N, 3–6 N, and above 6 N; and the bottom subfigure displays the time proportion within each force range. Red, green and blue colours in the middle and buttom graphs match with each other.

Analysis of the force sensor data distribution, as shown in the lower two graphs of Figure 19, reveals two prominent peaks around 3 N and 6 N, indicating that these force values are commonly maintained during the scanning process. Outside the optimal 3–6 N range, the distribution density decreases, suggesting that the system spends less time at force values beyond this interval. This pattern highlights the force control system's stability within the target range. Additionally, the pie chart shows the distribution proportions, with a 49.3% probability of the force being within the optimal 3–6 N range and a 93.7% probability of it remaining within the broader 1–8 N range. However, there are instances where the force spikes to higher values, such as around 21 s and 40 s. This occurs because the material was too rigid, making precise force control more challenging. In Sections 3.3.1 and 3.3.2, subsequent tests conducted on softer tissues, including the phantom with an added adipose layer and healthy volunteers, demonstrated smoother and more stable force readings.

To validate the path planning, we first scanned the phantom using RGB-D data, obtaining the abdominal phantom surface in the camera coordinate system. By tracking the AprilTag marker on the robot arm and applying the transformation matrix from the arm to the probe's face centre, we monitored the probe centre's movement in real time within the camera coordinate system. Overlaying the phantom surface and probe trajectory data on the same graph provided a clear visualization of the ultrasound probe's movement across the abdominal phantom, enabling a detailed assessment of path accuracy and coverage, as shown in Figure 20.

The probe follows the anticipated trajectory, moving back and forth between the upper and lower boundaries of the rectangular scan area to achieve full coverage. Although only the centre point of the probe's trajectory is plotted, the probe's width, exceeding 15 mm, ensures complete area coverage in practice. The probe paths were not perfectly parallel because slight tilting occurred due to friction between the probe and the phantom surface during movement, as well as adjustments made in response to real-time force readings.

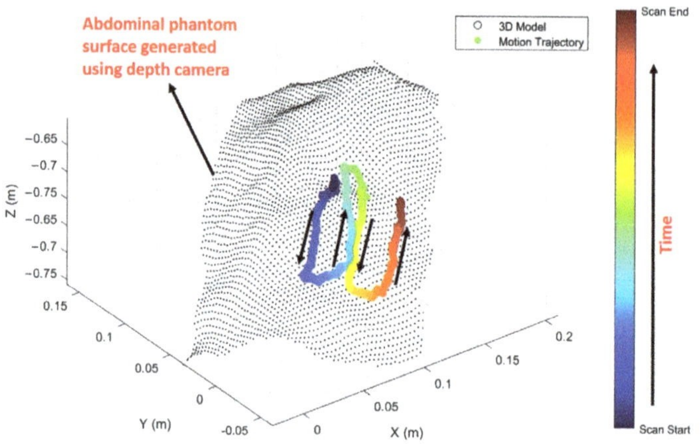

Figure 20. Probe's 3D trajectory and the phantom surface visualization in the real-world coordinate system. The abdominal phantom surface is represented as a black grid and the probe centre's 3D trajectory during the scanning process is colour-coded by time, transitioning from blue (start) to red (end).

3.3. *Automated Ultrasound Image Acquisition*

3.3.1. Phantom Tests

The image acquisition test followed a similar protocol to the standalone force-control experiments, so we used the same data analysis method to evaluate the force during the scan, as shown in the upper graph of Figure 21. In this experiment, 57.0% of the applied forces fall within the 3–6 N range, while 98% of forces remain within a ±2 N range (1–8 N). The standard deviation of the force readings is 1.59 N, which is lower compared to readings without the fat layer and ultrasound gel.

Figure 22 displays the ultrasound images acquired by the robot during the automatic sweep under varying force levels. Overall, the robot successfully generated clear abdominal aorta ultrasound images without human intervention. In these images, vessel walls appear as bright circular shapes, with water inside the circle and the surrounding area comprised of the phantom material (polyurethane). The smaller vein consistently appears clearer than the larger aorta, likely because it is closer to the phantom surface and, therefore, undergoes less attenuation.

We recorded the applied force for each ultrasound image and observed that the clearest images were obtained at force levels of 5 N and 6 N, while forces outside this range resulted in reduced image quality. The decline in quality at higher forces may be due to deformation of the phantom, impairing contact between the probe and the surface or compressing the embedded vessels—an issue unlikely to occur in real participants. In human subjects, signal strength and image clarity typically improve with increased force as the distance to the area of interest decreases and attenuation is reduced. However, excessive force can introduce artefacts and discomfort. To better understand this force–image relationship, we propose measuring the distance to the centre of the abdominal aorta (derived from the ultrasound depth scale) as a metric correlated with probe force and image quality in future human trials. These preliminary phantom tests demonstrate the robot's ability to autonomously acquire high-quality abdominal aorta images, laying a foundation for further studies to optimize the balance between image quality, patient comfort, and operator considerations in clinical settings.

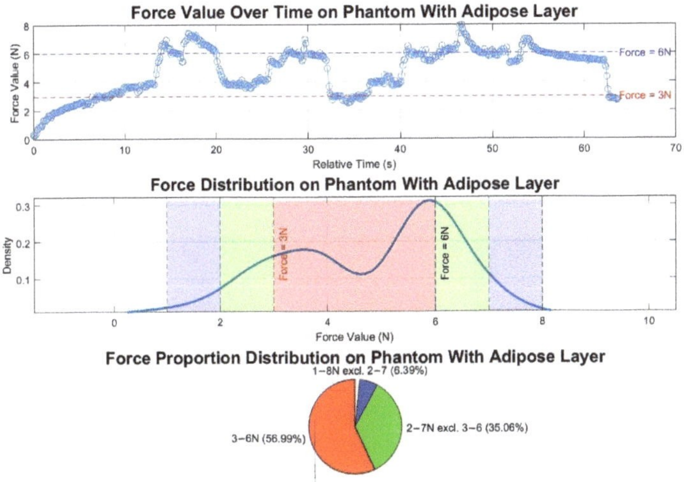

Figure 21. Force values and distributions in automated ultrasound image acquisition experiment on phantom with adipose layer. The top subfigure shows force variation over time, with 3 N and 6 N thresholds marked; the middle subfigure depicts the force value distribution, highlighting ranges below 3 N, 3–6 N, and above 6 N; and the bottom subfigure displays the time proportion within each force range. Red, green and blue colours in the middle and buttom graphs match with each other.

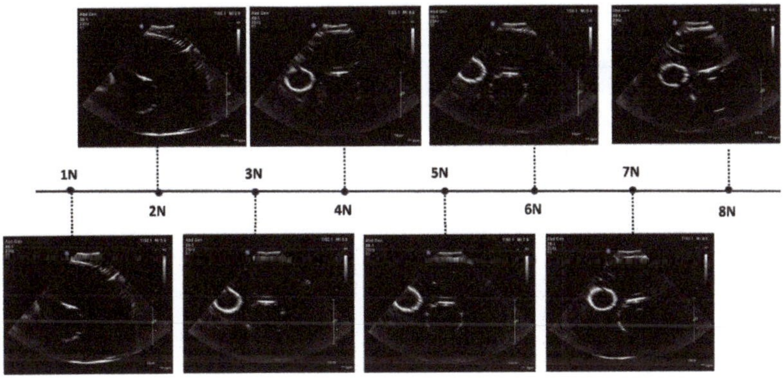

Figure 22. Ultrasound images acquired by robot at varying force levels on AAA phantom.

3.3.2. Volunteer Tests

The system successfully captured clear images of the abdominal aorta for both volunteers with different body shapes, which were reviewed by a registered clinical vascular scientist (HCPC CS20525) with 10 years of vascular ultrasound experience. Example images are presented in Figure 23.

Figure 24 illustrates the force readings and distributions recorded during automated ultrasound scans on two volunteers. The duration of each scan varies depending on the size of the selected region. The top graphs for both volunteers show a characteristic oscillatory pattern because of breathing motion, which was not observed in phantom tests. Regarding force distribution, Volunteer 1 demonstrated a lower proportion of forces within the desired range (39.6%) compared to Volunteer 2, where 65.1% of the forces fell within the target range. This difference can be explained by the higher BMI of Volunteer 2, where increased body fat and softer tissue facilitated smoother force control. The standard deviation of the forces further supports this observation, with Volunteer 1 exhibiting a higher standard deviation of 2.1 N compared to 1.6 N for Volunteer 2. The greater variability in Volunteer

1′s force readings is also attributed to the relatively firmer tissue, which made consistent force application more challenging. Nevertheless, the majority of force readings for both volunteers were within the ±2 N range, accounting for 93.1% and 96.7% of the forces for Volunteer 1 and Volunteer 2, respectively. The minor deviations outside this range primarily occurred at the beginning of the scan, as the robotic system transitioned from zero force to establishing surface contact. This is evident in the initial segments of the force value graphs. Throughout the scans, both volunteers reported no discomfort or pain.

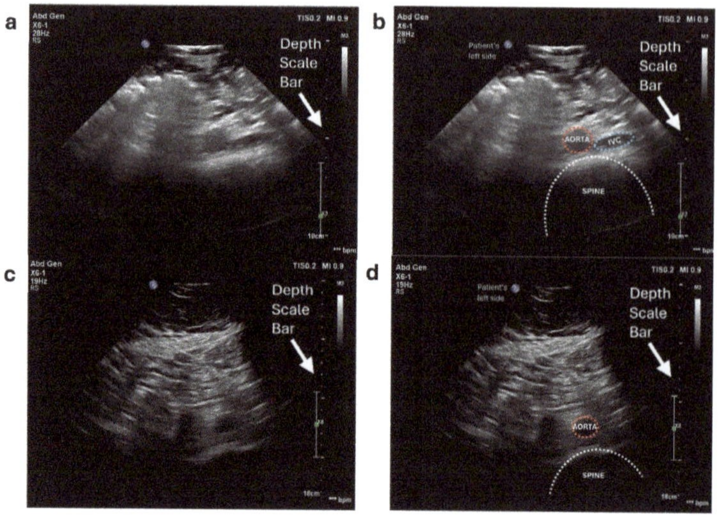

Figure 23. Abdominal aorta ultrasound images acquired by the robotic system: (**a**) Volunteer 1′s original image; (**b**) Volunteer 1′s images annotated by registered clinical vascular scientist; (**c**) Volunteer 2′s original image; (**d**) Volunteer 2′s images annotated by registered clinical vascular scientist.

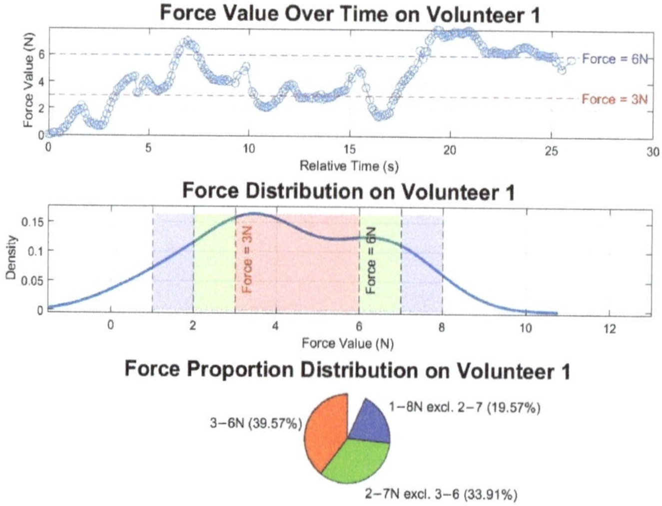

(**a**) Volunteer 1

Figure 24. *Cont.*

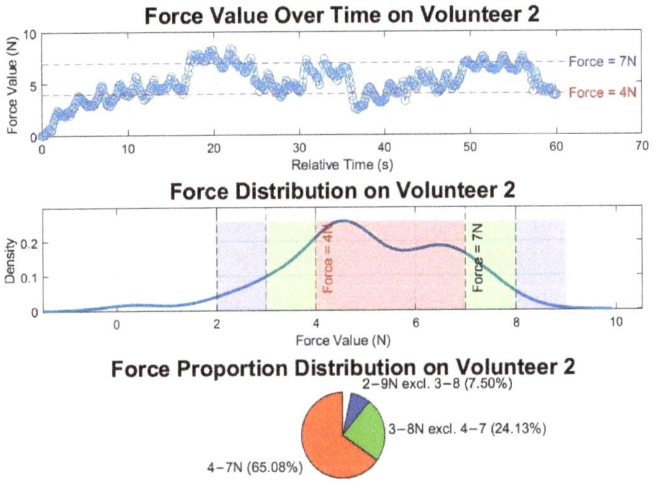

(**b**) Volunteer 2

Figure 24. Force values and distributions in automated ultrasound image acquisition experiment on two volunteers. Red, green and blue colours in the middle and buttom graphs match with each other.

4. Discussion

In this study, we developed a compact, single-axis ultrasound probe force sensor and implemented an automated ultrasound acquisition method for AAA using a custom-designed ultrasound robot. By combining path-planning with surface force-following control, the system enables full-coverage scanning within a defined area while maintaining a consistent force range, demonstrating promising potential for safe and effective robotic ultrasound scanning.

The custom-designed force sensor offers multiple advantages for robotic ultrasound applications. It is inexpensive and easy to manufacture with 3D printing, making it accessible for research and clinical use worldwide. Additionally, the sensor's round shape and probe-holding mechanism mimic the grip of a human hand, ensuring stable contact. Sensor performance tests revealed high linearity and accuracy, with minimal hysteresis, indicating that the sensor is well suited for abdominal scanning. Experiments demonstrated that the force-control algorithm effectively responds to platform movements, maintaining a consistent distance between the markers on the robot's wrist and the platform while dynamically adjusting the probe's position to ensure optimal force application.

Integrated experiments combining force-following control with path planning validated the system's ability to maintain a stable scanning trajectory while keeping contact force within safe thresholds, which is critical for patient comfort and safety. Additionally, it successfully acquired diagnostically usable abdominal aorta ultrasound images from two volunteers with different body shapes, demonstrating adaptability to variations in tissue properties and body composition. This adaptability marks a crucial step toward expanding the use of robotic ultrasound systems across diverse patient populations in healthcare settings.

Despite its advantages, the current sensor design has limitations. It measures force along only a single axis, limiting its ability to detect angular misalignments. To address this, future iterations should explore multi-axis force sensing, utilizing the four-load-cell design to enable three-axis measurements. This enhancement, coupled with an improved control algorithm, would allow for corrective tilting and more precise alignment with complex anatomical surfaces. Additionally, the current force-following control relies on a fixed proportional gain, which limits adaptability to sudden changes in surface position, such as those caused by patient movement or respiration.

In addition to addressing the current limitations, we are exploring the integration of image feedback into the system. This involves developing methods to automatically assess ultrasound image quality in real time and leveraging this feedback to optimize the probe's position and orientation. Furthermore, diagnostic scans for AAA, as specified by the National Abdominal Aortic Aneurysm Screening Programme (NAAASP) [20] guidelines, must encompass the region from the diaphragm to the iliac bifurcation, necessitating precise probe alignment and positioning. To ensure accurate measurements, it is crucial to avoid oblique angles when the aorta is not parallel to the probe. These advancements represent key areas for future development aimed at enhancing the system's diagnostic utility and clinical applicability.

5. Conclusions

In conclusion, our system offers a solution for safe human–robot interaction in extra-body ultrasound imaging acquisition. By employing multi-sensor fusion—including patient surface mapping, force sensing, ultrasound imaging, marker-based patient-to-robot registration, and scanning path planning—the system ensures both the safety and accuracy of sonographer-free, automated scanning. We hope that this work will contribute to the clinical acceptance of automatic and autonomous robotic ultrasound scanning systems, paving the way for broader adoption of such technologies in the future.

Author Contributions: Conceptualization, Y.Z., R.J.H., A.M. and K.R.; methodology, Y.Z., E.R., X.F., L.Y. and H.N.; software, Y.Z., H.N., W.W. and R.J.H.; validation, Y.Z., H.N., Z.X., W.W., S.M., P.K., K.R. and R.J.H.; formal analysis, Y.Z., H.N., A.G. and K.R.; data curation, Y.Z. and H.N.; writing—original draft preparation, Y.Z., E.R. and H.N.; writing—review and editing, A.G., R.J.H. and K.R.; supervision, L.L., R.J.H. and K.R. All authors have read and agreed to the published version of the manuscript.

Funding: This research was supported by core funding from the Wellcome/EPSRC Centre for Medical Engineering [WT203148/Z/16/Z] at King's College London, the British Heart Foundation (BHF) Centre of Excellence at King's College London, the Department of Health and Social Care (DHSC) through the National Institute for Health and Care Research (NIHR) MedTech and In Vitro Diagnostic Co-operative (MIC) award for Cardiovascular Diseases to Guy's & St Thomas' NHS Foundation Trust in partnership with King's College London, and the King's–China Scholarship Council PhD Scholarship Program (K-CSC).

Institutional Review Board Statement: This study obtained ethical approval from the King's College London local ethics committee to test our robotic ultrasound system on volunteers for general abdominal scans (study title: Investigating Robotic Abdominal Ultrasound Imaging, study reference: HR-22/23-5412).

Informed Consent Statement: Informed consent was obtained from all subjects involved in the study.

Data Availability Statement: Data are unavailable due to privacy or ethical restrictions.

Conflicts of Interest: The authors declare no conflicts of interest.

References

1. Harrison, G.; Harris, A. Work-related musculoskeletal disorders in ultrasound: Can you reduce risk? *Ultrasound* **2015**, *23*, 224–230. [CrossRef] [PubMed]
2. Berg, W.A.; Blume, J.D.; Cormack, J.B.; Mendelson, E.B. Operator dependence of physician-performed whole-breast US: Lesion detection and characterization. *Radiology* **2006**, *241*, 355–365. [CrossRef] [PubMed]
3. Kaminski, A.; Payne, A.; Roemer, S.; Ignatowski, D.; Khandheria, B.K. Answering to the call of critically ill patients: Limiting sonographer exposure to COVID-19 with focused protocols. *J. Am. Soc. Echocardiogr.* **2020**, *33*, 902–903. [CrossRef] [PubMed]

4. von Haxthausen, F.; Böttger, S.; Wulff, D.; Hagenah, J.; García-Vázquez, V.; Ipsen, S. Medical robotics for ultrasound imaging: current systems and future trends. *Curr. Robot. Rep.* **2021**, *2*, 55–71. [CrossRef]
5. Li, K.; Xu, Y.; Meng, M.Q.H. An overview of systems and techniques for autonomous robotic ultrasound acquisitions. *IEEE Trans. Med. Robot. Bionics* **2021**, *3*, 510–524. [CrossRef]
6. Merouche, S.; Allard, L.; Montagnon, E.; Soulez, G.; Bigras, P.; Cloutier, G. A Robotic Ultrasound Scanner for Automatic Vessel Tracking and Three-Dimensional Reconstruction of B-Mode Images. *IEEE Trans. Ultrason. Ferroelectr. Freq. Control.* **2016**, *63*, 35–46. [CrossRef] [PubMed]
7. Jiang, Z.; Grimm, M.; Zhou, M.; Hu, Y.; Esteban, J.; Navab, N. Automatic force-based probe positioning for precise robotic ultrasound acquisition. *IEEE Trans. Ind. Electron.* **2020**, *68*, 11200–11211. [CrossRef]
8. Kim, Y.J.; Seo, J.H.; Kim, H.R.; Kim, K.G. Development of a control algorithm for the ultrasound scanning robot (NCCUSR) using ultrasound image and force feedback. *Int. J. Med. Robot. Comput. Assist. Surg.* **2017**, *13*, e1756. [CrossRef] [PubMed]
9. Chen, M.; Huang, Y.; Chen, J.; Zhou, T.; Chen, J.; Liu, H. Fully Robotized 3D Ultrasound Image Acquisition for Artery. In Proceedings of the 2023 IEEE International Conference on Robotics and Automation (ICRA), London, UK, 29 May–2 June 2023; IEEE: Piscataway, NJ, USA, 2023; pp. 2690–2696.
10. Huang, Q.; Lan, J.; Li, X. Robotic arm based automatic ultrasound scanning for three-dimensional imaging. *IEEE Trans. Ind. Inform.* **2018**, *15*, 1173–1182. [CrossRef]
11. Noh, Y.; Lindenroth, L.; Wang, S.; Housden, R.J.; van Wingerden, A.S.; Li, W.; Rhode, K. A2-piece six-axis force/torque sensor capable of measuring loads applied to tools of complex shapes. In Proceedings of the 2019 IEEE/RSJ International Conference on Intelligent Robots and Systems (IROS), The Venetian Macao, Macau, 4–8 November 2019; IEEE: Piscataway, NJ, USA, 2019; pp. 7976–7981.
12. Goel, R.; Abhimanyu, F.; Patel, K.; Galeotti, J.; Choset, H. Autonomous ultrasound scanning using bayesian optimization and hybrid force control. In Proceedings of the 2022 International Conference on Robotics and Automation (ICRA), Philadelphia, PA, USA, 23–27 May 2022; IEEE: Piscataway, NJ, USA, 2022; pp. 8396–8402.
13. Suligoj, F.; Heunis, C.M.; Sikorski, J.; Misra, S. RobUSt–an autonomous robotic ultrasound system for medical imaging. *IEEE Access* **2021**, *9*, 67456–67465. [CrossRef]
14. Virga, S.; Zettinig, O.; Esposito, M.; Pfister, K.; Frisch, B.; Neff, T.; Navab, N.; Hennersperger, C. Automatic force-compliant robotic ultrasound screening of abdominal aortic aneurysms. In Proceedings of the 2016 IEEE/RSJ International Conference on Intelligent Robots and Systems (IROS), Daejeon, Republic of Korea, 9–14 October 2016; IEEE: Piscataway, NJ, USA, 2016; pp. 508–513.
15. Hennersperger, C.; Fuerst, B.; Virga, S.; Zettinig, O.; Frisch, B.; Neff, T.; Navab, N. Towards MRI-based autonomous robotic US acquisitions: a first feasibility study. *IEEE Trans. Med. Imaging* **2016**, *36*, 538–548. [CrossRef] [PubMed]
16. Housden, J.; Wang, S.; Bao, X.; Zheng, J.; Skelton, E.; Matthew, J.; Noh, Y.; Eltiraifi, O.; Singh, A.; Singh, D.; et al. Towards standardized acquisition with a dual-probe ultrasound robot for fetal imaging. *IEEE Robot. Autom. Lett.* **2021**, *6*, 1059–1065. [CrossRef] [PubMed]
17. Tsumura, R.; Iwata, H. Robotic fetal ultrasonography platform with a passive scan mechanism. *Int. J. Comput. Assist. Radiol. Surg.* **2020**, *15*, 1323–1333. [CrossRef] [PubMed]
18. Wang, S.; Housden, J.; Noh, Y.; Singh, D.; Singh, A.; Skelton, E.; Matthew, J.; Tan, C.; Back, J.; Lindenroth, L.; et al. Robotic-assisted ultrasound for fetal imaging: Evolution from single-arm to dual-arm system. In *Towards Autonomous Robotic Systems, Proceedings of the 20th Annual Conference, TAROS 2019, London, UK, 3–5 July 2019*; Proceedings, Part II 20; Springer: Berlin/Heidelberg, Germany, 2019; pp. 27–38.
19. Zheng, Y.; Wang, W.; Ferhusoglu, H.; Subramaniyam, T.; Xu, Z.; Rhode, K.; Housden, R.J. Fiducial Marker Based Patient-to-Robot Registration and Target Tracking for Automatic Extra-body Ultrasound Imaging. In Proceedings of the 2024 IEEE 20th International Conference on Automation Science and Engineering (CASE), Bari, Italy, 28 August–1 September 2024; pp. 2117–2122. [CrossRef]
20. Rankin, P. *Abdominal Aortic Aneurysm Screening: Clinical Guidance and Scope of Practice for Professionals Involved in the Provision of the Ultrasound Scan Within AAA Screening*; Technical Report; Public Health England: London, UK, 2016. Available online: https://assets.publishing.service.gov.uk/media/5a7f1ddc40f0b6230268d88f/NAAASP_scope_of_practice_document_V2.0_130916.pdf (accessed on 4 January 2025).

sensors

MDPI

Article

Self-Sensing with Hollow Cylindrical Transducers for Histotripsy-Enhanced Aspiration Mechanical Thrombectomy Applications

Li Gong [1,2,*], Alex R. Wright [2], Kullervo Hynynen [1,2,3] and David E. Goertz [1,2,*]

[1] Department of Medical Biophysics, University of Toronto, Toronto, ON M5G 1L7, Canada; khynynen@sri.utoronto.ca
[2] Physical Sciences Platform, Sunnybrook Research Institute, Toronto, ON M4N 3M5, Canada; alex.wright@sri.utoronto.ca
[3] Institute of Biomedical Engineering, University of Toronto, Toronto, ON M5S 3G9, Canada
* Correspondence: li.gong@mail.utoronto.ca (L.G.); david.goertz@sunnybrook.ca (D.E.G.)

Highlights:

What are the main findings?

- Miniature hollow cylindrical histotripsy transducers can detect intra-lumen cavitation without requiring external sensors.
- Voltage signals acquired across the transducer during and immediately after transmission provide insight into both the time and frequency domains.

What is the implication of the main finding?

- These self-sensing transducers offer a built-in method for cavitation monitoring, enabling feedback in therapeutic ultrasound applications such as histotripsy-mediated thrombectomy.

Abstract: Intravascular aspiration thrombectomy catheters are widely used to treat stroke, pulmonary embolism, and deep venous thrombosis. However, their performance is frequently compromised by clot material becoming lodged within the catheter tip. To address this, we develop a novel ultrasound-enhanced aspiration catheter approach that generates cavitation within the tip to mechanically degrade clots, with a view to facilitate extraction. The design employs hollow cylindrical transducers that produce inwardly propagating cylindrical waves to generate sufficiently high pressures to perform histotripsy. This study investigates the feasibility of self-sensing cavitation detection by analyzing voltage signals across the transducer during treatment. Experiments were conducted for two transmit pulse lengths at varying driving voltages with water or clot in the lumen. Cavitation clouds within the lumen were assessed using 40 MHz ultrasound imaging. Changes in the signal envelope during the pulse body and ringdown phases occurred above the cavitation threshold, the latter being associated with more rapid wave damping in the presence of bubble clouds within the lumen. In the frequency domain, voltage-dependent cavitation signals—subharmonics, ultra-harmonics, and broadband—emerged alongside transmit pulses. This work demonstrates a highly sensitive, sensor-free method for detecting cavitation within the lumen, enabling feedback control to further improve histotripsy-assisted aspiration.

Keywords: histotripsy; sonothrombolysis; cavitation; aspiration thrombectomy; ultrasound; transducer

1. Introduction

Thrombotic large-vessel occlusions are a significant contributor to global mortality and morbidity, particularly in cases of myocardial infarction [1–3], ischemic stroke [4], pulmonary embolism (PE) [5], and deep venous thrombosis (DVT) [6]. Catheter-based mechanical thrombectomy techniques have become increasingly common in standard clinical practice to remove acute thrombotic occlusions in large vessels. Mechanical thrombectomy procedures are employed in the treatment of stroke [7–11], pulmonary embolism [5,12–14], and peripheral vessels [6,15,16]. One of the most widely used mechanical thrombectomy methods is aspiration, which operates by using suction to extract clot material through the lumen of a hollow catheter. In the most common form of this procedure—direct contact aspiration (DA)—the catheter tip is navigated be in contact with the proximal end of the clot and an external vacuum is applied. While this approach has significantly impacted the treatment of patients, in many circumstances it is unsuccessful or performs sub-optimally [17,18] due to thrombus clogging within the catheter or 'corking' at the catheter tip [19,20]. Therefore, there is a recognized need to improve the performance of aspiration mechanical thrombectomy devices, particularly their ability to ingest more challenging clots (e.g., stiffer or larger volume clots). We are developing a novel ultrasound-enhanced aspiration approach that involves the placement of a hollow cylindrical transducer (HCT) at the catheter tip. The premise is that established clinical DA procedures will be followed, but, upon the application of suction, histotripsy will be performed on clot material as it enters the HCT lumen to degrade its mechanical integrity and thereby facilitate deformation and ingestion further into the catheter. Histotripsy is an established ultrasound method that employs cavitation clouds to fractionate targeted regions of tissue into an acellular liquid [21–24]. The approach involves using very-high-pressure ultrasound to generate violent bubble clouds within the HCT focal region. There are several general classes of histotripsy, determined by the pulse lengths employed: extrinsic histotripsy typically uses ~1–2 μs pulses and requires the highest pressure levels; shock scattering histotripsy uses 5–20 μs pulses; and 'boiling' histotripsy involves 1–100 ms pulses and requires the lowest pressure level due to pulse-scale thermal elevations. With few exceptions [25–27], large aperture extracorporeal spherically focused transducer configurations are used to achieve the pressure levels required to create cavitation clouds.

With the proposed aspiration technique, sufficiently high pressures must be achieved within the lumen of HCTs that are of a scale compatible with intravascular catheters. In our prior work [28], we demonstrated that cavitation clouds could be generated within a radially polarized HCT (3.3/2.5 mm outer/inner diameter) when operating within its 'thickness' mode resonant frequency bandwidth. The associated transducer vibrations produce predominantly cylindrical waves and a high-pressure zone along the HCT lumen axis of symmetry. In particular, when operating at a standing wave frequency, constructive interference can increase the effective focal gain, resulting in increased internal pressures (estimated > 20 MPa). In this work, cavitation within the lumen was confirmed with a hydrophone situated external to the HCT lumen and high-frequency (40 MHz) ultrasound imaging. The pulse lengths employed (10 and 100 μs) were sufficiently long to permit the development of standing wave. This was supported by hydrophone data that revealed pressure buildup at the outset of electrical pulse stimulation that was associated with the development of the interference pattern, and a decay (ringdown) following the end of the pulse associated with the dissipation of internal reflections. In a more recent work [29], we demonstrated the feasibility of performing histotripsy within clots situated in an HCT lumen.

An important element of any cavitation-mediated therapeutic ultrasound approach is monitoring and controlling the level and characteristics of cavitation. Cavitation monitoring and control methods have been widely investigated in the setting of microbubble-mediated drug delivery [30–33]. Generally, these involve using receive transducers, separate from the transmit transducers, to passively detect and, in some cases, spatially map cavitation emissions arising from the therapy pulses. In histotripsy, the primary approach to cavitation monitoring to date has involved situating an ultrasound imaging transducer within the therapy transducer aperture. B-scan ultrasound imaging is then used to 'actively' (pulse-echo) detect bubble clouds that form during the treatment process. The imaging information is used for targeting purposes, confirming the formation and extent of lesions, and potentially adjusting the transmit conditions to ensure sufficient cavitation activity. More recently, efforts are underway to develop transmit–receiver arrays compatible with short-pulse (extrinsic) histotripsy. In [34,35], a subset of elements was configured with both transmit and receive capabilities, enabled by the short duration of the transmitted pulses relative to the propagation time to regions of interest, resulting in time separation between the transmit and receive signals.

For the proposed ultrasound-enhanced aspiration approach, as with conventional histotripsy, the ability to detect and monitor induced cavitation will be necessary to ensure that effective treatments are occurring and to potentially adjust transmit parameters as necessary. As the intended application areas are in the setting of stroke, PE, and the peripheral vasculature, using a separate B-scan imaging transducer to monitor cavitation within the HCT lumen is not viable. Furthermore, due to the spatial lengths of the pulses employed relative to the lumen diameter, there will not be temporal separation between the transmit and receive signals [28]. In the present study, we will investigate the use of voltage signals measured across the transmit transducer to monitor cavitation induced within the lumen of the HCT. In the setting of boiling histotripsy conditions, using a single-element large aperture spherically focused transducer, there have been several reports of fluctuations in the voltage or power across the transducer after the onset of cavitation within the focal region [36,37]. These fluctuations result from the signals across the transducer being a superposition of the applied transmit voltage and the arrival of emissions and reflections/scattering from the cavitation cloud at the focal region. Outside the biomedical ultrasound field, particularly in the area of sonochemistry, there have been numerous reports of examining measured transducer signals (voltage and current) to monitor cavitation with long-duration pulses (>100 ms) [38–41]. This work has been carried out for acoustic horn or larger-scale cylindrical focusing configurations. It has been shown that the measured signals can detect the presence of inertial cavitation, harmonics, and subharmonics [42,43]. This is often referred to as 'self-sensing' due to the dual use of the transducer as a transmitter and receiver, with the target cavitation signals being concurrent with the transmission stimulation signal.

Here, we investigate the use of self-sensing approaches to detect and monitor cavitation within an intravascular scale HCT. Following the conditions employed in recent work (between 10 and 100 μs pulses) [28], signals measured across the HCT will be assessed over a range of applied voltages from sub-cavitation to cavitation levels. The temporal (pulse-body and ringdown) and spectral aspects of the pulses will be assessed. The presence of cavitation will be confirmed using high-frequency ultrasound imaging. The majority of experiments will be conducted with water-filled lumens, followed by a proof-of-principle experiment with thrombus-filled lumens.

2. Materials and Methods

2.1. Transducer Configuration

The transducer used for experiments was a radially polarized HCT with an outer diameter of 3.3 mm and an inner diameter of 2.5 mm, with a length of 2.5 mm, as employed in [28]. The transducer material was DL-47 (DeL Piezo Specialties, West Palm Beach, FL, USA), a hard PZT ceramic suitable for high-power applications due to its high mechanical quality factor, relatively high dielectric constant, and low dielectric loss.

The transducer had a micro-coaxial wire attached to the inner signal and outer ground electrodes using silver epoxy (Epotek H20E, Epoxy Technology Inc., Billerica, MA, USA) and a catheter liner (PTFE Sub-Lite-Wall™ Liner 0.095″ ID 0.0015″ wall thickness, Zeus Company Inc., San Jose, SC, USA) attached to its inner wall using epoxy (Epotek 301, Epoxy Technology Inc., Billerica, MA, USA). This design is capable of achieving high internal pressures and generating cavitation clouds within the HCT lumen, as reported in previous work [28]. Finite element simulations of the HCT, including the liner and epoxy layers, were performed in OnScale™, as per the methods previously described in [28] (material properties in Supplementary Table S1). Figure 1A,B depict the simulated normalized pressure map and lateral pressure distribution within the lumen for the transducer using 10 μs pulses at 6.17 MHz in water. These results show a narrow high-pressure region along the central axis of the cylinder and the presence of side lobes that decay in amplitude towards the HCT lumen walls. Figure 1C shows fiber-optic hydrophone (10 μm active region, Precision Acoustics Inc., Dorset, UK)-based measurements of the internal pressure field, and Figure 1D highlights a matching detailed view of the simulation for comparison, emphasizing the similarities between the two results. Absolute pressures are not measured here due to the directional sensitivity of the hydrophone, coupled with the angle of incidence of the wavefronts being perpendicular to the hydrophone axis, as discussed in [28]. It was previously established that sufficiently high pressure levels can be achieved to initiate cavitation within the lumen, and the primary purpose of the present study is to examine self-sensing-based methods to detect this form of cavitation. Supplementary Figure S1 shows a comparison of simulated profiles for 10 μs and 100 μs pulses. They exhibit similar spatial distributions to peak pressures, which is expected, as the luminal standing-wave pattern is fully developed for the particular HCT employed with pulse lengths of 10 μs or greater [28].

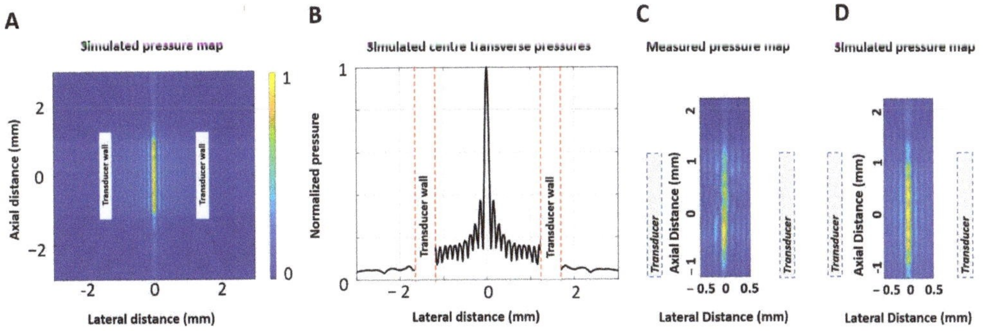

Figure 1. Simulated sagittal 2D pressure map (**A**) and pressure across the HCT transverse axis of symmetry (**B**) operating at 6.17 MHz in water. Detailed views of measured (**C**) and simulated (**D**) pressures within the HCT lumen show qualitatively consistent profiles.

2.2. Cavitation Observation with High-Speed Ultrasound Imaging

2.2.1. Experimental Configuration

The experimental setup is shown in Figure 2, where an HCT and imaging transducer were situated in a tank of gas-equilibrated deionized water held at 30 °C. A 2.5mm diameter PTFE liner (Sub-Lite-Wall™ Liner 0.095″ ID 0.0015″ wall thickness, Zeus Company Inc., Orangeburg, SC, USA) was epoxied to the lumen surface of HCT to mimic the eventual intended catheter configuration, as well as facilitate positioning within the tank. A Vevo

2100 imaging system with a 40 MHz probe (MS550D, FUJIFILM VisualSonics Inc., Toronto, ON, Canada) was positioned to image the central longitudinal cross-section of the HCT lumen. The framerate was set to 916 Hz, which was the system's maximum for a field of view that captured the entirety of the lumen diameter across the full transducer length. The imaging system was synchronized with treatment pulses using a pulse delay generator (Model 575, Berkeley Nucleonics Corp., San Rafael, CA, USA) to ensure that every treatment pulse occurred at the same time point during each frame. The transducer was driven by a function generator (AFG3102, Tektronix, Beaverton, OR, USA) and an RF amplifier (A150, E&I, Jericho, NY, USA). The voltage across the HCT was measured using a 100:1 voltage probe and a digitizer PicoScope 5242B (Pico Technology, England, UK) at a sample rate of 125 MS/s (14 bits). The applied voltage amplitudes reported are the average steady-state peak voltages (V_p) measured across the transducer.

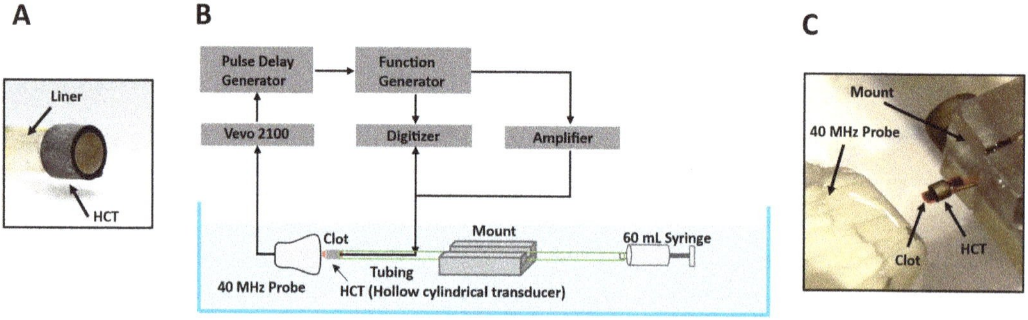

Figure 2. (**A**) Image of the constructed transducer with attached liner. (**B**) Schematic overview of experimental configuration. (**C**) Photo of experimental setup with a clot present in the HCT lumen.

2.2.2. Transmit Frequency Selections

The impedance of HCT in water was measured from 300 Hz to 8 MHz in 5 kHz steps using a network analyzer (AA-30.ZERO, RigExpert, Kyiv, Ukraine) with both water and a clot in the HCT lumen. Within the thickness mode bandwidth of the HCT geometry, there are a number of prominent minima. These are associated with standing wave frequencies within the lumen, and were previously shown to correspond to local (frequency) pressure maxima [28]. As in [28], the transmit frequency was selected by observing the HCT lumen under B-mode imaging while sending 'treatment' pulses (82 V_p voltage, 10 μs pulses) at each frequency, corresponding to a minimum within the thickness mode bandwidth. The minima frequency that resulted in the largest cavitation cloud on B-scan imaging was then employed as the transmit frequency for the experiments. This process was carried out for both water-filled and clot-filled lumens to determine the most appropriate operating frequency for each case as the impedance minima differed, as described below.

2.2.3. Experimental Protocol

For experiments with water in the lumen, the transducer was operated at 6.17 MHz, and pulse lengths of 10 μs and 100 μs were tested. A total of 100 pulses were sent per sequence, with a pulse interval of 10.92 ms, such that one pulse was imaged every 10 frames for 1000 frames. A total of 26 applied voltages ranging from 19 to 82 V_p were tested. Each 100-pulse sequence was repeated three times at each voltage and pulse length, resulting in 300 pulses per voltage condition. We note that the maximum voltage applied here corresponds closely to that reported in [28] (82 V_p), which resulted in high levels of cavitation and did not result in any change in the performance of the transducer over time.

The lowest tested voltage (19 V_p) did not produce cavitation in any configuration, again consistent with previous work [28].

For experiments with a clot in the lumen, a frequency of 6.43 MHz was used. A pulse length of 10 μs and a pulse interval of 1.09 ms were used for 916 pulses, resulting in a pulse every frame for 916 frames. A total of 5 clots were tested in this manner. For each clot, the treatment was performed 5 times at 19 V_p (sub-cavitation threshold) and then once at 82 V_p. After treatment, the clots were extracted and bisected for optical imaging under a microscope.

2.3. *Cavitation Threshold*

B-mode imaging data was used to determine whether cavitation occurred on a given pulse. Each frame, corresponding to a specific treatment pulse, was manually assessed for the presence or absence of visible cavitation, which manifests as hyperechoic regions within the transducer lumen. To determine cavitation probability in water, for a given applied voltage, the ratio of the number of pulses which had visible cavitation activity to the number of total pulses was calculated. This probability, as a function of applied voltage, was fitted to a sigmoid curve. The voltage value at a probability of 0.5 was defined as the cavitation threshold voltage.

2.4. *Self-Sensing Signals Analysis*

Signal analysis was performed using MATLAB R2021a (MathWorks, Natick, MA, USA) and GraphPad Prism 10 (GraphPad Software, San Diego, CA, USA). For each pulse, time domain voltage signals acquired from the transducer can be divided into three segments: pre-pulse (noise only), 'pulse body' (during transmit), and 'ringdown' (post-transmit). In this study, ringdown refers to the period of decay following the end of the transmit pulse (applied voltage), which is primarily associated with the dissipation of internal reflections within the HCT lumen. Note that such reflections are distinct from transducer ringdown, which arises from residual mechanical vibrations in the transducer material itself.

2.4.1. Time Domain Analysis

For time domain signal analysis, two temporal windows were examined: a pulse body window and a ringdown window. The pulse body window (duration 1.2 μs for water and 1.67 μs for clot) began 3 cycles after the start of the transmit pulse (applied voltage) to allow for a steady state to be reached. The ringdown window was 2 μs in duration, beginning after 3 cycles following the cessation of the transmit pulse. Note that there was a range of window durations that was evaluated and produced relatively similar results. For each of these windows, the envelope of the signal was squared and then summed over the window duration to obtain an estimate of power. The ratio of these resulting metrics (ringdown to pulse body) was then calculated to provide a measure of the ringdown signal level relative to the pulse body amplitude, which we define here as the *ringdown ratio*. Employing this ratio effectively normalizes the results to the different driving voltages tested.

For the water-in-lumen case, to provide an indication of the impact of cavitation on the pulse-to-pulse variability in the ringdown ratio, the variance of this metric was calculated over the 300-pulse sets at each applied voltage.

2.4.2. Spectral Analysis

Frequency domain analysis was conducted on the pulse body portion, during which active cavitation could occur. The 10 μs Blackman windows were used for 10 μs and 100 μs Tukey windows with a cosine fraction of 0.25 were used for 100 μs pulses. For 10 μs pulses, zero padding was employed in order to match the frequency resolution of the 100 μs windows. Fast Fourier transforms were performed on each windowed signal, and the spectra were power averaged for each applied voltage. The same process was applied to segments of signal acquired pre-trigger prior to the arrival of the transmit pulse in order to generate baseline noise-only spectra.

As a means of quantifying cavitation, the sum of the power was calculated over the frequency windows surrounding the fundamental frequency (i.e., 6.03–6.34 MHz for 10 μs pulses and 6.15–6.20 MHz for 100 μs pulses in water), ultra-harmonic frequency (i.e., 9.21–9.31 MHz for both pulse lengths in water and 9.55–9.67 MHz with a clot), and a frequency range corresponding to broadband signals (i.e., 19.30–19.70 MHz for 10 μs pulses in water; 5.13–5.40 and 7.00–7.30 MHz for 100 μs pulses in water; 5.45–5.60 and 7.45–7.70 MHz with a clot).

2.5. Blood Clot Preparation

Clots were prepared in a manner similar to previous sonothrombolysis studies [29,43–48]. Clots were prepared from blood extracted from the femoral vein of a pig and were captured in 3.2% sodium citrate vacutainers (BD Vacutainer™, Thermo Fisher Scientific Inc., Waltham, MA, USA). Clots were formed in 2 mL borosilicate pipettes (Fisherbrand™ Disposable Borosilicate Glass Pasteur Pipets, Thermo Fisher Scientific Inc., Waltham, MA, USA) after mixing anticoagulated blood with 100 mMol/L $CaCl_2$ at a ratio of 1 mL of anticoagulated blood to 200 μL of $CaCl_2$. The sealed pipettes were incubated in a water bath at 37 °C for 3 h and then transferred to a fridge at 4 °C for 3 days. The resulting retracted clots were extracted and cut into homogenous 5 mm lengths prior to the experiments.

3. Results

3.1. Impedance and Frequency Selection

The impedance as a function of frequency for the water-immersed HCT is shown in Figure 3A. Local minima associated with the fundamental length mode and its third harmonic appear at 660 kHz and 1.93 MHz, along with a more complex series of peaks and troughs near the thickness mode bandwidth (4.9–7 MHz). Figure 3B shows measurements for both water- and clot-filled lumens within the thickness mode bandwidth. The minima are each associated with a standing-wave frequency [28]. Having a clot in the lumen shifts the frequencies of these minima higher relative to the water case.

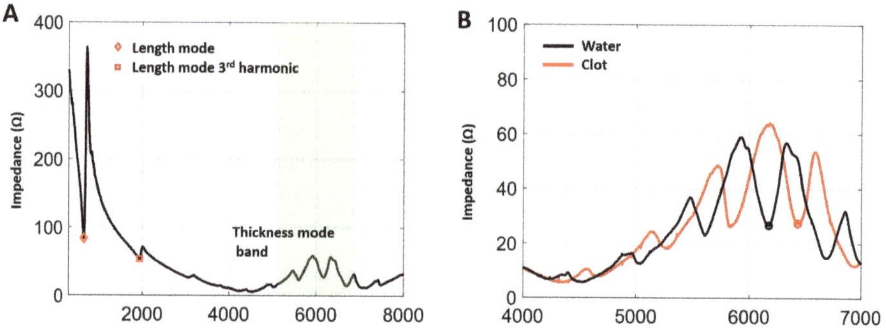

Figure 3. The measured electrical impedances as a function of frequency in a range of 300–8000 kHz (**A**) and 4000–7000 kHz (**B**), with either water or clot occupying the transducer lumen. Minima at 660 kHz and 1.93 MHz correspond to the length mode and its third harmonic. The transmit frequencies used for clot treatments and treatments with water in the lumen are highlighted with circle markers in (**B**).

Following the previously described methods for transmit frequency selection, the thickness mode minima identified above were evaluated for their capacity to generate cavitation. The frequencies selected for the water-in-lumen and clot-in-lumen cases were 6.17 and 6.43 MHz, respectively. The difference between these two transmit frequencies is consistent with the minima shift in the impedance curve.

3.2. Cavitation Imaging and Cavitation Probability with Water-Filled Lumen

Figure 4 displays example B-mode imaging frames of the HCT operating in water. At low voltages, the lumen is hypo-echogenic, consistent with the absence of cavitation [28]. As the voltage increases, cavitation clouds begin to appear, and, at high voltages, there is strong cavitation activity, although it remains confined within the transducer lumen. This trend holds for both the 10 and 100 μs pulses, although the 100 μs pulses have larger cavitation clouds at higher voltages. The cavitation probability in water is shown in Figure 5. No cavitation was observed below 54 V_p for either the 10 or 100 μs pulses. For both pulse lengths, above 65 V_p cavitation was observed for every pulse.

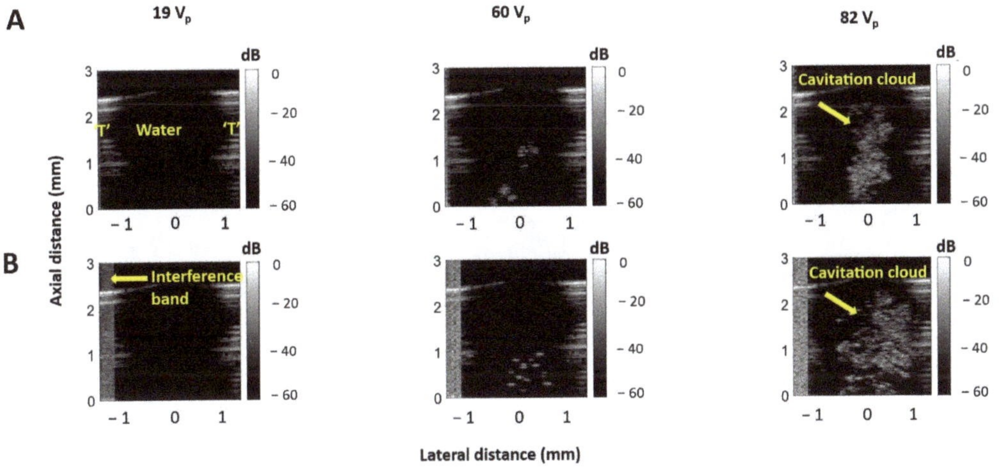

Figure 4. B-mode imaging frames of the lumen of an HCT in water with 10 μs (**A**) and 100 μs pulses (**B**). 'T' indicates HCT wall.

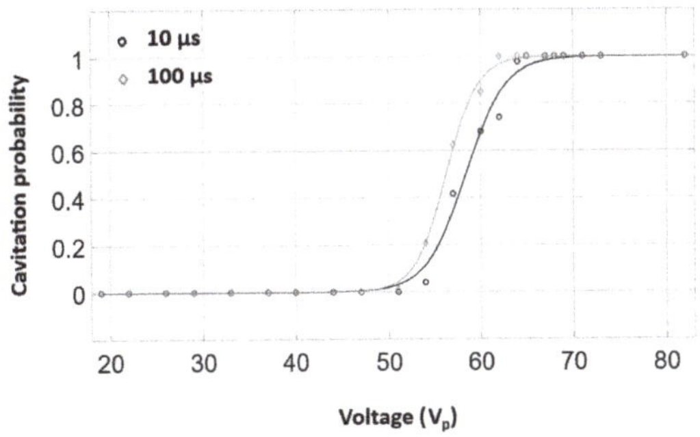

Figure 5. Cavitation probability in gas-equilibrated water as a function of applied voltage, as measured using B-mode imaging for both 10 and 100 μs pulses.

3.3. *Self-Sensing Signals in Water-Filled Lumens*

Figure 6 presents example time domain voltage signals for both 10 and 100 μs pulses and at 19, 60, and 82 V_p in water. An amount of 19 V was shown to be well below the cavitation threshold (Figure 5), and no cavitation is observed with B-mode imaging (Figure 4), while 60 V_p is around the cavitation threshold voltage range and 82 V_p is substantially above the cavitation threshold. Comparing the envelopes of the voltages at each applied voltage reveals differences in the ratios of the amplitude of the envelopes of the ringdown portions to the pulse bodies. In particular, the relative ringdown ratios are progressively lower at the higher voltages—where cavitation is present—relative to the 19 V_p case (sub-cavitation threshold). In addition, it can be seen that fluctuations are present in the envelope of the pulse body in the presence of cavitation. This is consistent for both the 10 and 100 μs pulses.

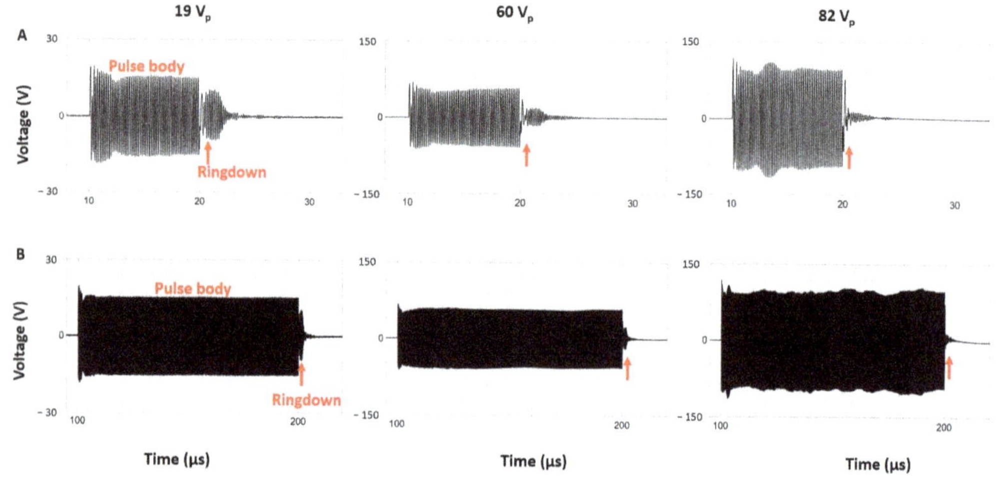

Figure 6. Voltages measured across the transducer during (**A**) 10 μs and (**B**) 100 μs pulses at different applied voltages in water. The red arrows indicate the ringdown portion of the signals.

Figure 7A,B shows a box plot of the ringdown ratio as a function of applied voltage. For both 10 and 100 μs pulses, at lower voltages—below the cavitation threshold—the ratio remains constant. As the cavitation threshold voltage is approached, the ratio begins to decrease with increasing applied voltage. Figure 7C,D shows the variance in the ringdown ratio as a function of voltage. The variance increases at the onset of cavitation before plateauing at approximately 64 V_p in the 10 μs case and 67 V_p in the 100 μs case.

Pulse body spectra at select voltages are shown in Figure 8. Sub-harmonic, ultra-harmonic, and broadband signals emerge for voltages around and above the cavitation threshold voltage for both the 10 and 100 μs pulses.

The quantification of selected frequency bands as a function of voltage is shown in Figure 9. For 10 μs pulses, the fundamental frequency signal increases monotonically as a function of applied voltage. An increase in broadband signal becomes apparent from 62 V_p and continues to rise with applied voltage and begins to plateau around 69 V_p. The ultra-harmonic signal exhibits a similar pattern, but with a lower voltage onset (47 V_p). The sub-harmonic signal follows a more complex pattern, but also suggests the presence of an onset voltage and eventual plateau. This complexity may be due to overlap with contributions from thickness mode frequencies. The overall trends for the 100 μs pulses share many of these characteristics. Notable differences include lower standard deviations below the cavitation threshold, which results in a more distinct cavitation threshold curve.

Interestingly, both the ultra- and sub-harmonic signals have more similar curve features and also undergo a reduction at the highest voltage levels.

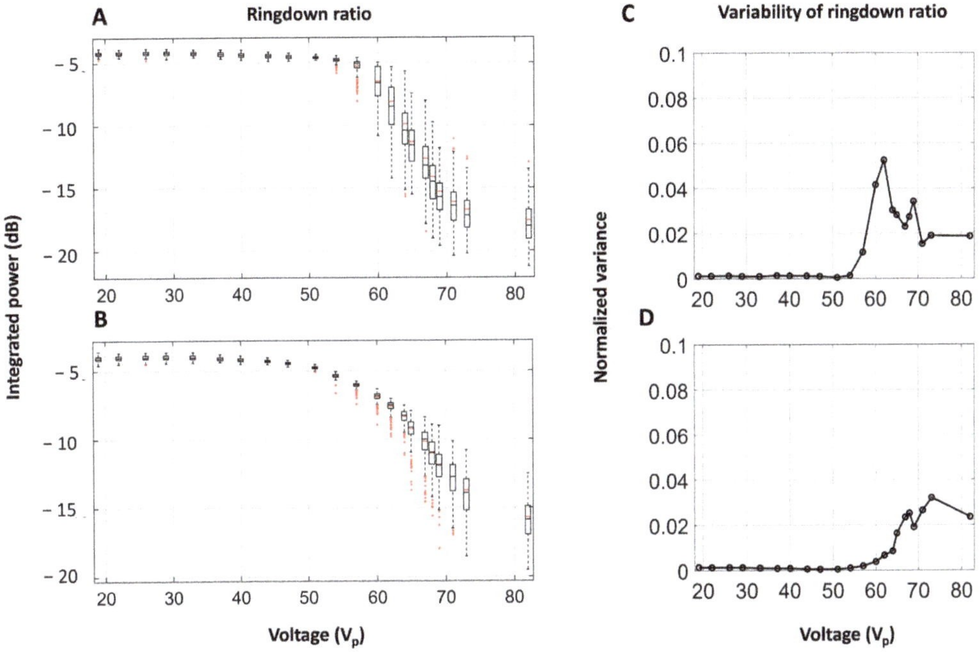

Figure 7. Box plots of the ringdown ratio as a function of voltage for (**A**) 10 µs and (**B**) 100 µs pulses in water. The results are normalized to the mean ringdown ratios at 19 V for each dataset, and the box plots identify the median as the central marks, the 25th and 75th percentiles as the box edges, the most extreme non-outlier datapoints as the whiskers, and individual outliers as red markers. (**C**) and (**D**) show the variance of these ringdown ratios as a function of voltage for the 10 µs and 100 µs pulses, respectively.

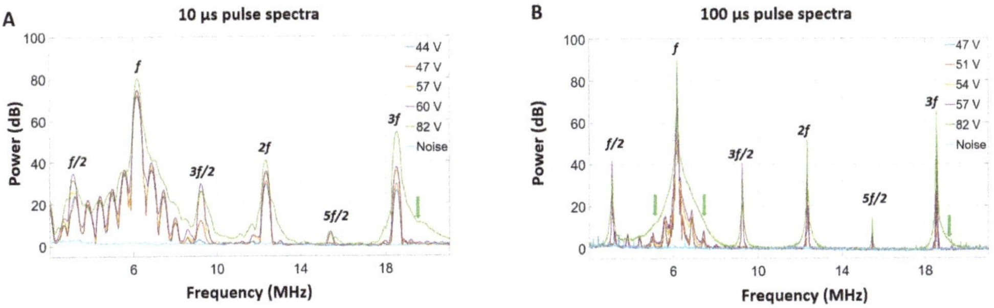

Figure 8. The spectra of (**A**) 10 µs and (**B**) 100 µs pulses at different voltages in water. The green arrows point to regions of broadband emissions which are visible at 82 V_p; f indicates the fundamental frequency of transmit pulses (6.17 MHz).

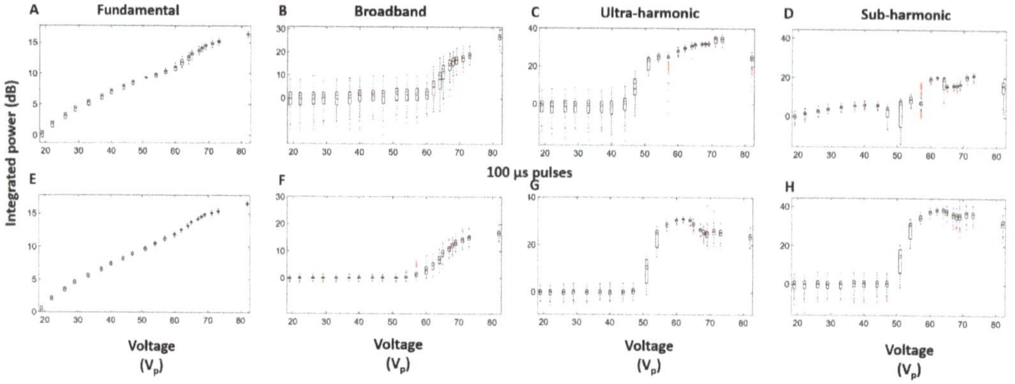

Figure 9. Box plots of integrated power of specific frequency bands of the transducer voltage spectra as a function of applied voltage in water. The broadband (**B**,**F**), ultra-harmonic (**C**,**G**), and sub-harmonic (**D**,**H**) for both 10 μs and 100 μs pulses show a threshold behavior, while the fundamental (**A**,**E**) show a monotonically increasing behavior. The integrated powers are normalized to the mean values at 19 V_p within each dataset.

3.4. Self-Sensing Signals During Clot Treatments

Figure 10 shows representative B-mode images of the treatment (82 V_p) of a clot in the lumen of the HCT. A central cavitation cloud is visible from the beginning of the treatment, and, post-treatment, a lesion in the same location is visible. The bisected extracted clot shows the lesion hole under optical imaging.

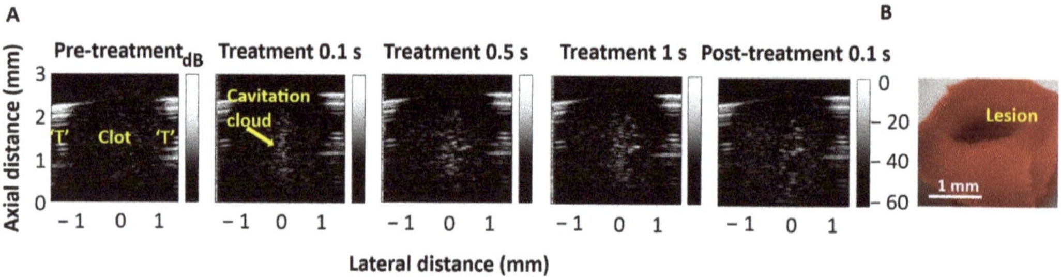

Figure 10. (**A**) Representative B-mode imaging of a cross-section of a clot situated in the transducer before, during, and after treatment. 'T' represents HCT walls. Treatment parameters were 10 μs pulse length, 1000 Hz PRF, and 1 s treatment duration. Before treatment, the clot is visible as uniform speckle bounded by the side walls of the transducer. During treatment, the cavitation cloud is visible as a hyperechoic region along the central axis of the HCT. (**B**) Immediately post-treatment, the hypoechoic region is visible over the same area of the cavitation cloud, indicating the generation of a lesion, which is confirmed by the bisected clot.

Figure 11A shows representative time domain voltage signals during the exposure of a clot (1 s duration) at applied voltages of 19 (sub-cavitation) or 82 V_p, along with their associated frequency domain spectra. Similarly to the case for the water-filled HCT lumen, the ringdown ratio decreased significantly from 19 to 82 V_p. The spectra shown in Figure 11B also indicate that sub-harmonic and ultra-harmonic signals are present at 82 V_p which are not detected at 19 V_p.

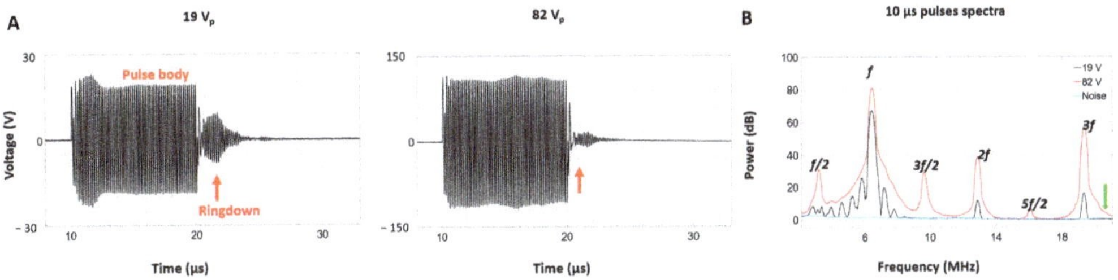

Figure 11. Voltage signals across the HCT with a clot in the lumen at (**A**) 19 V_p and (**B**) 82 V_p, and (**C**) the spectra of these signals. The red arrows on the time domain traces indicate the ringdown portions of the signals, and the green arrows on the spectra indicate broadband emissions at 82 V_p.

The quantification of these features across the five clots tested is presented in Figure 12 as violin plots. Metrics using ringdown ratio, broadband signals, sub-harmonic, and ultra-harmonic signals all demonstrate significant ($p < 0.0001$) differences in integrated power between 19 and 82 V_p.

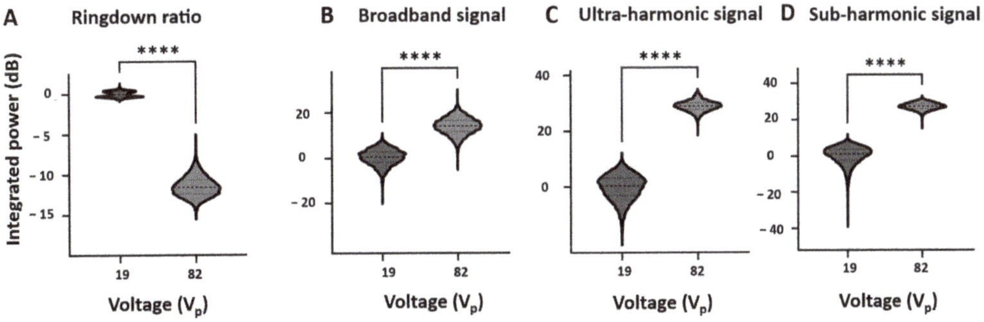

Figure 12. Violin plots comparing the integrated power for (**A**) the ringdown ratio in the time domain, and the broadband, ultra-harmonic, and sub-harmonic (**B–D**) of specific frequency bands of the HCT voltage signal spectra at both 19 and 82 V_p with a clot in the lumen. The integrated power is normalized to the mean value at 19 V_p. Dashed lines represent the median and quantiles. **** indicates statistical significance between groups with $p < 0.0001$ using a two-tailed unpaired *t*-test.

4. Discussion

In conventional extracorporeal histotripsy systems, the detection and monitoring of cavitation is an important element of the treatment process. Active and passive cavitation detection approaches continue to be investigated to facilitate this process. In the setting of the ultrasound (histotripsy)-enhanced aspiration approach that we are developing, cavitation detection also has potential to provide valuable information for treatment monitoring and control. The approach taken here is to examine the use of the transmit transducer for this purpose, which, for a disposable catheter, would have notable advantages, including simplicity and cost-effectiveness. The results collectively demonstrate the use of measured voltage signals across an HCT to detect cavitation induced within its lumen. The measured signals underwent both temporal and spectral analyses, which were conducted by separating the signals into pulse body and ringdown segments. The cavitation-related experiments were supplemented by electrical impedance measurements across the HCT.

For water-filled lumens, fluctuations in the envelope of the pulse body became evident following the onset of cavitation, which is consistent with previous reports for spherically focused transducers in the setting of boiling histotripsy conditions [37,49]. These fluctuations arise from the superposition of signals associated with cavitation (emissions, reflections/scattering) and the applied transmit signal voltage.

To extend these observations beyond time domain fluctuations, pulse body spectral analysis was conducted. At lower voltages—below the imaging-detected cavitation threshold—the spectra feature a prominent peak at the transmit frequency alongside a number of smaller adjacent peaks/lobes within the thickness mode bandwidth. The side lobes may be associated with standing wave frequencies, as well as Fourier analysis artifacts. Multiple harmonics (second, third) are also present, which can be associated with nonlinear bubble behavior, as well as contributions from nonlinear propagation. The onset of detected sub-harmonic and ultra-harmonic signals (1/2, 3/2, and 5/2) at higher voltages is consistent with pressure threshold-dependent bubble behavior. The emergence of broadband signals followed a similar pattern, albeit with a higher voltage onset and variance. In interpreting the broadband signal, it should be noted that the frequency ranges selected for quantification were within the thickness mode range (between the local peaks) and adjacent to the third harmonic signal peak, as shown in Supplementary Figure S2. In [28], hydrophone measurements performed adjacent to the lumen at higher voltages demonstrated a broadband appearance across a wide range of frequencies. This was not observed in the self-sensing HCT spectra, which is attributed to the frequency response of the transducer when exposed to emissions impinging on its lumen surface. As the frequency ranges employed were in proximity to the fundamental and third harmonic peaks, there may be contributions from spectral broadening of these peaks in addition to general broadband noise that can be associated with inertial cavitation. This effect may have been more prominent for the 10 μs relative to the 100 μs case due to the impact of Fourier transform-related spectral broadening. Overall, the results shown in Figure 9 indicate that the selected sub-harmonic, ultra-harmonic, and broadband signals are substantially elevated over baseline values in the presence of cavitation. This is further supported by the analysis of Supplementary Figure S3, which shows that these signals are significantly elevated at high voltages relative to lower sub-cavitation voltages (e.g., 82 V_p vs. 19 V_p).

It was also observed that the ringdown ratio remained relatively flat at lower voltages (~<50 V_p), but, after the onset of detected cavitation, it began to drop precipitously. In [28], the simulated and hydrophone-measured pressure waveforms external to the transducer exhibited a stepped decay over time following the cessation of the stimulating pulse, which was attributed to the dissipation of internally reflected cylindrical waves. The measured ringdown signals are therefore interpreted to be predominantly associated with internally reflecting waves that dissipate over time. The progressive reduction in amplitude of the ringdown ratio with increasing levels of cavitation is hypothesized to result from interactions with the internally reflected waves with the bubble clouds. That is, it can be expected that waves encountering clouds will be scattered and reflected and lose energy through absorption processes. Unlike self-sensing based cavitation detection in the context of spherically focused transducers, the reduced ringdown ratio is a feature that is distinct to the HCT configuration, where internally reflected waves can be generated. It is also notable that cavitation detection via the ringdown analysis presented here has been previously reported with large-scale HCTs (>10 cm typically), such as those employed in sonochemistry [38–41]. Indeed, in sonochemistry configurations, continuous waves or very long pulses are employed and the emphasis is to assess cavitation signatures in the frequency domain. In the application being investigated here, which employs shorter pulses, ringdown analysis appears to robustly detect the presence of cavitation within the lumen.

Impedance spectra measurements show a pattern of minima within the thickness mode bandwidth for both water- and clot-filled lumens. These were previously shown in [28] (for water-filled lumens) to be associated with standing-wave frequencies, which produced higher pressure levels within the lumen due to the effects of constructive interference. As in [28], the operating frequency was selected based on the frequency that produced the highest level of cavitation, based on B-mode imaging. The locations of impedance minima were found to differ between water- and clot-filled lumens (Figure 2B). This can be attributed to differences in the compressional wave propagation velocity: clots have a higher wave speed than water [50], which, in turn, shifts the frequency at which minima occur. In a clinical setting, clots can have a range of compositions and ages, which can be expected to give rise to variations in the propagation velocity. Although a dedicated network analyzer was used in this study to measure electrical impedance, an alternate approach could involve monitoring the current, in addition to the voltage, to derive the impedance.

The proof-of-principle experiments to examine self-sensing signals in the context of clot erosion were conducted at two voltages: one below the cavitation threshold and one at the maximum voltage. The high-voltage exposures resulted in the production of a lesion within the clots, as observed in our recent study [29]. The cavitation and lesions were preferentially initiated in the central portion of the lumen, where pressure levels are highest (Figure 1). This is consistent with the results reported in [28,29]. No cavitation was observed lateral to the transducer due to the low-pressure levels outside the transducer lumen. The average spectra during the course of the treatment had similar features to the water-filled lumens. A comparison of the ringdown, sub-harmonic, ultra-harmonic, and broadband signals between the two voltages showed highly significant differences. This supports the notion that these are candidate metrics to detect cavitation during clot treatments.

The results of this work suggest that both electrical impedance and voltage signals of an intravascular-scale HCT have the potential to provide information that is relevant to the monitoring and control of ultrasound-enhanced aspiration techniques that are being developed. For example, the impedance spectra could be employed to identify candidate transmit frequencies (e.g., at minima within the thickness bandwidth) when the clot has entered the lumen. The candidate frequencies could then be tested to evaluate their capacity to produce cavitation. As B-mode imaging of the lumen cannot be performed in a clinical setting, the degree of cavitation at the candidate frequencies could be assessed by self-sensing cavitation metrics, such as those evaluated in this study, to select the operating frequency. Furthermore, an examination of the evolution of these signals over time during treatments—both for 'static' and aspirating clots—is necessary to determine their possible role in dynamically adapting exposures during the course of treatments. Work is underway in relation to examining the impact of clot erosion on aspiration success rate of in vitro clots.

The implementation of self-sensing signal-based monitoring and control methods would necessarily have to undergo pre-clinical experiments in relevant animal models and, eventually, clinical testing. While valuable, benchtop experiments cannot recapitulate all relevant physiologic and anatomic conditions that are present in vivo. As the signals are derived predominantly from within the catheter lumen, it is not anticipated that the presence of a vessel wall or relative motion between the catheter and wall would significantly impact their characteristics. The presence of blood may alter the signals, as there are differences in the properties (e.g., attenuation and speed of sound [50]) between blood, clots, and saline. There may also be different pressure thresholds for cavitation in vivo, although this is not expected to significantly impact the methods described above. Ultimately, this work must be conducted using fully constructed catheters.

Catheter-based ultrasound systems have been developed for treating large-vessel occlusions, most notably the clinically approved EKOS catheter used for DVT [51,52] and PE [53]. The EKOS catheter must be inserted directly into the clot, where it delivers thrombolytic drugs through side ports and applies low-intensity ultrasound to improve enzymatic activity, typically over several hours. The use of microbubbles to enhance the performance of the EKOS catheter is also being investigated [54,55]. There have also been research efforts to develop forward-looking ultrasound catheters that apply energy at the proximal clot face, generally in conjunction with cavitation agents such as microbubbles and nanodroplets [55–61]. These approaches rely on the progressive in situ erosion of thrombus to restore flow, assuming the downstream clearance of clot fragments. In contrast, the objective of ultrasound-enhanced aspiration is to use histotripsy improve the performance of an established and widely used mechanical thrombectomy procedure that restores blood flow by removing thrombotic vascular occlusions.

Supplementary Materials: The following supporting information can be downloaded at: https://www.mdpi.com/article/10.3390/s25175417/s1, Figure S1: Simulated pressure maps using 10 μs (left) and 100 μs (right) pulses at 6.17 MHz in water. Pressures are normalized to the peak pressure within the field. Figure S2: Box plots of integrated power of specific frequency bands of the transducer voltage spectra as a function of applied voltage in water for both 10 μs and 100 μs pulses. The 2nd harmonic (A,D) increases in a linear behavior. The 3rd harmonic (B,E) signal exhibits an inflection at a sub-cavitation voltage, while the 4th harmonic (C,F) starts to increase at voltage over the cavitation threshold. The integrated powers are normalized to the mean value at 19 V_p within each dataset. Figure S3: Violin plots for both 10 μs and 100 μs pulses comparing the integrated power for (A,E) the ringdown ratio in time domain and the broadband (B,F), ultra-harmonic (C,G), sub-harmonic (D,H) of specific frequency bands of the HCT voltage signal spectra at both 19 and 82 V_p with water in the lumen. The integrated powers are normalized to the mean values at 19 V_p within each dataset. Dashed lines represent the median and quantiles. **** indicates statistical significance between groups with $p < 0.0001$ using two-tailed, unpaired *t*-test. Table S1: Material properties of HCT (DL-47) used in simulations.

Author Contributions: Conceptualization, L.G. and D.E.G.; methodology, L.G., A.R.W. and D.E.G.; software, L.G.; validation, L.G.; formal analysis, L.G.; investigation, L.G.; resources, D.E.G.; data curation, L.G.; writing—original draft preparation, L.G.; writing—review and editing, L.G., A.R.W., K.H. and D.E.G.; visualization, L.G.; supervision, D.E.G.; project administration, D.E.G.; funding acquisition, D.E.G. All authors have read and agreed to the published version of the manuscript.

Funding: This work was funded by NSERC Discovery (2019-07132) and CIHR (PRJ165849 and MOP353808) grants.

Institutional Review Board Statement: Not applicable.

Informed Consent Statement: Not applicable.

Data Availability Statement: The original contributions presented in this study are included in the article/Supplementary Material. Further inquiries can be directed to the corresponding authors.

Acknowledgments: The authors would like to thank Jennifer Barry, Shawna Rideout, Stephanie Furdas, and Dallan McMahon for performing animal blood collection.

Conflicts of Interest: The authors declare no conflicts of interest.

References

1. Johansson, S.; Rosengren, A.; Young, K.; Jennings, E. Mortality and Morbidity Trends after the First Year in Survivors of Acute Myocardial Infarction: A Systematic Review. *BMC Cardiovasc. Disord.* **2017**, *17*, 53. [CrossRef]
2. Ye, Q.; Zhang, J.; Ma, L. Predictors of All-Cause 1-Year Mortality in Myocardial Infarction Patients. *Medicine* **2020**, *99*, e21288. [CrossRef]
3. Ko, D.T.; Ahmed, T.; Austin, P.C.; Cantor, W.J.; Dorian, P.; Goldfarb, M.; Gong, Y.; Graham, M.M.; Gu, J.; Hawkins, N.M.; et al. Development of Acute Myocardial Infarction Mortality and Readmission Models for Public Reporting on Hospital Performance in Canada. *CJC Open* **2021**, *3*, 1051–1059. [CrossRef]

4. Tsao, C.W.; Aday, A.W.; Almarzooq, Z.I.; Anderson, C.A.M.; Arora, P.; Avery, C.L.; Baker-Smith, C.M.; Beaton, A.Z.; Boehme, A.K.; Buxton, A.E.; et al. Heart Disease and Stroke Statistics—2023 Update: A Report From the American Heart Association. *Circulation* **2023**, *147*, e93–e621. [CrossRef]
5. Tapson, V.F. Acute Pulmonary Embolism. *N. Engl. J. Med.* **2008**, *358*, 1037–1052. [CrossRef]
6. Sailer, A.; Revzin, M.V.; Pollak, J.; Ayyagari, R.; Mojibian, H.R.; Nezami, N.; Pellerito, J.S.; Marino, A.G. Deep Vein Thrombosis: Update on Mechanical Thrombectomy and Intravascular US. *Radiographics* **2022**, *42*, E184–E185. [CrossRef]
7. Powers, W.J.; Rabinstein, A.A.; Ackerson, T.; Adeoye, O.M.; Bambakidis, N.C.; Becker, K.; Biller, J.; Brown, M.; Demaerschalk, B.M.; Hoh, B.; et al. Guidelines for the Early Management of Patients With Acute Ischemic Stroke: 2019 Update to the 2018 Guidelines for the Early Management of Acute Ischemic Stroke: A Guideline for Healthcare Professionals From the American Heart Association/American Stroke Association. *Stroke* **2019**, *50*, e344–e418. [CrossRef] [PubMed]
8. Goyal, M.; Menon, B.K.; Zwam, W.H.v.; Dippel, D.W.J.; Mitchell, P.J.; Demchuk, A.M.; Dávalos, A.; Majoie, C.B.L.M.; van der Lugt, A.; de Miquel, M.A.; et al. Endovascular Thrombectomy after Large-Vessel Ischaemic Stroke: A Meta-Analysis of Individual Patient Data from Five Randomised Trials. *Lancet* **2016**, *387*, 1723–1731. [CrossRef]
9. Qiu, K.; Zhao, L.-B.; Xu, X.-Q.; Wang, Y.; Liu, J.; Liu, S.; Shi, H.-B.; Zu, Q.-Q. Acute Embolic Stroke with Large-Vessel Occlusion: Does Contact Aspiration Thrombectomy Show Superiority? *Clin. Radiol.* **2022**, *77*, 577–583. [CrossRef] [PubMed]
10. Kaneko, N.; Ghovvati, M.; Komuro, Y.; Guo, L.; Khatibi, K.; Mejia, L.L.P.; Saber, H.; Annabi, N.; Tateshima, S. A New Aspiration Device Equipped with a Hydro-Separator for Acute Ischemic Stroke Due to Challenging Soft and Stiff Clots. *Interv. Neuroradiol.* **2021**, *28*, 43–49. [CrossRef] [PubMed]
11. Amireh, A.O.; Kuybu, O.; Adeeb, N.; Kelley, R.E.; Javalkar, V.; Cuellar, H.; Sharma, P. Utilization of the Large-Bore Penumbra JET 7 Reperfusion Catheter in Thrombectomy for Acute Ischemic Stroke: A Single-Center Experience. *Interv. Neuroradiol.* **2020**, *27*, 99–106. [CrossRef]
12. Sedhom, R.; Abdelmaseeh, P.; Haroun, M.; Megaly, M.; Narayanan, M.A.; Syed, M.; Ambrosia, A.M.; Kalra, S.; George, J.C.; Jaber, W.A. Complications of Penumbra Indigo Aspiration Device in Pulmonary Embolism: Insights From MAUDE Database. *Cardiovasc. Revasc. Med.* **2022**, *39*, 97–100. [CrossRef]
13. Avgerinos, E.D.; Ali, A.A.; Toma, C.; Wu, B.; Saadeddin, Z.; McDaniel, B.; Al-Khoury, G.; Chaer, R.A. Catheter-Directed Thrombolysis versus Suction Thrombectomy in the Management of Acute Pulmonary Embolism. *J. Vasc. Surg. Venous Lymphat. Disord.* **2019**, *7*, 623–628. [CrossRef] [PubMed]
14. Sista, A.K.; Horowitz, J.M.; Tapson, V.F.; Rosenberg, M.; Elder, M.D.; Schiro, B.J.; Dohad, S.; Amoroso, N.E.; Dexter, D.J.; Loh, C.T.; et al. Indigo Aspiration System for Treatment of Pulmonary Embolism Results of the EXTRACT-PE Trial. *JACC Cardiovasc. Interv.* **2021**, *14*, 319–329. [CrossRef] [PubMed]
15. Loffroy, R.; Falvo, N.; Guillen, K.; Galland, C.; Baudot, X.; Demaistre, E.; Fréchier, L.; Ledan, F.; Midulla, M.; Chevallier, O. Single-Session Percutaneous Mechanical Thrombectomy Using the Aspirex®S Device Plus Stenting for Acute Iliofemoral Deep Vein Thrombosis: Safety, Efficacy, and Mid-Term Outcomes. *Diagnostics* **2020**, *10*, 544. [CrossRef] [PubMed]
16. Shah, K.J.; Roy, T.L. Catheter-Directed Interventions for the Treatment of Lower Extremity Deep Vein Thrombosis. *Life* **2022**, *12*, 1984. [CrossRef]
17. Yeo, L.L.L.; Bhogal, P.; Gopinathan, A.; Cunli, Y.; Tan, B.; Andersson, T. Why Does Mechanical Thrombectomy in Large Vessel Occlusion Sometimes Fail? *Clin. Neuroradiol.* **2019**, *29*, 401–414. [CrossRef]
18. Kaesmacher, J.; Maegerlein, C.; Kaesmacher, M.; Zimmer, C.; Poppert, H.; Friedrich, B.; Boeckh-Behrens, T.; Kleine, J.F. Thrombus Migration in the Middle Cerebral Artery: Incidence, Imaging Signs, and Impact on Success of Endovascular Thrombectomy. *J. Am. Heart Assoc.* **2017**, *6*, e005149. [CrossRef]
19. Nogueira, R.G.; Thornton, J.; Connolly, K.; Mullins, L.; Fitzgerald, S. Investigation of Current and Super-Bore 088″ Treatment Strategies of Soft and Stiff Clots at Internal Carotid Artery and Middle Cerebral Artery Occlusion Sites in an In Vitro Thrombectomy Model. *Stroke Vasc. Interv. Neurol.* **2021**, *2*, e000240. [CrossRef]
20. Yuki, I.; Kan, I.; Vinters, H.V.; Kim, R.H.; Golshan, A.; Vinuela, F.A.; Sayre, J.W.; Murayama, Y.; Vinuela, F. The Impact of Thromboemboli Histology on the Performance of a Mechanical Thrombectomy Device. *Am. J. Neuroradiol.* **2012**, *33*, 643–648. [CrossRef]
21. Xu, Z.; Ludomirsky, A.; Eun, L.Y.; Hall, T.L.; Tran, B.C.; Fowlkes, B.; Cain, C.A. Controlled Ultrasound Tissue Erosion. *IEEE Trans. Ultrason. Ferroelectr. Freq. Control* **2004**, *51*, 726–736. [CrossRef] [PubMed]
22. Vidal-Jove, J.; Serres, X.; Vlaisavljevich, E.; Cannata, J.; Duryea, A.; Miller, R.; Merino, X.; Velat, M.; Kam, Y.; Bolduan, R.; et al. First-in-Man Histotripsy of Hepatic Tumors: The THERESA Trial, a Feasibility Study. *Int. J. Hyperth.* **2022**, *39*, 1115–1123. [CrossRef] [PubMed]
23. Khokhlova, V.A.; Fowlkes, J.B.; Roberts, W.W.; Schade, G.R.; Xu, Z.; Khokhlova, T.D.; Hall, T.L.; Maxwell, A.D.; Wang, Y.-N.; Cain, C.A. Histotripsy Methods in Mechanical Disintegration of Tissue: Towards Clinical Applications. *Int. J. Hyperth.* **2015**, *31*, 145–162. [CrossRef] [PubMed]

24. Xu, Z.; Hall, T.L.; Vlaisavljevich, E.; Lee, F.T., Jr. Histotripsy: The First Noninvasive, Non-Ionizing, Non-Thermal Ablation Technique Based on Ultrasound. *Int. J. Hyperth.* **2021**, *38*, 561–575. [CrossRef]
25. Mallay, M.G.; Woodacre, J.K.; Landry, T.G.; Campbell, N.A.; Brown, J.A.; Mallay, M. A Dual-Frequency Lens-Focused Endoscopic Histotripsy Transducer. *IEEE Trans. Ultrason. Ferroelectr. Freq. Control* **2021**, *68*, 2906–2916. [CrossRef]
26. Landry, T.G.; Gannon, J.; Vlaisavljevich, E.; Mallay, M.G.; Woodacre, J.K.; Croul, S.; Fawcett, J.P.; Brown, J.A. Endoscopic Coregistered Ultrasound Imaging and Precision Histotripsy: Initial In Vivo Evaluation. *BME Front.* **2022**, *2022*, 9794321. [CrossRef]
27. Dadgar, M.M.; Hynynen, K. High-Power Phased-Array Transducer Module for the Construction of a System for the Treatment of Deep Vein Thrombosis. *IEEE Trans. Ultrason. Ferroelectr. Freq. Control* **2020**, *67*, 2710–2716. [CrossRef]
28. Gong, L.; Wright, A.R.; Hynynen, K.; Goertz, D.E. Inducing Cavitation Within Hollow Cylindrical Radially Polarized Transducers for Intravascular Applications. *Ultrasonics* **2024**, *138*, 107223. [CrossRef]
29. Gong, L.; Wright, A.R.; Hynynen, K.; Goertz, D.E. Histotripsy of Blood Clots Within a Hollow Cylindrical Transducer for Aspiration Thrombectomy Applications. *Ultrasound Med. Biol.* **2025**. *submitted*.
30. Jones, R.M.; Deng, L.; Leung, K.; McMahon, D.; O'Reilly, M.A.; Hynynen, K. Three-Dimensional Transcranial Microbubble Imaging for Guiding Volumetric Ultrasound-Mediated Blood-Brain Barrier Opening. *Theranostics* **2018**, *8*, 2909–2926. [CrossRef]
31. Coussios, C.C.; Roy, R.A. Applications of Acoustics and Cavitation to Noninvasive Therapy and Drug Delivery. *Annu. Rev. Fluid Mech.* **2008**, *40*, 395–420. [CrossRef]
32. Coussios, C.C.; Farny, C.H.; Haar, G.T.; Roy, R.A. Role of Acoustic Cavitation in the Delivery and Monitoring of Cancer Treatment by High-Intensity Focused Ultrasound (HIFU). *Int. J. Hyperth.* **2007**, *23*, 105–120. [CrossRef] [PubMed]
33. Haworth, K.J.; Salgaonkar, V.A.; Corregan, N.M.; Holland, C.K.; Mast, T.D. Using Passive Cavitation Images to Classify High-Intensity Focused Ultrasound Lesions. *Ultrasound Med. Biol.* **2015**, *41*, 2420–2434. [CrossRef]
34. Sukovich, J.R.; Hall, T.L.; Komaiha, M.; Haskell, S.; Lu, N.; Stocker, G.; Xu, Z. Acoustic Cavitation Localization during Histotripsy Using Transmit-Receive Capable Arrays. *J. Acoust. Soc. Am.* **2023**, *153*, A316. [CrossRef]
35. Sukovich, J.R.; Macoskey, J.J.; Lundt, J.E.; Gerhardson, T.I.; Hall, T.L.; Xu, Z. Real-Time Transcranial Histotripsy Treatment Localization and Mapping Using Acoustic Cavitation Emission Feedback. *IEEE Trans. Ultrason. Ferroelectr. Freq. Control* **2020**, *67*, 1178–1191. [CrossRef]
36. McLaughlan, J.; Rivens, I.; Leighton, T.; Haar, G.T. A Study of Bubble Activity Generated in Ex Vivo Tissue by High Intensity Focused Ultrasound. *Ultrasound Med. Biol.* **2010**, *36*, 1327–1344. [CrossRef] [PubMed]
37. Canney, M.S.; Khokhlova, T.D.; Khokhlova, V.A.; Bailey, M.R.; Hwang, J.H.; Crum, L.A.; Hynynen, K.; Souquet, J. Tissue Erosion Using Shock Wave Heating and Millisecond Boiling in HIFU Fields. *AIP Conf. Proc.* **2010**, *1215*, 36–39. [CrossRef]
38. Saalbach, K.-A.; Twiefel, J.; Wallaschek, J. Self-Sensing Cavitation Detection in Ultrasound-Induced Acoustic Cavitation. *Ultrasonics* **2019**, *94*, 401–410. [CrossRef]
39. Haller, J.; Wilkens, V.; Shaw, A. Determination of Acoustic Cavitation Probabilities and Thresholds Using a Single Focusing Transducer to Induce and Detect Acoustic Cavitation Events: I. Method and Terminology. *Ultrasound Med. Biol.* **2018**, *44*, 377–396. [CrossRef]
40. Samah, D.; Tierce, P.; Rouchon, J.-F. Influence of Cavitation on Ultrasonic Piezoelectric Transducers Impedance: Modelling and Experimentation. In Proceedings of the IEEE International Workshop of Electronics, Control, Measurement, Signals and Their Application to Mechatronics, Toulouse, France, 24–26 June 2013; Volume 1, pp. 1–6. [CrossRef]
41. Bornmann, P.; Hemsel, T.; Sextro, W.; Memoli, G.; Hodnett, M.; Zeqiri, B. Self-Sensing Ultrasound Transducer for Cavitation Detection. In Proceedings of the 2014 IEEE International Ultrasonics Symposium, Chicago, IL, USA, 3–6 September 2014; pp. 663–666. [CrossRef]
42. Lauterborn, W.; Ohl, C.-D. Cavitation Bubble Dynamics. *Ultrason. Sonochem.* **1997**, *4*, 65–75. [CrossRef]
43. Lauterborn, W.; Cramer, E. Subharmonic Route to Chaos Observed in Acoustics. *Phys. Rev. Lett.* **1981**, *47*, 1445–1448. [CrossRef]
44. Zhang, X.; Owens, G.E.; Cain, C.A.; Gurm, H.S.; Macoskey, J.; Xu, Z. Histotripsy Thrombolysis on Retracted Clots. *Ultrasound Med. Biol.* **2016**, *42*, 1903–1918. [CrossRef] [PubMed]
45. Hendley, S.A.; Bollen, V.; Anthony, G.J.; Paul, J.D.; Bader, K.B. In Vitro Assessment of Stiffness-Dependent Histotripsy Bubble Cloud Activity in Gel Phantoms and Blood Clots. *Phys. Med. Biol.* **2019**, *64*, 145019. [CrossRef]
46. Maxwell, A.D.; Haworth, K.J.; Holland, C.K.; Hendley, S.A.; Kreider, W.; Bader, K.B. Design and Characterization of an Ultrasound Transducer for Combined Histotripsy-Thrombolytic Therapy. *IEEE Trans. Ultrason. Ferroelectr. Freq. Control* **2022**, *69*, 156–165. [CrossRef]
47. Sutton, J.T.; Ivancevich, N.M.; Perrin, S.R.; Vela, D.C.; Holland, C.K. Clot Retraction Affects the Extent of Ultrasound-Enhanced Thrombolysis in an Ex Vivo Porcine Thrombosis Model. *Ultrasound Med. Biol.* **2013**, *39*, 813–824. [CrossRef]
48. Hendley, S.A.; Bhargava, A.; Holland, C.K.; Wool, G.D.; Ahmed, O.; Paul, J.D.; Bader, K.B. (More than) Doubling down: Effective Fibrinolysis at a Reduced Rt-PA Dose for Catheter-Directed Thrombolysis Combined with Histotripsy. *PLoS ONE* **2022**, *17*, e0261567. [CrossRef]

49. Canney, M.S.; Khokhlova, V.A.; Bessonova, O.V.; Bailey, M.R.; Crum, L.A. Shock-Induced Heating and Millisecond Boiling in Gels and Tissue Due to High Intensity Focused Ultrasound. *Ultrasound Med. Biol.* **2010**, *36*, 250–267. [CrossRef]
50. Shung, K.K.; Fei, D.Y.; Ballard, J.O. Further Studies on Ultrasonic Properties of Blood Clots. *J. Clin. Ultrasound* **1986**, *14*, 269–275. [CrossRef]
51. Owens, C. Ultrasound-Enhanced Thrombolysis: EKOS EndoWave Infusion Catheter System. *Semin. Intervent. Radiol.* **2008**, *25*, 37–41. [CrossRef]
52. Dumantepe, M.; Tarhan, I.A.; Ozler, A. Treatment of Chronic Deep Vein Thrombosis Using Ultrasound Accelerated Catheter-Directed Thrombolysis. *Eur. J. Vasc. Endovasc. Surg.* **2013**, *46*, 366–371. [CrossRef]
53. Piazza, G.; Hohlfelder, B.; Jaff, M.R.; Ouriel, K.; Engelhardt, T.C.; Sterling, K.M.; Jones, N.J.; Gurley, J.C.; Bhatheja, R.; Kennedy, R.J.; et al. A Prospective, Single-Arm, Multicenter Trial of Ultrasound-Facilitated, Catheter-Directed, Low-Dose Fibrinolysis for Acute Massive and Submassive Pulmonary Embolism The SEATTLE II Study. *JACC Cardiovasc. Interv.* **2015**, *8*, 1382–1392. [CrossRef]
54. Kennedy, S.R.; Lafond, M.; Haworth, K.J.; Escudero, D.S.; Ionascu, D.; Frierson, B.; Huang, S.; Klegerman, M.E.; Peng, T.; McPherson, D.D.; et al. Initiating and Imaging Cavitation from Infused Echo Contrast Agents through the EkoSonic Catheter. *Sci. Rep.* **2023**, *13*, 6191. [CrossRef] [PubMed]
55. Lafond, M.; Salido, N.G.; Haworth, K.J.; Hannah, A.S.; Macke, G.P.; Genstler, C.; Holland, C.K. Cavitation Emissions Nucleated by Definity Infused through an EkoSonic Catheter in a Flow Phantom. *Ultrasound Med. Biol.* **2021**, *47*, 693–709. [CrossRef] [PubMed]
56. Kim, J.; Lindsey, B.D.; Chang, W.-Y.; Dai, X.; Stavas, J.M.; Dayton, P.A.; Jiang, X. Intravascular Forward-Looking Ultrasound Transducers for Microbubble-Mediated Sonothrombolysis. *Sci. Rep.* **2017**, *7*, 3454. [CrossRef] [PubMed]
57. Goel, L.; Wu, H.; Zhang, B.; Kim, J.; Dayton, P.A.; Xu, Z.; Jiang, X. Nanodroplet-Mediated Catheter-Directed Sonothrombolysis of Retracted Blood Clots. *Microsyst. Nanoeng.* **2021**, *7*, 3. [CrossRef]
58. Goel, L.; Wu, H.; Zhang, B.; Kim, J.; Dayton, P.A.; Xu, Z.; Jiang, X. Safety Evaluation of a Forward-Viewing Intravascular Transducer for Sonothrombolysis: An in Vitro and Ex Vivo Study. *Ultrasound Med. Biol.* **2021**, *47*, 3231–3239. [CrossRef]
59. Zhang, B.; Wu, H.; Kim, J.; Dayton, P.; Xu, Z.; Jiang, X. Integration of Forward-Viewing and Side-Viewing Ultrasound Transducers in an Intravascular Sonothrombolysis Catheter. In Proceedings of the 2022 IEEE International Ultrasonics Symposium (IUS), Venice, Italy, 10–13 October 2022; pp. 1–4. [CrossRef]
60. Wu, H.; Kim, J.; Zhang, B.; Owens, G.; Stocker, G.; Chen, M.; Kreager, B.C.; Cornett, A.; Bautista, K.; Kaovasia, T.; et al. Rotational Intravascular Multidirectional Ultrasound Catheter for Sonothrombolysis of Retracted Clots: An in Vitro and in Vivo Study. *Engineering* **2024**, *42*, 235–243. [CrossRef]
61. Vu, P.T.; Rojas, S.S.; Ott, C.C.; Lindsey, B.D. A 9-Fr Endovascular Therapy Transducer With an Acoustic Metamaterial Lens for Rapid Stroke Thrombectomy. *IEEE Trans. Ultrason. Ferroelectr. Freq. Control* **2024**, *71*, 1627–1640. [CrossRef]

sensors

MDPI

Article

Imaging Through Scattering Tissue Based on NIR Multispectral Image Fusion Technique

Nisan Atiya [1], Amir Shemer [1,*], Ariel Schwarz [1], Yevgeny Beiderman [2] and Yossef Danan [1]

[1] Department of Electrical and Electronics Engineering, Azrieli College of Engineering, Jerusalem 9103501, Israel; nisanat@post.jce.ac.il (N.A.); arielsc@jce.ac.il (A.S.); yossefda@jce.ac.il (Y.D.)
[2] Faculty of Electrical and Electronics Engineering, Holon Institute of Technology, Holon 5810201, Israel; beidermany@hit.ac.il
* Correspondence: amirse@jce.ac.il

Abstract: Non-invasive diagnostics play a crucial role in medicine, and they ensure both contamination safety and patient comfort. The proposed study integrates hyperspectral imaging with advanced image fusion, enabling non-invasive, diagnostic procedure within tissue. It utilizes near-infrared (NIR) wavelength vision that is suitable for reflections from objects within a dispersive layer, enabling the reconstruction of internal tissue layers images. It can detect objects, including cancerous tumors (presented as phantoms), inside human tissue. This involves processing data from multiple images taken in different NIR bands and merging them through image fusion techniques. Our research demonstrates evident data about objects within the diffusive media, visible only in the reconstructed images. The experimental results demonstrate a significant correlation with the samples employed in the study's experimental design.

Keywords: multispectral imaging; image fusion; scattering tissue imaging; pyramid decomposition

1. Introduction

The identification of clear tumor margins is a critical factor in the complete removal of tumors and reducing the risk of their recurrence [1]. This is particularly important when the tumor is located near neurological structures, as removing too much tissue can be dangerous and harmful to vital organs or nerve tissue [2]. Several methods are aimed at improving tumor margin visualization, including MRI, CT, and targeted fluorescence imaging [1,3–9]. However, these methods have limitations such as reduced spatial resolution due to tissue shifting during surgery in MRI and CT, dependence on photo-physics and photochemistry in targeted fluorescence imaging, as well as issues such as phototoxicity and photobleaching [10–16]. Moreover, MRI and CT devices are costly and less accessible to common patients. The CT technique can also be harmful to the human body due to its radiative detection nature, and a major concern with this medical imaging is the risk from ionizing radiation, such as X-rays and gamma rays, which can increase the likelihood of DNA damage and cancer.

Utilizing a hyperspectral camera within the near-infrared spectrum introduces a novel approach aimed at identifying reflections originating from objects situated within a dispersive layer. By processing data from images captured in various NIR bands and utilizing image fusion techniques [17], one can produce a non-invasive and non-harmful tumor detection method. This technique holds promise for detecting tumors located inside human tissue layers.

The human body's tissues, rich in fluids, exhibit dispersion, contrasting with the dense composition of tumor tissue. When directing halogen light sources with a black-body radiation spectrum toward the body, the heightened density of tumors yields significantly more reflections compared to a typical healthy tissue [18,19]. Conventional silicon-based cameras primarily function within the visible spectrum range of 400–700 [nm] and are much less sensitive to NIR spectrum [20]. Yet, detecting objects obscured by diffuse tissue in this spectrum proves challenging. This limitation arises from the scattering or absorption of much of the visible light within the tissue, preventing deep penetration. A hyperspectral camera in the NIR range can be used to sense light coming from tissue at a depth of up to 2 cm; thus, cancerous tumors and blood vessels can be detected [21,22].

In the proposed work, the illumination for the hyperspectral imaging is achieved using a halogen light, which is an incandescent lamp (that generates light by heating a solid body to a high temperature) [23]. Halogen illumination was chosen because it has wide-spectrum illumination, with most of the energy emitted lying in the NIR regions of the spectrum. Only 15–20% of the light falls into the visible range (400–700 nm), and less than 1% is ultraviolet light [24,25]. As a source of black-body radiation, a halogen bulb can be considered a rough approximation of a black-body radiator. A black-body radiator is a theoretical object that absorbs all the electromagnetic radiation that falls on it and emits radiation according to Planck's law [26]. The main innovation lies in combining NIR and multispectral imaging with advanced image fusion techniques. In detail, Different wavelengths penetrate tissue with different paths to different depths, each providing specific information about tissue structure. By extracting valuable data from each image, the fused result offers enhanced detailed image.

2. Theoretical Explanation

Exposure fusion is a technique for synthesizing a single image from a series of input images by retaining only the regions corresponding to the highest-quality parameters defined in the model [17,27–30]. The process involves the computation of weight maps for each input image, where higher weights indicate that a pixel should contribute more prominently to the final composite image. These weights are designed to reflect desired image attributes, such as high contrast and optimal exposure. Unlike many other fusion techniques, exposure fusion does not depend on generating a high-dynamic-range (HDR) image [28]. This approach offers faster computational performance and is particularly advantageous for visualization on displays that do not support HDR format.

The quality parameters used for the weight calculations are as follows:

a. Contrast: Contrast is typically measured using Michelson contrast [31] or RMS contrast [32]. However, in this application, a Laplacian filter was applied to the grayscale version of each image. The absolute value of the filter's response was then calculated for each pixel. This method tends to assign higher weights to significant image features, such as edges and textures. This feature is denoted as C (the contrast weights) and is computed separately for each pixel in the image

b. Exposure and Illumination power: Examining the unprocessed intensities within a channel allows us to assess the exposure quality of a pixel. Our objective is to retain intensities that are not close to zero (indicating underexposure) or one (indicating overexposure). We assign a weight to each intensity, denoted as i, based on its proximity to the pixel-normalized intensity middle value, 0.5, which is represented by employing a Gaussian curve:

$$E = \exp(-\frac{(i-0.5)^2}{2\sigma^2}) \tag{1}$$

Here, E denotes the image exposedness per pixel, and σ (the standard deviation) controls the width of the Gaussian intensity distribution at each pixel. In our method, $\sigma = 2$.

Now we establish a weighted image parameter, W, in the k-th image defined as follows:

$$W_k = C_k \cdot E_k \tag{2}$$

where k is the index of the image in a sequence. Furthermore, a weighted image, $\widetilde{W}$, over the sequence of images is defined as:

$$\widetilde{W} = \sum_{n=1}^{N} W_n \tag{3}$$

where N is the number of images in the sequence. Then, the normalized weight per k-th image per each pixel is defined as follows:

$$\hat{W}_k = \frac{W_k}{\widetilde{W}} \tag{4}$$

The resulting fused image, R, can be defined as follows, calculated pixel-wise:

$$\hat{R} = \sum_{n=1}^{N} \hat{W}_n \cdot I_n \tag{5}$$

where I_n is the n-th image in the sequence.

Currently, the resulting image is of inadequate quality to properly depict the outcome. This issue arose because of the visible stitch lines, a consequence of rapid fluctuations in the weights. To overcome this issue, Laplacian pyramid decomposition was applied [33,34]. The Laplacian pyramid decomposition is a multi-scale image representation technique that decomposes an image into layers, each representing different levels of detail. It is built by iteratively applying a Gaussian filter to the image and down-sampling it by a factor of two, creating a sequence of images at progressively lower resolutions (known as the Gaussian pyramid). At each level, the higher-resolution image is reconstructed from the lower-resolution version using interpolation, and the difference (residuals) between the original and the reconstructed image is stored. These residuals form the layers of the Laplacian pyramid, which captures fine details at each scale. Figure 1 shows the concept of the Gaussian and Laplacian pyramid decomposition in the proposed research.

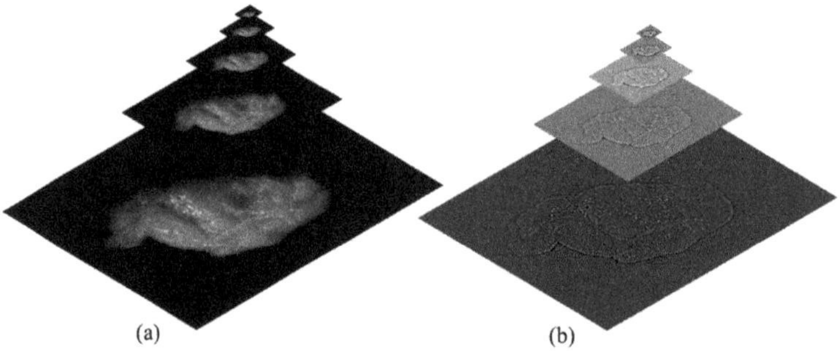

Figure 1. (**a**) Gaussian pyramid decomposition and (**b**) Laplacian pyramid decomposition.

Following the Gaussian and Laplacian decompositions, there are N images along with N normalized weight maps functioning as alpha masks. The l-th level in a Laplacian pyramid decomposition of image I is represented as $L\{I\}^l$, while a Gaussian pyramid of image I is denoted as $G\{I\}^l$. The Laplacian pyramid of the image and the Gaussian pyramid of the normalized weights maps is pixel-wise multiplexed, as described in Equation (5):

$$L\{R\}^l = \sum_{n=1}^{N} G\{\hat{W}\}_n^l L\{I\}_n^l \tag{6}$$

where n is the n-th image out of N images in the sequence. The computation of each level "l" within the resulting Laplacian pyramid involves a weighted average of the original

Laplacian decompositions for that level. The weights are determined by the l-th level of the pyramid of the corresponding weight map. Finally, the pyramid, $L\{R\}$, collapses to derive R, which is the final fused image.

Our experiment involves two stages for capturing the initial sets of images: First, we capture ten images at a specific wavelength with varying exposures (ranging from underexposed to overexposed, evenly distributed). This procedure is subsequently repeated for ten distinct wavelengths. Consequently, in our experiment, N equals 100 (resulting from 10 wavelengths multiplied by 10 exposures each).

3. Experimental Setup

The experimental setup was designed to capture images of back-reflection light coming from a sample situated under a halogen lamp using a hyperspectral camera with a wavelength range of 713 nm–920 nm (Monarch II hyperspectral, Unispectral, Ramat Gan, Israel, 1280 × 1024 Pixels, H-FOV (H, V, D) 31.5°, 25.5°, 39.8°, 10 CWL bands (±10 nm) 713, 736, 759, 782, 805, 828, 851, 874, 897, 920 nm). In Figure 2, one can see a schematic sketch of the setup. The camera was positioned perpendicular (90°) to the tissue with the light source angled at 70° towards an object that resembles the human body and with a denser material placed underneath (a piece of chicken breast with a plastic disk of 175 mm dia. underneath it) to simulate a tumor. The disk size in camera pixels was about 125 pixels in diameter. Both the object and denser material were placed on a non-reflective surface. The hyperspectral camera was positioned above the object (40 cm above the sample area), and images were taken in a dark environment. To ensure a uniform distribution of gray levels across the frame, we captured an image of a gray uniform patch before proceeding with the samples. The subsequent step involved acquiring the sequence of images. Then, a weight map was generated after considering the quality measures of each pixel in the images within the sequence. The final image was produced by consolidating the stack of images through weighted blending, with each image making a proportional contribution to the result determined by its weight in the weight map (Equations (1)–(6)).

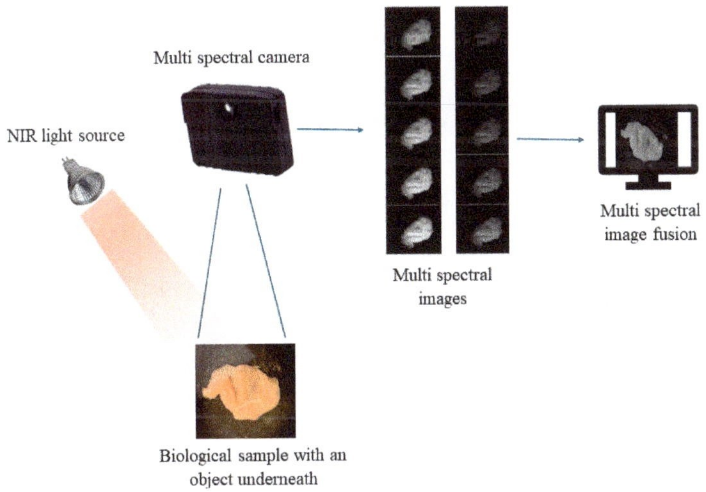

Figure 2. The schematic sketch of the setup.

4. Experimental Results and Discussion

A standard CMOS camera, which records light wavelengths below the near-infrared (NIR) spectrum, would not be able to distinguish the plasticine underneath the chicken tissue. Consequently, a hyperspectral camera was used to acquire 10 sets of images at different wavelength bandwidths, with different 10 exposures per wavelength (100 images in total). The different exposures were taken to compensate for the uneven light intensities per wavelength of the light source. In Figure 3, one can see experimental images: hyperspectral images taken in a sequence. The object appears brighter at lower wavelengths because of detector sensitivity and the illumination spectrum. The corresponding filter wavelength of a specific image is mentioned below each one. The FWHM (Full-Width Half Maxima) of all images is 40 nm per bandwidth, with a spectral band accuracy of +/− 2 nm and a 100 ms exposure time. All the images presented in Figure 3 were taken at the same exposure time. One can see the different intensities due to the light source as mentioned above.

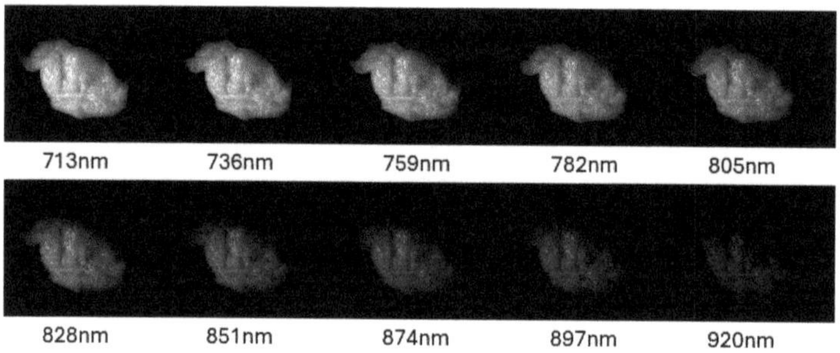

Figure 3. Experimental images: Hyperspectral images taken in a sequence with the same exposure of 100 ms. The corresponding filter wavelength of a specific image is shown below each one.

Subsequently, the image fusion technique was employed to merge all the images, aiming to generate a clearer and more harmonized composite. Following the merging of exposures, we iteratively applied the image fusion procedure to refine the detection of the plasticine area beneath the examined tissue, thereby facilitating a clearer delineation of its position. Furthermore, the illumination penetrating deeper into the region of interest (ROI) primarily comprises longer wavelengths within the near-infrared spectrum, undetectable by conventional silicon-based cameras.

To overcome this challenge, we utilized a camera equipped with near-infrared (NIR) detection capabilities. Leveraging this specialized camera, we captured reflections penetrating deeply into the tissue, thus revealing obscured objects. We acquired ten images from the hyperspectral camera for each exposure, spanning a wavelength range of 705–920 [nm], adequately suited for our objectives.

The purpose of the following step is to prepare all the images for the final stage of the fusion. After demonstrating that the camera can be used to detect objects behind scattering tissue, we wanted to improve its detection capability. In the first step, we captured the object at 10 different wavelengths as in the previous step (Figure 3). However, this time, we took sets of images with different exposures ranging from underexposed to overexposed (10 sets of images). The image fusion method was used to combine all the wavelengths into one image (as described before in Equations (1)–(6)). This method helped us extract the relevant information from each image and ignore any irrelevant information using the different interactions of each wavelength with the tissue. Since each wavelength reflection provides slightly different information about the area where the object is located, the combination of the images yielded a single image with more information, resulting in a clearer identification.

In Figure 4, one can see the experimental results. In Figure 4a, a white-light-illumination image (RGB) is shown, and no signs of the internal phantom can be seen. In Figure 4b, a one-shot of the NIR image is shown, resolving some internal phantom image, and the resulting exposure fusion image is shown in Figure 4c, with a clear resolution of the internal tissue (a black spot in the middle of the image). Note that the gray-colored images are due to the monochromatic nature of the images.

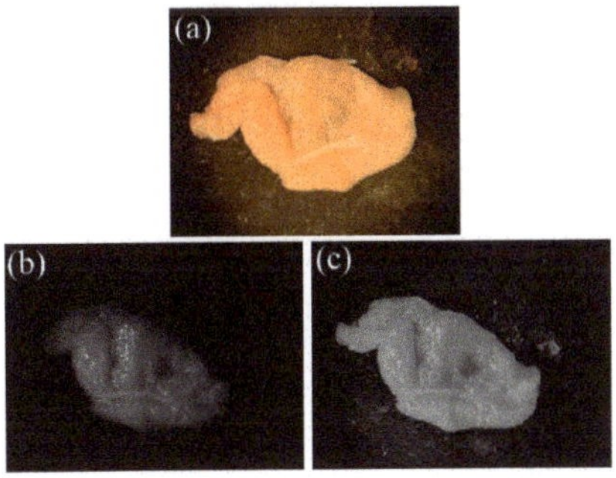

Figure 4. Experimental results: (**a**) white-light-illuminated image (RGB), (**b**) one-shot of NIR image, (**c**) resulting exposure fusion image.

In Figure 5, one can see cross-sections of the fusion image that clearly show a black spot (tumor location) indicated on the phantom. Figure 5a,c show the fusion image with dashed lines indicating the X- and Y-axis cross-section locations, respectively. Figure 5b,d show the intensity profile along the X- and Y-axis at the black spot location (marked by the dashed lines). The green line denotes the horizontal axis intensity profile, while the red line denotes the vertical axis.

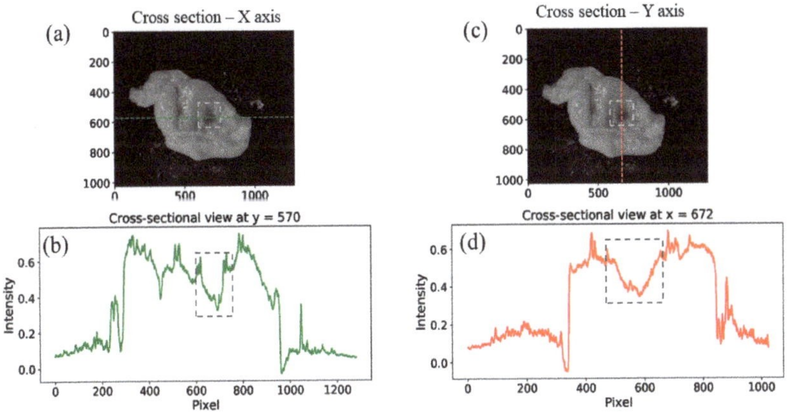

Figure 5. Images (**a**,**c**) are the fusion images, with green and red dashed lines indicating the cross-section's locations. (**b**,**d**) Intensity profile along the X- and Y-axis at the black spot location along the dashed lines.

The resulting cross-sectional graphs presented enable us to clearly distinguish the presence of a high-density body beneath the diffusing tissue with a high level of certainty.

In Figure 6, one can see a zoom-in of the fused image with the X- and Y-axis cross-section intensity profile (Figure 6a). The zoomed-in location is marked with dashed white lines, as can be seen in Figure 5.

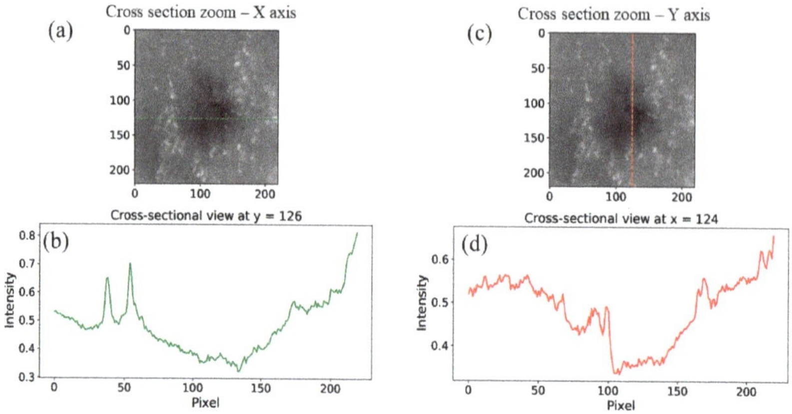

Figure 6. Experimental results: (**a**,**c**) zoomed version of the white dashed line areas in Figure 5 of the fusion image with the intensity profile along the Y-axis at the black spot location, (**b**,**d**) cross-section intensity profiles of the zoomed-in area.

The cross-sectional analysis presented in these images demonstrates the effectiveness of multispectral imaging and image fusion in detecting subsurface structures within diffusing biological tissues. The significant variations in intensity profiles along both the X- and Y-axis confirm the presence of a high-density object beneath the diffusing tissue.

Hyperspectral imaging with exposure fusion provides a powerful non-invasive method to visualize subsurface features in tissue. It is done by capturing a full reflectance spectrum at each pixel and combining multiple NIR wavelength images. It greatly enhances contrast and reveals hidden objects (e.g., tumors) that are indiscernible with conventional cameras. The proposed approach exploits the rich spectral and spatial information to differentiate materials by their optical signatures without any prior knowledge of the object, yielding superior delineation of the target in comparison with the background. However, the method is also characterized by high complexity and cost. It requires specialized NIR hyperspectral sensors, multi-exposure image acquisition and intensive data processing, resulting in large multi-dimensional datasets. Its performance is inherently limited to low tissue depths due to light attenuation, which collectively constrains its practical applicability. Our next step involves using deep convolutional neural networks to achieve the same results but using fewer images. Thus, we will need less computational and hyperspectral resources.

5. Conclusions

In this study, we demonstrated the feasibility of detecting reflections from objects situated within a dispersive layer using the near-infrared (NIR) range. Employing a hyperspectral camera operating within the NIR spectrum coupled with advanced image fusion techniques can enable the detection of cancerous tumors (phantoms) with improved resolution and clarity. By processing data from images captured across various bands of the near-IR range and merging them using image fusion methods, we devised a non-invasive and safe approach to tumor detection. This research lays the groundwork for future endeavors aimed at detecting cancerous tumors and blood vessels buried within tissue depths of the human body.

Author Contributions: Conceptualization, Y.D.; methodology, Y.D.; software, N.A. and Y.D.; validation, N.A. and Y.D.; formal analysis, N.A. and Y.D.; investigation, N.A. and Y.D.; resources, A.S. (Amir Shemer); data curation, Y.D.; writing—original draft preparation, N.A., Y.D., and A.S. (Amir Shemer); writing—review and editing, Y.B., A.S. (Ariel Schwarz), and A.S. (Amir Shemer); visualization, Y.B., A.S. (Ariel Schwarz), A.S. (Amir Shemer), and Y.D.; supervision, Y.D. and A.S. (Amir Shemer); project administration, A.S. (Amir Shemer); funding acquisition, A.S. (Amir Shemer). All authors have read and agreed to the published version of the manuscript.

Funding: This research was partially supported by the RAMA 302-2024 grant provided by MOST Science Accelerators and Azrieli internal fund.

Institutional Review Board Statement: Not applicable.

Data Availability Statement: Data is unavailable due to privacy or ethical restrictions.

Conflicts of Interest: The authors declare no conflicts of interest.

Abbreviations

The following abbreviations are used in this manuscript:

NIR	Near Infrared
MRI	Magnetic Resonance Imaging
CT	Computed Tomography
FWHM	Full-Width Half Maxima
RGB	Red, Green, Blue

References

1. Danan, Y.; Yariv, I.; Zalevsky, Z.; Sinvani, M. Improved Margins Detection of Regions Enriched with Gold Nanoparticles inside Biological Phantom. *Materials* **2017**, *10*, 203. [CrossRef] [PubMed]
2. Unger, J.; Hebisch, C.; Phipps, J.E.; Lagarto, J.L.; Kim, H.; Darrow, M.A.; Bold, R.J.; Marcu, L. Real-time Diagnosis and Visualization of Tumor Margins in Excised Breast Specimens using Fluorescence Lifetime Imaging and Machine Learning. *Biomed. Opt. Exp.* **2020**, *11*, 1216–1230. [CrossRef] [PubMed]
3. Baba, A.I.; Cătoi, C. Chapter 14, Nervous System Tumors. In *Comparative Oncology*; The Publishing House of the Romanian Academy: Bucharest, Romania, 2007.
4. Heidkamp, J.; Scholte, M.; Rosman, C.; Manohar, S.; Fütterer, J.J.; Rovers, M.M. Novel Imaging Techniques for Intraoperative Margin Assessment in Surgical Oncology, A Systematic Review. *Int. J. Cancer* **2021**, *149*, 635–645. [CrossRef] [PubMed]
5. Arica, S.; Altuntas, T.S.; Erbay, G. On Visualization and Quantification of Lesion Margin in CT Liver Images. In Proceedings of the Medical Technologies Congress (TIPTEKNO), Antalya, Turkey, 19–20 November 2020; pp. 1–4.
6. Veluponnar, D.; Dashtbozorg, B.; Jong, L.-J.S.; Geldof, F.; Guimaraes, M.D.S.; Peeters, M.-J.T.F.D.V.; van Duijnhoven, F.; Sterenborg, H.J.C.M.; Ruers, T.J.M.; de Boer, L.L. Diffuse Reflectance Spectroscopy for Accurate Margin Assessment in Breast-Conserving Surgeries: Importance of an Optimal Number of Fibers. *Biomed. Opt. Express* **2023**, *14*, 4017. [CrossRef]
7. Veluponnar, D.; de Boer, L.L.; Geldof, F.; Jong, L.-J.S.; Guimaraes, M.D.S.; Peeters, M.-J.T.F.D.V.; van Duijnhoven, F.; Ruers, T.; Dashtbozorg, B. Toward Intraoperative Margin Assessment using a Deep Learning-based Approach for Automatic Tumor Segmentation in Breast Lumpectomy Ultrasound Images. *Cancers* **2023**, *15*, 1652. [CrossRef]
8. Veluponnar, D.; de Boer, L.L.; Dashtbozorg, B.; Jong, L.-J.S.; Geldof, F.; Guimaraes, M.D.S.; Sterenborg, H.J.C.M.; Vrancken-Peeters, M.-J.T.F.D.; van Duijnhoven, F.; Ruers, T. Margin Assessment During Breast Conserving Surgery using Diffuse Reflectance Spectroscopy. *J. Biomed. Opt.* **2024**, *29*, 045006. [CrossRef]
9. Jong, L.J.S.; Veluponnar, D.; Geldof, F.; Sanders, J.; Guimaraes, M.D.S.; Peeters, M.-J.T.F.D.V.; van Duijnhoven, F.; Sterenborg, H.J.C.M.; Dashtbozorg, B.; Ruers, T.J.M. Toward Real-time Margin Assessment in Breast-conserving Surgery with Hyperspectral Imaging. *Sci. Rep.* **2025**, *15*, 9556. [CrossRef]
10. Yang, J.; Li, K.; Deng, H.; Feng, J.; Fei, Y.; Jin, Y.; Liao, C.; Li, Q. CT Cinematic Rendering for Pelvic Primary Tumor Photorealistic Visualization. *Quant. Imaging Med. Surg.* **2018**, *8*, 804–818. [CrossRef]
11. Kim, K.; Park, H.; Lim, K.M. Phototoxicity: Its Mechanism and Animal Alternative Test Methods. *Toxicol. Res.* **2015**, *31*, 97–104. [CrossRef]
12. Bernas, T.; Robinson, J.P.; Asem, E.K.; Rajwa, B. Loss of Image Quality in Photobleaching During Microscopic Imaging of Fluorescent Probes Bound to Chromatin. *J. Biomed. Opt.* **2005**, *10*, 064015. [CrossRef]

13. Berglund, A.J. Nonexponential Statistics of Fluorescence Photobleaching. *J. Chem. Phys.* **2004**, *121*, 2899–2903. [CrossRef] [PubMed]
14. Pertzborn, D.; Nguyen, H.N.; Hüttmann, K.; Prengel, J.; Ernst, G.; Guntinas-Lichius, O.; von Eggeling, F.; Hoffmann, F. Intraoperative Assessment of Tumor Margins in Tissue Sections with Hyperspectral Imaging and Machine Learning. *Cancers* **2023**, *15*, 213. [CrossRef]
15. Zhang, L.; Liao, J.; Wang, H.; Zhang, M.; Liu, Y.; Jiang, C.; Han, D.; Jia, Z.; Qin, C.; Niu, S.; et al. Near-Infrared II Hyperspectral Imaging Improves the Accuracy of Pathological Sampling of Multiple Cancer Types. *Lab Investig.* **2023**, *103*, 100212. [CrossRef]
16. Parasca, S.V.; Calin, M.A.; Manea, D.; Radvan, R. Hyperspectral Imaging with Machine Learning for in Vivo Skin Carcinoma Margin Assessment: A Preliminary Study. *Phys. Eng. Sci. Med.* **2024**, *47*, 1141. [CrossRef]
17. Mertens, T.; Kautz, J.; Van Reeth, F. Exposure Fusion. In Proceedings of the 15th Pacific Conference on Computer Graphics and Applications (PG'07), Maui, HI, USA, 29 October–2 November 2007; pp. 382–390. [CrossRef]
18. Jung, K.Y.; Cho, S.W.; Kim, Y.A.; Kim, D.; Oh, B.-C.; Park, D.J.; Park, Y.J. Cancers with Higher Density of Tumor-Associated Macrophages Were Associated with Poor Survival Rates. *J. Pathol. Transl. Med.* **2015**, *49*, 318–324. [CrossRef]
19. Heusmann, H.; Kölzer, J.; Otto, J.; Puls, R.; Friedrich, T.; Heywang-Koebrunner, S.; Zinth, W. *Photon Transport in Highly Scattering Tissue*; SPIE: Bellingham, WA, USA, 1995; Volume 2326, p. 370.
20. Mangold, K.; Shaw, J.; Vollmer, M. The physics of near-infrared photography. *Eur. J. Phys.* **2013**, *34*, 51. [CrossRef]
21. Nilsson, A.M.K.; Heinrich, D.; Olajos, J.; Andersson-Engels, S. Near Infrared Diffuse Reflection and Laser-induced Fluorescence Spectroscopy for Myocardial Tissue Characterisation. *Spectrochim. Acta Part A Mol. Biomol. Spectrosc.* **1997**, *53*, 1901–1912. [CrossRef] [PubMed]
22. Henderson, T.A.; Morries, L.D. Near-infrared Photonic Energy Penetration: Can Infrared Phototherapy Effectively Reach the Human Brain? *Neuropsychiatr. Dis. Treat.* **2015**, *11*, 2191–2208. [CrossRef]
23. Van Bommel, W. Halogen Lamp. In *Encyclopedia of Color Science and Technology*; Luo, R., Ed.; Springer: New York, NY, USA, 2012. [CrossRef]
24. Ojanen, M.; Kärhä, P.; Ikonen, E. Spectral Irradiance Model for Tungsten Halogen Lamps in 340–850 nm Wavelength Range. *Appl. Opt.* **2010**, *49*, 880–886. [CrossRef]
25. Ohmi, M.; Haruna, M. Ultra-High Resolution Optical Coherence Tomography (OCT) Using a Halogen Lamp as the Light Source. *Opt. Rev.* **2003**, *10*, 478–481. [CrossRef]
26. Ribbing, C.G. Chapter 8.2. Blackbody radiation. In *Optical Thin Films and Coatings*; Woodhead Publishing Limited: Cambridge, UK, 2013.
27. Hessel, C. An Implementation of the Exposure Fusion Algorithm. *Image Process. Line* **2018**, *8*, 369–387. [CrossRef]
28. Xu, F.; Liu, J.; Song, Y.; Sun, H.; Wang, X. Multi-Exposure Image Fusion Techniques: A Comprehensive Review. *Remote Sens.* **2022**, *14*, 771. [CrossRef]
29. Li, S.; Kang, X.; Fang, L.; Hu, J.; Yin, H. Pixel-level Imageaw Fusion: A Survey of the State of the Art. *Inf. Fusion* **2017**, *33*, 100–112. [CrossRef]
30. Michelson, A. *Studies in Optics*; University of Chicago Press: Chicago, IL, USA, 1927.
31. Peli, E. Contrast in Complex Images. *J. Opt. Soc. Am. A* **1990**, *7*, 2032–2040. [CrossRef]
32. Burt, P.J.; Hanna, K.; Kolczynski, R.J. Enhanced image capture through fusion. In Proceedings of the Workshop on Augmented Visual Display Research, Berlin, Germany, 11–14 May 1993; pp. 207–224.
33. Ogden, J.M.; Adelson, E.H.; Bergen, J.R.; Burt, P.J. Pyramid-based computer graphics. *RCA Eng.* **1985**, *30*, 4–15.
34. Tet, A. Hierarchical image fusion. *Mach. Vis. Appl.* **1990**, *3*, 1–11. [CrossRef]

sensors

MDPI

Article

Real-Time Multi-Channel Epileptic Seizure Detection Exploiting an Ultra-Low-Complexity Algorithm–Hardware Co-Design Approach

Andrea Vittimberga, Giovanni Nicolini and Giuseppe Scotti *

Department of Information, Electronics and Telecommunication Engineering, Sapienza University of Rome, 00184 Roma, Italy; andrea.vittimberga@uniroma1.it (A.V.); giovanni.nicolini@uniroma1.it (G.N.)
* Correspondence: giuseppe.scotti@uniroma1.it

Abstract: This paper presents an automated threshold-based multi-channel epileptic seizure detection algorithm designed for low-complexity hardware implementations. The algorithm relies on two discriminative, computationally simple time-domain features, based on power and amplitude variations, that enable accurate and timely detections due to their rapid adaptiveness to fluctuations in neural activity. To ensure long-term functionality and high sensitivity, system thresholds are optimized through an offline calibration process that exploits the statistical analysis of patient-specific inter-ictal and ictal periods. The novelty of the approach lies in its multi-channel decision-making strategy, which enhances reliability against false alarms. The proposed algorithm is tested on multiple datasets to assess its adaptability to different recording conditions, achieving roughly 98% accuracy and over 98% sensitivity on both the EEG CHB-MIT dataset and the iEEG SWEC-ETHZ dataset, with average latencies of 3.37 s and 7.84 s, respectively. These results are comparable to, and in some cases outperform, several published machine-learning-based approaches. On the hardware side, FPGA synthesis highlights the minimal and scalable resource requirements of the proposed architecture, achieved through Time-Division Multiplexing (TDM) of both filtering and feature extraction. When compared to state-of-the-art proposals, the system emerges as an ideal candidate for real-time, resource-constrained hardware implementations.

Keywords: electroencephalogram; intracranial EEG; adaptive neurostimulation; epilepsy; seizure onset detection; FPGA

1. Introduction

Epilepsy remains one of the most prevalent neurological disorders worldwide, and while pharmacological treatments are effective in many cases, approximately a quarter of the affected individuals fail to respond adequately to conventional medications [1,2]. As a consequence, recent advancements have focused on the development of automated device-based approaches to address the challenge of refractory epilepsy. Among these, techniques like Vagus Nerve Stimulation (VNS) and Deep Brain Stimulation (DBS) rely on the localized delivery of electrical stimuli to ameliorate the effects of the ongoing seizure in a controlled manner [3–5].

In its conventional, open-loop form, DBS delivers continuous electrical pulses to targeted brain regions according to pre-set parameters, without taking into account the patient's current neural state. Adaptive variations in DBS, however, enact neuromodulation based on real-time monitoring of the subject's neural responses, dynamically adjusting stimulation to optimize therapeutic outcomes and reduce side effects [6,7]. For this reason, automated seizure detection is becoming increasingly prevalent as an ideal candidate to regulate the stimulation protocol.

Traditional seizure detection has typically relied on visual inspection of neural recordings, a process that is inherently subjective, time-consuming, and therefore ill-suited not only for automated detection but also for deployment in implantable devices. Developing objective, efficient, and reliable methods for automated seizure detection presents a critical hurdle due to the strong variability of pre-ictal and inter-ictal patterns both within individual patients and across different patients, as well as the non-stationary nature of scalp EEG and intracranial EEG (iEEG) signals. This high variability complicates the design of generalized detection algorithms and introduces an inherent trade-off between sensitivity and accuracy, as systems must balance the minimization of both false positives and false negatives. These challenges make it difficult to develop algorithms capable of maintaining robust performance across diverse individuals and recording conditions.

A key task in seizure detection involves identifying optimal representations from neural signals within the feature space. From this perspective, a plethora of signal descriptors and metrics has been proposed and evaluated in terms of their discriminative capabilities. In this context, features may be extracted from the time domain, including peak density, signal power, or Hjorth parameters [8,9]. Furthermore, non-linear features like Approximate Entropy (ApEn) have seen increased usage, adding complementary information to traditional time-based metrics [10]. In frequency analysis, Cross-Frequency Coupling (CFC) stands out in particular, with Phase-Amplitude Coupling (PAC) being widely employed to quantify the relationship between the amplitude of high-frequency oscillations and the phase of slower frequency bands [11].

More recently, machine-learning and deep learning strategies have taken center stage: models such as Support Vector Machines (SVM), Convolutional Neural Networks (CNN), and other architectures are able to combine and process large sets of features, achieving high levels of accuracy in seizure onset detection and prediction [12,13]. Despite their strong performance, these methods often involve considerable computational overhead, which limits their practical use in implantable systems. Such devices must function under strict constraints on area and power consumption, conditions that remain difficult to meet with current state-of-the-art algorithms. The study in [14] proposes a patient-specific, FPGA-implemented, threshold-based detection algorithm that leverages eleven time-domain features. A feature ranking strategy is employed to identify the most discriminative subset of features, while a channel selection mechanism is integrated to reduce data dimensionality. However, the initial number of extracted features is relatively high, and the resources required for their computation, as well as for the associated pre-processing steps, are non-trivial. On that topic, the work presented in [15] aims to reduce the computational load by restricting the set of features to a single time-domain Local Binary Pattern (LBP) code extracted from the recorded iEEG signal. While effective, the algorithm has not been implemented on hardware platforms and presents significant latency. Additionally, the work proposed in [16] employs a combination of under-sampling and boosting techniques to improve the performance of a classifier, which in turn makes use of statistical metrics extracted from pre-processed EEG signals. Although the original paper does not report implementation details, it can be reasonably assumed that calculating higher-order metrics, such as kurtosis or skewness, could limit the applicability of the approach on portable or resource-constrained hardware. In [17], the authors proposed an adaptive distance-based seizure detector which makes use of both statistical and morphological metrics to elaborate signals that undergo principal component analysis and common spatial patterns to enhance the available data. Similar potential limitations as [16] also apply in this case.

Regarding more specific implementations that target hardware realization on FPGA, the system proposed in [18] implements a mixed-signal SoC featuring an analog front-end for EEG acquisition and a digital detection core in 55 nm CMOS technology. The detection unit performs on-chip feature extraction based on four eigenvalues derived from time-domain features. Hardware modules handle feature calculation, threshold update, and detection logic. The design was prototyped on FPGA and later realized as an ASIC.

In addition, the system presented in [19], implemented on an Artix-7 FPGA, realizes energy-efficient, real-time seizure detection. The hardware consists of dedicated modules for feature extraction, calculation of Hjorth mobility and non-linear energy in parallel pipelines, and on-chip SVM or QDA classifiers. The design includes optimized memory buffers and control logic to manage data flow between extraction and classification units, achieving minimal dynamic power (approximately 0.057 mW for SVM). In both of the aforementioned cases, there remains margin for improvement in terms of resource allocation.

In this work, we propose a novel framework to analyze and detect abnormal patterns related to epileptic seizure events, based on the power fluctuations of the EEG signal, and applied through the use of two discriminative time-domain features of simple derivation. The proposed algorithm is threshold-based and, as such, it is much simpler in terms of hardware resources with respect to the aforementioned solutions, yet it is able to achieve state-of-the-art comparable performance regarding accuracy, sensitivity, and latency. The novelty of this work lies in the combination of a minimal feature set, consisting of Line Length (*LL*) and instantaneous power difference (*PD*) between consecutive samples, which enables a simple and low-complexity algorithm. The *PD* feature benefits from being compatible with partial LL computation, as both require only two samples at a time, reducing hardware requirements.

In addition, instead of implementing a channel reduction approach, a full-channel majority-voting mechanism leverages all recording channels to improve true positive detection while limiting false alarms, made possible by the low-complexity algorithm and minimal hardware usage. Finally, a hardware-oriented Time-Division Multiplexing (TDM) architecture executes filtering and partial feature extraction within the TDM framework, allowing scalable operation across a large number of channels and making the system suitable for implantable applications.

The paper is organized as follows: Section 2 is devoted to the presentation of the proposed work, from the employed set of features to the detailed description of the algorithm's functionality in its distinct steps, including the pre-processing stage. Section 3 delves into the details of the hardware implementation of the proposed seizure detector, showing its functionality with simulations conducted across different software environments. Section 4 reports the simulation results obtained using EEG and iEEG datasets as benchmarks for system validation, providing an extensive description of the statistical metrics employed to assess the algorithm and tune its parameters. In addition, this section hosts a comparative analysis that weighs the algorithm's performance and synthesis on FPGA against current state-of-the-art proposals. Finally, Section 5 concludes the work.

2. Proposed Work

This section details the main components of the proposed seizure detection system, starting with a description of the double feature extraction stage that provides the basis for the algorithm. Consequently, the multi-channel signal processing flow and threshold-based decision rule used to detect the seizure onset are thoroughly explained.

2.1. Features

Feature extraction is a critical step in seizure detection algorithms, as it transforms the filtered neural signals into representations that highlight the more relevant patterns pertaining to the ictal activity. In this work, two time-domain features were selected based on their discriminative capability, computational efficiency, and ability to follow variations associated with seizure onsets with relatively low latency. In limiting the feature set to two, the overall complexity of the extraction stage is greatly reduced, making the system suitable for resource-constrained hardware implementations. The specific features to be extracted in the proposed algorithm are introduced and discussed in the following.

2.1.1. Power Difference

As seizure events occur, they manifest as sudden, abnormal bursts of synchronized neural activity, often observable through sharp increases in amplitude and shifts in frequency. To characterize rapid changes in signal dynamics, the first of the two considered features quantifies the variation in instantaneous power by computing the difference between the squared values of consecutive samples. Such power difference (PD) can be analytically expressed as

$$PD = x[n]^2 - x[n-1]^2 \quad (1)$$

where $x[n]$ denotes the amplitude of the n-th sample. The PD feature is effective in capturing instantaneous power and frequency fluctuations with reliable consistency, as it emphasizes localized shifts in amplitude, allowing it to distinguish between pre-ictal and ictal states.

2.1.2. Line Length

Line length is a time-domain feature widely used in EEG/iEEG signal analysis due to its high sensitivity to variations in both amplitude and frequency [20,21]. It is computed as the cumulative sum of the absolute differences between consecutive signal samples within a given time window of N samples, as defined in [22]:

$$LL = \sum_{n=1}^{N} |x[n] - x[n-1]| \quad (2)$$

In the initial stages of the ictal state, neural signals often exhibit subtle changes, such as low-amplitude rhythmic oscillations or minor waveform distortions. These early alterations may not be efficiently captured by the power difference feature alone, which, being quadratic in nature, responds more prominently to abrupt, localized increases in amplitude while tending to suppress smaller fluctuations, potentially resulting in delayed seizure onset detection. In contrast, the line length feature offers greater sensitivity to these early dynamics and, through its window-based cumulative process, generally yields a smoother, monotonic increase in value that reflects the progression of the seizure event.

2.2. Proposed Algorithm

The proposed seizure detection algorithm, which is outlined in Figure 1, is based on a thresholding principle that fully harnesses information provided by multi-channel neural recordings to achieve a consistent and responsive discrimination. Particularly, since seizure events are typically characterized by pronounced fluctuations, the proposed algorithm leverages the LL and PD features to track and locally enhance such variations with respect to inter-ictal neural dynamics. The combination of these two extracted features raises the probability of correctly distinguishing false alarms from true epileptic events. To that end, an initial offline calibration stage is employed to determine the optimal detection thresholds, thereby accounting for patient-specific variability and recording conditions, as well as ensuring that the algorithm is able to operate reliably in a real-time scenario.

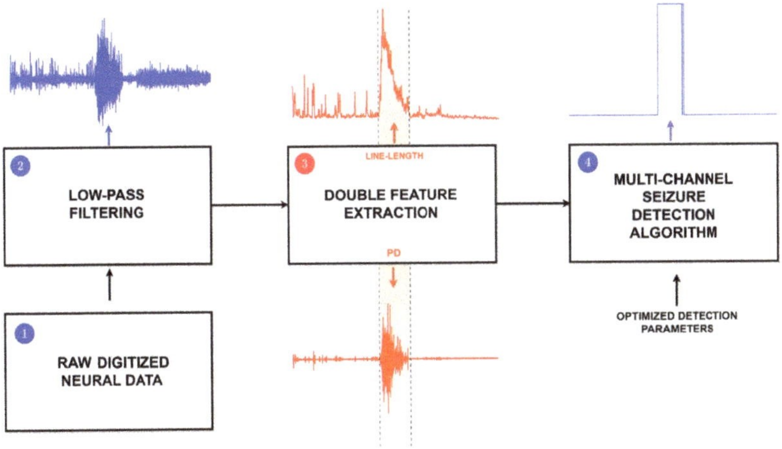

Figure 1. General diagram of the proposed algorithm: (1) neural data are digitized by the analog front-end, and (2) low-pass filtered for pre-processing. (3) Main time-domain features are extracted, and (4) aggregated information is processed to detect seizure-related activity.

A window-based processing scheme is used to improve the temporal resolution of detection and to organize the processing pipeline in a structured, stepwise manner. With an appropriately chosen window length *N*, this approach also enhances the consistency of the analysis. Indeed, windows that are too large can reduce the algorithm's discriminative ability and cause increases in latency, whereas excessively short windows may fail to capture the characteristic fluctuations between ictal and inter-ictal states or make the system more susceptible to false detections. For each window, the algorithm first extracts the *PD* and *LL* features after low-pass filtering of the signal is performed. Next, seizure discrimination is carried out by monitoring whether these features cross the detection thresholds. The remainder of the section presents a detailed description of the general processing chain involved in the proposed methodology.

2.2.1. Data Pre-Processing

Pre-processing is deliberately kept to a minimum in order to maintain a lightweight algorithm that is well-suited for resource-constrained applications. In particular, advanced artifact-removal techniques such as Independent Component Analysis (ICA) are not employed, as they introduce additional computational complexity and risk discarding signal components that may still contain relevant seizure-related information.

A 64th-order low-pass Finite Impulse Response (FIR) filter, implemented using TDM across recording channels, is applied to the incoming neural data. The FIR structure ensures a linear phase response, which preserves the temporal coherence of the EEG signals. The stop-band frequency is set to 50 Hz (Figure 2), effectively suppressing high-frequency artifacts and noise while retaining the spectral bands most relevant for seizure detection according to our study, namely Delta, Theta, Alpha, and Beta. Furthermore, the selected order guarantees an attenuation of at least 60 dB at 50 Hz, ensuring robust rejection of power-line interference.

This filter is included in the processing chain regardless of any pre-processing already applied to the test datasets, providing a complete system that can operate directly on in vivo neural data and remain adaptable for real-time implementation.

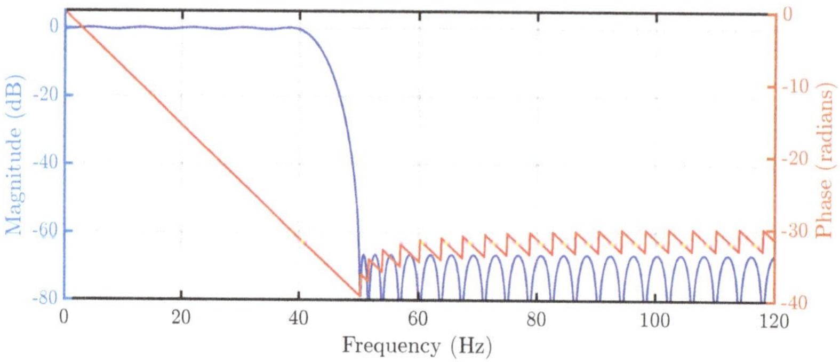

Figure 2. 64-taps FIR filter magnitude and phase response.

2.2.2. Seizure Detection Algorithm

The overall algorithm is summarized in the high-level flow diagram shown in Figure 3. Its operation can be divided into two main stages: single-channel processing and multi-channel decision-making. The first stage focuses on the analysis of each EEG (or iEEG) channel independently, while the second stage integrates information across multiple channels to provide a final decision on seizure occurrence.

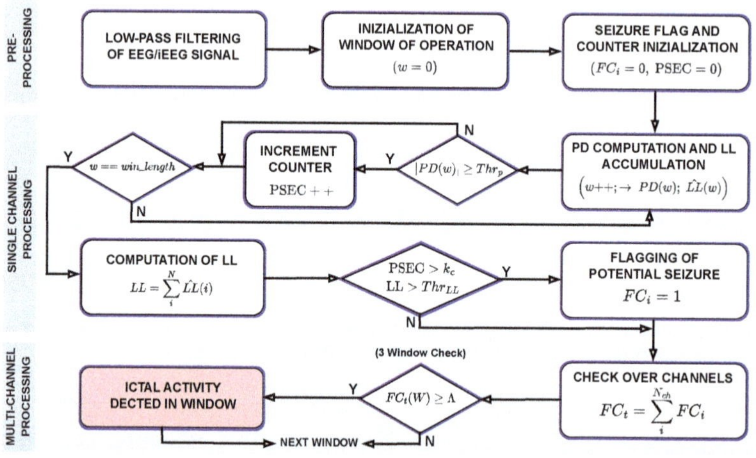

Figure 3. Pseudo-algorithmic flow-diagram of the proposed seizure detection algorithm. The steps are grouped by their functionality within the processing chain.

As a preliminary step, the acquired neural signals undergo low-pass filtering to suppress high-frequency noise and reduce the influence of potential artifacts, as described in Section 2.2.1. Following this pre-processing step, the algorithm proceeds in a window-based fashion. At the beginning of the analysis, corresponding to the first sample of the first window, all relevant parameters are initialized to zero. Specifically, the window sample-related variable w is set to 0, and for each recording channel two dedicated variables are initialized: (i) the Potential Seizure Event Counter (PSEC), which tracks the number of PD threshold crossings within the window under analysis for a given channel, and (ii) the Flag Counter (FC), which is later used in the decision stage to identify the presence of epileptic seizure evidence shared among channels.

At this point, the single-channel processing phase is triggered. In particular, following a first increment of the sample variable w, the PD is computed according to Equation (1).

Meanwhile, the algorithm updates the running accumulation of the LL feature, with the partial value being defined as $\hat{LL}(w)$. Since LL is defined as an accumulated measure, only this intermediate value is provided during window progression, whereas the final value of the accumulation, given in Equation (2), becomes available only at the end of the window under analysis.

The first condition in the detection process evaluates the absolute value of the extracted PD at each sample against a predefined threshold, Thr_p. Such threshold is obtained from an offline statistical analysis of a sufficiently large dataset to ensure proper tuning, accounting for the trade-off between detection sensitivity, latency, and overall accuracy. Whenever the threshold is exceeded, it indicates a potential seizure event for the corresponding channel, resulting in an increment of PSEC.

$$\text{PSEC} = \begin{cases} \text{PSEC} + 1, & \text{if } |PD(w)| \geq Thr_p, \\ \text{PSEC}, & \text{otherwise.} \end{cases} \tag{3}$$

This process is reiterated for each sample in the current analysis window until all samples have been processed. Once the last sample is evaluated, the window is complete, and the algorithm can proceed to the next state of the flow diagram. Here, the LL is available to be used for further evaluations. In particular, a double condition is checked in order to potentially raise the seizure flag FC, and it is defined as follows:

$$\begin{cases} \text{PSEC} \geq k_c, \\ LL \geq Thr_{LL} \end{cases} \tag{4}$$

where k_c represents an empirical factor that quantifies the peak density of the PD feature with respect to its threshold within the analysis window, while Thr_{LL} denotes the threshold applied to the LL accumulated over the same window. Both parameters, similar to Thr_p, are determined offline through a calibration process aimed at optimizing system performance. The system of equations in Equation (4) implements a logical OR operation, meaning that if either condition is satisfied the samples within the current window are flagged as abnormal, and the flag counter FC_i is set to 1, where i denotes the i-th channel.

Although the use of a logical OR operation makes the detection framework inherently more permissive compared to a stricter AND condition, thereby potentially increasing vulnerability to false positives, the actual impact on the False Positive Rate (FPR) remains limited when the calibration process is properly carried out. In practice, the thresholds are tuned in a way that balances the contribution of the PD and LL features, allowing them to complement one another. This compensation mechanism ensures that the system can achieve an improved trade-off between sensitivity and latency without a significant degradation in accuracy.

The final stage of the algorithm integrates a multi-channel voting strategy to determine the effective presence of ictal activity. Specifically, the seizure flags from all channels are aggregated, and the percentage of marked channels is compared against a predefined threshold, denoted as Λ. A seizure is declared only if the number of flagged channels exceeds Λ, thus enhancing robustness against spurious detections on individual channels:

$$FC_t = \sum_{i}^{N_{ch}} FC_i \geq \Lambda \tag{5}$$

with FC_t being the sum of the flags FC_i across all processing channels. To further improve the system's reliability, the condition presented in Equation (5) is required to be satisfied over three consecutive windows, ensuring that the detected activity corresponds to a sustained ictal pattern rather than a non-ictal fluctuation.

3. Hardware Implementation

The following section provides a detailed account of the hardware implementation of the various functional blocks that compose the digital processing chain underlying the proposed epilepsy detection algorithm. In that regard, the FPGA platform AMD Artix-7 was considered for the design's realization. Specifically, the proposed processing chain was implemented as a hardware-oriented simulation model in Simulink, using the AMD proprietary System Generator libraries, integrated within the Vitis Model Composer tool.

This methodology allowed the proposed system to be thoroughly tested and validated within a controlled software environment, while simultaneously supporting the automatic generation of synthesizable HDL code. As such, the resulting system was simulated both within the Simulink environment and the Vivado Design Suite using the CHB-MIT scalp EEG dataset [23], which will be described in more detail in Section 4.1.

3.1. FIR Filter

As previously mentioned, the processing chain includes a 64th-order low-pass FIR filter, implemented using a TDM approach. The choice of a FIR structure over an Infinite Impulse Response (IIR) alternative ensures a linear phase response, thereby preserving the temporal integrity of the EEG signals and avoiding distortions that could compromise feature extraction. To optimize resources usage, the TDM approach enables the reuse of a single DSP unit across the multiple input channels, which significantly reduces hardware complexity while still meeting the required throughput.

The filter is designed to operate at a sufficiently high clock frequency to handle all available recording channels in real time. Specifically, the operating frequency is defined as

$$f_{FIR} \geq f_s \cdot N_{ch}, \quad (6)$$

where f_s is the input sampling frequency, and N_{ch} represents the number of channels. This relation ensures that the filter can sequentially process the i-th sample of each channel across all coefficients before the next input sample arrives. The high-level functional timing diagram of the FIR's operation is shown in Figure 4. For illustrative purposes, the diagram refers to a simplified case study with 16 input channels.

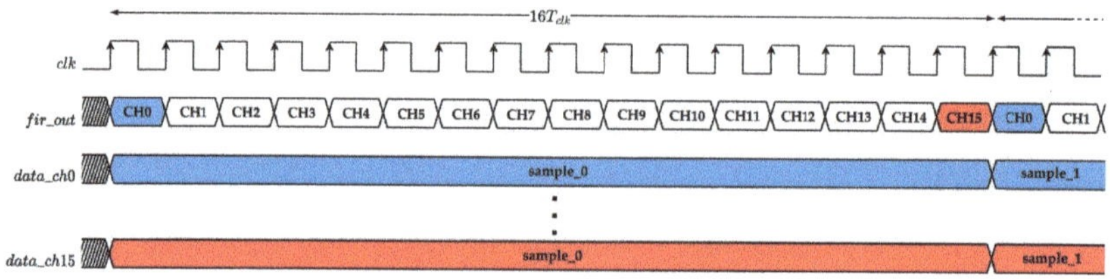

Figure 4. Timing diagram of the FIR filter operation, illustrating the sequential time-division multiplexing process for a generic case with 16 channels.

The filter coefficients were generated with MATLAB's Filter Design and Analysis (FDA) tool and provided as input to the Vivado FIR Compiler, which handles coefficient quantization and hardware mapping. In this implementation, the coefficients are represented with a 16-bit word length, which ensures that the realized frequency response closely matches the ideal design. The underlying architecture adopted for the FIR is a systolic Multiply-Accumulate (MAC) structure, well suited for FPGA-based realizations

(Figure 5). In this architecture, data flows through an array of interconnected processing elements, each performing a partial multiplication and accumulation. This pipelined organization improves resource utilization, allowing the filter to sustain real-time operation across multiple channels while maintaining scalability and predictable timing behavior. For data alignment purposes, a series of First-In First-Out (FIFO) elements is used to ensure proper TDM operation, while dedicated registers are inserted in the processing chain to break critical paths.

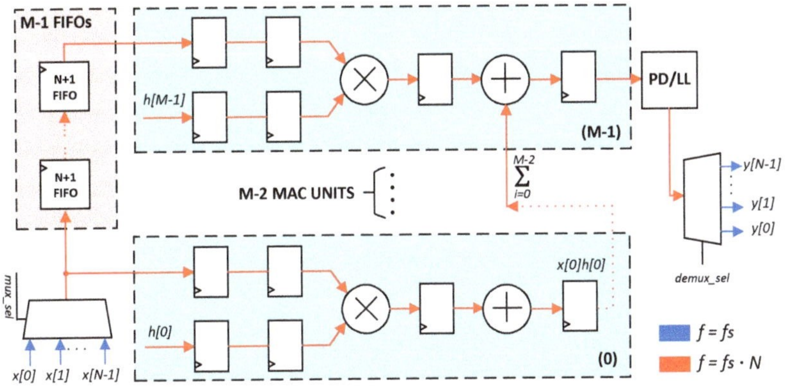

Figure 5. DSP slice of a multi-channel FIR filter operating in TDM.

To further optimize resources usage, part of the FIR FIFO used to delay the input data is implemented using Block RAM (BRAM) elements instead of flip-flops, thereby reducing the overall register count and improving area efficiency on the FPGA.

3.2. Double Feature Extractor

The hardware implementation of the double feature extracting core benefits from the TDM architecture as well, specifically with regard to resource minimization (Figure 6). The input data stream coming from the FIR filter is first demultiplexed, to account for the sequential logic required by the definition of the *PD* and *LL* feature, and then multiplexed again to perform the TDM operation. This approach enables the reuse of combinational logic across channels, significantly reducing hardware overhead. Consequently, the *PD* feature is implemented in TDM using two multipliers and a single subtractor.

The same multiplexed processing is applied to the computation of the absolute difference between consecutive samples required for the *LL* feature. However, unlike the *PD*, the accumulation over the full window cannot be shared across channels, as each channel requires an independent sum of its differences. For this reason, the *LL* implementation employs a dedicated accumulator per channel, while keeping the absolute-difference computation time-shared across the multiplexed data stream.

3.3. Detection Unit

After filtering and processing by the TDM double feature extractor, the data stream is sent to a demultiplexer and, subsequently, to the detection units, which include counters, comparators, and a deglitch module apiece. Each *PD* sample triggers a comparator when the condition in Equation (3) is satisfied, enabling an accumulator that increments by one unit and is reset at the end of the operational window through dedicated logic. A second comparator evaluates the peak density within the window, while, in parallel, a third one assesses the *LL* at the end of the same temporal interval. The outputs of the two comparators are combined using a logical OR and drive specific logic that triggers a flag whenever either the PSEC or the accumulated line length exceeds their respective thresholds.

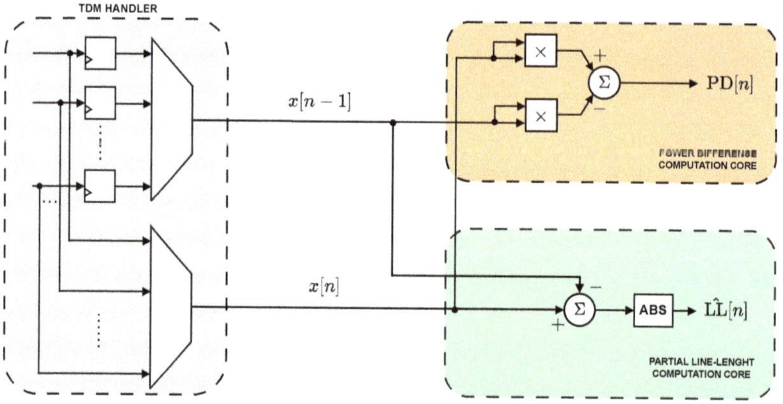

Figure 6. Hardware architecture of the TDM double feature extractor.

A dedicated deglitch unit, which is a temporal filter, stabilizes the output over a defined time interval, with its operation being controlled by an edge detector that monitors the comparators' outputs.

Regarding the multi-channel voting, after proper type conversion, the single detectors' outputs are fed to a tree of adders, which sum the contributions from each channel to determine the percentage that have flagged potential ictal activity. The output of the summing operation is evaluated by a comparator against the threshold Λ, implementing the condition described in Equation (5). Once again, a deglitch unit is exploited to stabilize the comparator's output, which is subject to a final check by dedicated logic over three consecutive windows.

3.4. Architecture-Level Hardware Simulation

Shown in Figure 7a is the model of the hardware architecture implementing a 20-channel version of the proposed algorithm, emphasizing the scalability of the system. The choice of twenty channels was deemed sufficient to meaningfully exploit the multi-channel voting mechanism, while still maintaining a realistic hardware configuration. As shown at the top of the figure, the 20 EEG inputs logged from the MATLAB environment are collected and passed through a dedicated time-multiplexing block, which combines them into a single interleaved stream. This stream, operating at the frequency specified in Equation (6), is directed to the FIR filter stage, whose output signal is then delivered to the dual feature-extraction unit, responsible for producing the *PD* and partial *LL*. After said extraction, the TDM streams are demultiplexed to reconstruct 20 parallel signals per feature at frequency f_s, which are finally routed to the individual detection units, where the initial stage of the decision process is carried out independently on each channel (Figure 7b).

The final element of the system incorporates the multi-channel voting block that was previously described in detail. To demonstrate the proper functionality of the system at hardware-level, the output of the detector when fed with a seizure-affected sample is presented in Figure 8.

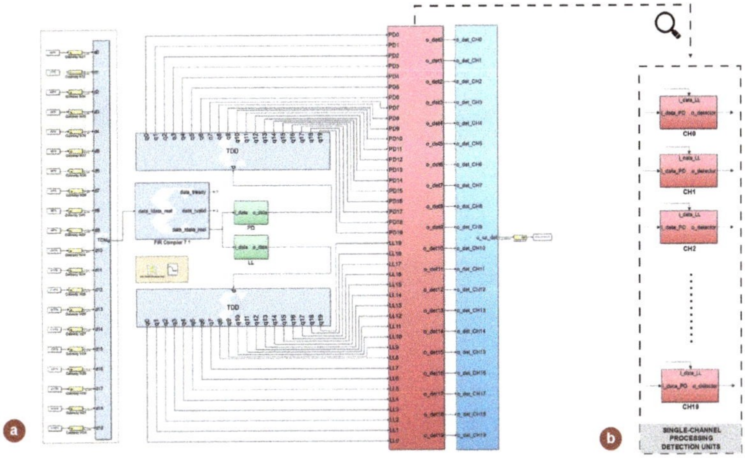

Figure 7. (**a**) Simulink model of a 20 channel implementation of the proposed algorithm. (**b**) Focus on the single-channel detection blocks that comprise the overall detection unit.

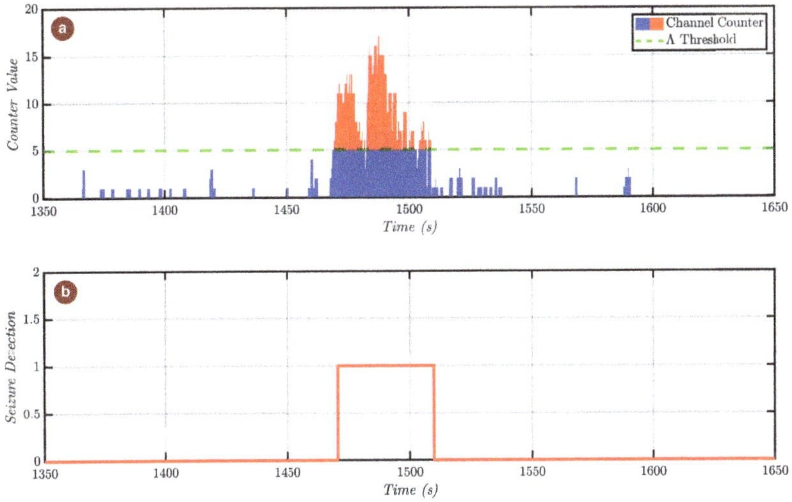

Figure 8. Detector's output simulated in Simulink. Depicted are (**a**) the output of the channel voting unit, along with (**b**) the actual seizure detection result.

Notably, it can be observed that, as the ictal state begins, the number of channels that flag the presence of seizure-related activity (Figure 8a) starts incrementing rapidly, as a consequence of both the increased density of *PD* peaks and the elevated *LL* values observed across multiple channels. The multi-channel voting mechanism continuously monitors this count, and once the latter exceeds the predefined threshold Λ, set to 5 in this example, for at least three consecutive windows, the detector output is activated, as illustrated in Figure 8b.

The system was then imported into the Vivado Suite to be synthesized and to perform a functional simulation. Figure 9 shows the result of the detection when fed with EEG data capturing a segment that includes the transition from non-ictal to ictal activity. For visual clarity, only three channels are present in the waveform window. In this case, it is evident that the system reacts rapidly to the presence of abnormal seizure-related patterns.

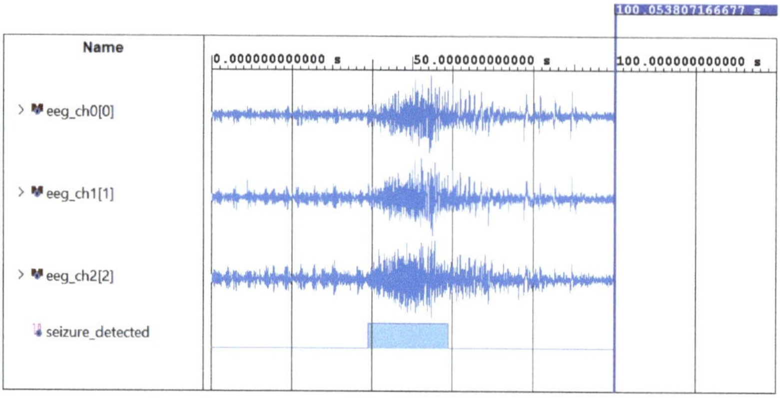

Figure 9. Vivado simulation reporting key waveforms of the algorithm's RTL implementation. Depicted are the three input EEG data taken from the CHB dataset, and the output of the detector.

3.5. FPGA Synthesis Results

To validate the simplicity of the proposed work, the overall architecture was synthesized on an FPGA platform, namely the AMD Artix-7 XC7A100T FPGA. Without loss of generality, a 4-channel version was prepared for the sake of the synthesis, in order to be reasonably comparable with state-of-the-art single-channel implementations while keeping the multi-channel facet peculiar to the proposed algorithm. The resulting report is compiled in Table 1, showing the exact unit utilization, correlated with the percentage of the same with respect to the available resources on the chosen platform. As can be observed, the reported results corroborate the low complexity and efficiency of the proposed architecture, confirming its suitability for real-time implementation on resource-constrained hardware.

Table 1. Synthesis results showing resource utilization on the AMD Artix-7 XC7A100T FPGA with a 4 input channel configuration.

Resources	Utilization	Available	Utilization (%)
LUT	192	63,400	0.30
FF	500	126,800	0.39
DSP	3	240	1.25
BRAM	0.5	135	0.37

4. Experimental Results and Comparative Analysis

This section describes the simulation framework, with a focus on the datasets used for tuning and validating the algorithm in the MATLAB 2021a environment. Key figures of merit employed to assess the detector are also presented. Finally, the results of applying the algorithm to the CHB-MIT and SWEC-ETHZ datasets are reported and analyzed in detail and compared against current state-of-the-art proposals. Additional comparisons will also address the hardware resource allocation with respect to other implemented designs.

4.1. Simulation Setup and Datasets

The following sub-section describes the datasets employed to evaluate the proposed algorithm. For a proper assessment, both a scalp EEG dataset and an iEEG dataset were considered. The latter includes short-term sessions, focused on few minutes centered around the ictal activity, and long-term recordings, which provide realistic conditions of extended monitoring periods.

4.1.1. CHB Dataset

For validation, the proposed algorithm was tested on the publicly available CHB-MIT Scalp EEG dataset, collected at Boston Children's Hospital in collaboration with the Massachusetts Institute of Technology [23]. The dataset contains long-term EEG recordings from 23 pediatric patients with epilepsy, comprising 9 to 40 sessions per subject and totaling over 900 h. Signals were acquired with the international 10–20 electrode system, sampled at 256 Hz with 16-bit resolution, and generally included 23 recording channels. Each recording is accompanied by expert-annotated seizure onset and offset times, providing a benchmark for the statistical analysis of the detector. Information pertaining to the CHB dataset is annotated in Table 2.

Table 2. Summary of the CHB dataset, including the number of patients, their age and gender, seizure counts, and average seizure time per patient.

Metric	chb01	chb02	chb03	chb04	chb05	chb06	chb07	chb08	chb09	chb10	chb11	chb12	chb13	chb14	chb15	chb16	chb17	chb18	chb19	chb20	chb21	chb22	chb23
Sex	F	M	F	M	F	F	F	M	F	M	F	F	F	F	M	F	F	F	F	F	F	F	F
Age (years)	11	11	14	22	7	1	14	3	10	3	12	2	3	9	16	7	12	18	19	6	13	9	6
N° Seizures	7	3	7	4	5	10	3	5	4	7	3	35	9	9	21	10	3	6	3	8	4	3	7
Avg. Time (s)	63	57	57	94	112	15	108	183	69	64	268	35	38	19	103	8	98	53	79	37	50	68	61

4.1.2. SWEC-ETHZ iEEG Database

The long-term SWEC-ETHZ iEEG dataset comprises 2656 h of continuous recordings from 18 patients with pharmaco-resistant epilepsy, collected during pre-surgical evaluation at the Sleep-Wake-Epilepsy-Center (SWEC) of the University Department of Neurology, Inselspital Bern, in collaboration with the Integrated Systems Laboratory of ETH Zurich. The dataset includes 116 annotated seizure events, acquired with strip, grid, and depth electrodes. Signals were median-referenced, band-pass filtered between 0.5 Hz and 120 Hz using a fourth-order Butterworth filter, and digitized at 512 Hz or 1024 Hz with 16-bit resolution [24].

As with the previously mentioned dataset, the recordings are accompanied by annotations that label the beginning and conclusion of seizure events, determined through visual inspection by clinical experts. The overall number of seizures per patient, as well as the average seizure time, have been reported in Table 3. Contrary to the CHB-MIT dataset, it can be observed that the number of channels per patient varies greatly, ranging from 24 to over 100.

Table 3. Summary of the long-term SWEC-ETHZ iEEG dataset, including the number of channels, seizure counts, and average seizure time per patient.

Metric	ID01	ID02	ID03	ID04	ID05	ID06	ID07	ID08	ID09	ID10	ID11	ID12	ID13	ID14	ID15	ID16	ID17	ID18
N° Ch.	88	66	64	32	128	32	75	61	48	32	32	56	64	24	98	34	60	42
N° Seizures	2	2	4	14	4	8	4	70	27	17	2	9	7	60	2	5	2	5
Avg. Time (s)	602	88	65	42	35	43	53	22	34	42	43	53	50	34	35	43	39	44

To further test the adaptability of the algorithm, the short-term SWEC-ETHZ iEEG dataset was also taken into consideration (Table 4). Such collection of samples includes 100 recordings taken from 16 patients. Each recording contains 3 min of pre-ictal activity, the ictal segment, and 3 min of post-ictal activity [25].

Table 4. Summary of the short-term SWEC-ETHZ iEEG dataset, including the number of patients, average seizure time, seizure counts, and variability in the number of channels per patient.

Metric	ID01	ID02	ID03	ID04	ID05	ID06	ID07	ID08	ID09	ID10	ID11	ID12	ID13	ID14	ID15	ID16
N° Ch.	47	42	98	62	54	64	36	59	56	100	64	49	92	74	61	59
N° Seizures	13	4	2	14	10	4	2	2	9	5	2	10	7	7	3	6
Avg. Time (s)	71	223	99	153	98	146	15	56	144	14	109	45	81	602	121	88

Results from the simulation of the algorithm on the aforementioned datasets are thoroughly analyzed in the coming section.

4.2. *Performance of the Proposed Algorithm*

This section presents the metrics used to calibrate and assess the performance of the proposed algorithm. The subsequent results summarize its application across the selected datasets, with all simulations conducted in the MATLAB environment.

4.2.1. Evaluation Metrics

Before delving into the simulation results, it is necessary to introduce the statistical metrics used to both tune the thresholds and assess the algorithm's efficiency on test data. First, the definition of accuracy adopted in this work, as defined in [15], is based on the ratio of correctly classified samples to the total number of samples, thus providing a comprehensive performance indicator that accounts for both correct detections and misclassifications. As such, it can be expressed as:

$$\text{Accuracy} = \frac{\text{TP} + \text{TN}}{\text{TP} + \text{TN} + \text{FN} + \text{FP}} \times 100. \quad (7)$$

Here, TP denotes the number of true positives, corresponding to correctly detected seizure samples, while FN represents the amount of false negatives, that is, missed detections. FP indicates the portion of false positives, interpreted as incorrect detections during normal neural activity. Finally, TN refers to the number of samples correctly identified as non-seizure related (Figure 10).

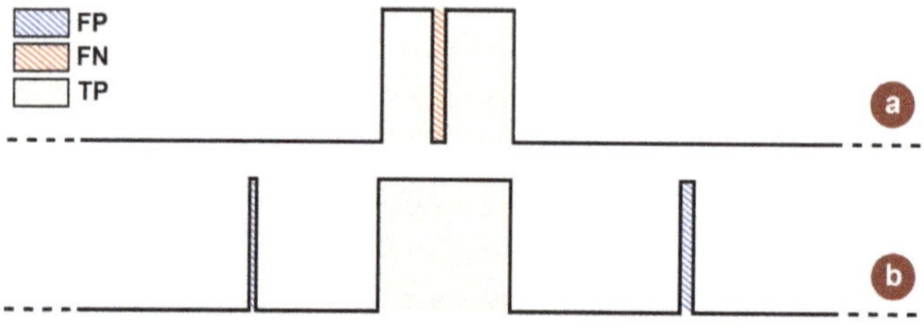

Figure 10. Explanatory illustration of (**a**) false negatives, and (**b**) false positives.

In addition, sensitivity is introduced as a metric that quantifies the algorithm's ability to correctly detect epileptic seizures. It is defined as the ratio between true positives and the sum of true positives and false negatives, thereby measuring the percentage of seizures that are successfully identified. As such, this metric is not influenced by the quantity of true negatives, nor by the number of false positives, as it interests solely the classification of the actual seizure event. The formula for sensitivity is defined as follows:

$$\text{Sensitivity} = \frac{\text{TP}}{\text{TP} + \text{FN}} \times 100. \quad (8)$$

In practice, it depends not only on whether a seizure is detected, but also on how quickly the detection occurs, as delayed detections increase the number of unidentified positives and thus lower the sensitivity.

To quantify the algorithm's responsiveness, latency has also to be taken into account. The latter refers to the time interval, typically measured in seconds, between the actual onset of a seizure and its detection performed by the algorithm. In this study, where testing is performed on recorded datasets rather than live subjects, latency is calculated as the difference between the detection time and the seizure onset as indicated by the clinical annotation provided with each sample.

The expression for latency adopted throughout the development of this work is therefore equal to:

$$\text{Latency} = T_{onset} - T_{detection}, \tag{9}$$

where T_{onset} denotes the timestamp of the seizure's beginning provided with the test samples, while $T_{detection}$ represents the instant in which the detection first occurs. Specifically, $T_{detection}$ is identified as the earliest time instant within the annotated ictal window from which the algorithm consistently flags seizure-related activity across channels. It must be noted that the metric defined in Equation (9) carries a degree of uncertainty due to the subjective nature of the annotations, with the marked onsets possibly not aligning with the physiological genesis of the epileptic seizures.

Figure 11 shows a practical example of the delay described above, illustrating how the inherent properties of the threshold-based algorithm can cause its identification of physiological changes in neural patterns to diverge from the annotated seizure onset due to uncertainty in the onset timing.

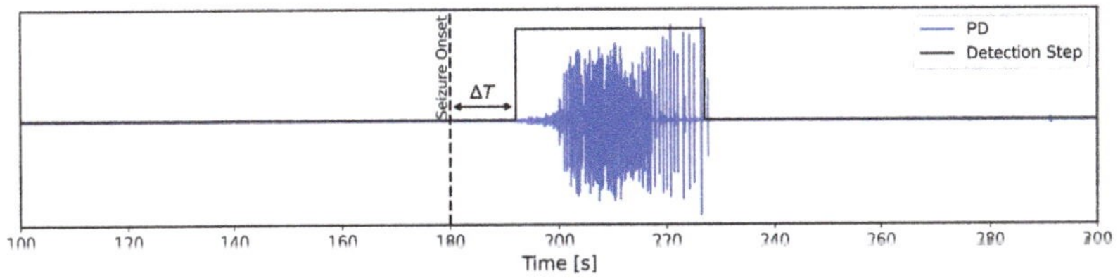

Figure 11. Example case illustrating a detection delay with the threshold-based algorithm, showing the seizure being identified slightly after the annotated onset.

4.2.2. Simulation Results

Simulations were first conducted on the CHB-MIT dataset, selecting five patients as representative examples. All available seizure samples for these patients were used to evaluate the algorithm. Since each sample is 60 min long, they provided sufficient ictal and inter-ictal data to perform threshold calibration using only the first seizure-containing sample per subject. Such tuning aimed to achieve an optimal trade-off between accuracy and sensitivity. Subsequent test samples were then evaluated using the fixed thresholds obtained from the initial tuning.

The calibration itself was performed individually for each patient in the CHB-MIT dataset and reiterated for those in the other datasets. The process involved iteratively running the algorithm on the sample of data reserved for tuning. Initial thresholds for the *LL* and *PD* features were defined as scaled versions of their standard deviations and progressively adjusted, together with the counter variable k_c, to maximize detection accuracy. Once satisfactory accuracy was reached, Thr_p and Thr_{LL} were gradually lowered to achieve at least 90% sensitivity, or preferably higher, while maintaining accuracy above a target value. Detection latency showed a strong dependence on sensitivity, as higher sensitivity generally reduced missed detections and thus latency. The parameter Λ was contextually optimized based on the number of channels, corresponding to the number of electrodes used to record the neural data, and was set to represent a fixed percentage of the total. The results reported for the CHB dataset, as well as the SWEC-ETHZ datasets presented later, reflect this patient-specific calibration procedure.

For simulations of the algorithm conducted on the CHB dataset, as well as the intracranial SWEC-ETHZ datasets, a duration of 1 s was established for the processing windows, corresponding to f_s samples. This configuration was identified as optimal after evaluating multiple window sizes and observing superior overall detection performance and consistency across datasets.

Regarding the scalp EEG CHB-MIT dataset, results reported in Table 5 demonstrate that the proposed algorithm achieves excellent performance across all tested subjects. Average accuracy per patient ranges from 95.95 % (CHB07) to 98.23 % (CHB05), while sensitivity remains consistently high, often approaching 100 %. Detection delays are low, with an overall mean of 3.37 s, and false positive rates are well contained, averaging 2.26 % per hour, supporting the method's reliability in distinguishing seizure and non-seizure activity. Since the threshold is kept constant across all test samples, there is a slight, though not majorly significant, decrease in accuracy for later seizure events, suggesting potential for improvement through an adaptive thresholding or autocalibration strategy to maintain uniformly high performance over time.

Table 5. Detection metrics per seizure and patient-level averages. CHB Dataset.

Patient	Test Seizure	Accuracy (%)	Sens. (%)	Delay (s)	FPR/h (%)
CHB01	#1	98.91	99.99	0	1.09
	#2	96.90	99.99	0	3.10
	#3	98.36	100	0	1.64
	#4	99.77	100	2	0.23
	#5	97.73	96.47	4	2.18
	#6	96.62	100	3	3.38
	Avg	**98.04**	**99.40**	**1.5**	**1.94**
CHB05	#1	99.40	98.05	8	0.55
	#2	99.22	92.39	5	0.57
	#3	99.20	97.41	4	0.72
	#4	95.11	100	13	4.89
	Avg	**98.23**	**96.96**	**7.5**	**1.68**
CHB07	#1	96.64	97.77	6	3.30
	#2	95.26	97.65	7	4.66
	Avg	**95.95**	**97.71**	**6.5**	**3.98**
CHB09	#1	98.51	99.99	3	1.49
	#2	96.78	99.99	2	3.22
	Avg	**97.64**	**99.99**	**2.5**	**2.35**
CHB23	#1	96.93	99.99	2	3.07
	#2	98.00	99.99	−5	2.04
	Avg	**97.46**	**99.99**	**−1.5**	**2.55**
AVERAGE		**97.71**	**98.73**	**3.37**	**2.26**

Concerning the short-term SWEC iEEG dataset, threshold calibration required a different approach due to the limited inter-ictal data in each sample. Approximately half of the seizure-marked recordings were used for training to properly tune the thresholds, while the remaining half were reserved for testing. It must be noted that, given the short duration of the samples, particularly in terms of inter-ictal activity, the identification of true positives carries more weight, as there are fewer true negatives available.

The results reported in Table 6 indicate that the proposed algorithm performs very well under these conditions. Accuracy for individual seizures is high, with patient-level averages exceeding 98 % in the examples reported. Sensitivity remains consistently close to 100 %, demonstrating reliable detection of ictal activity. Detection delays are generally low, ranging from 5 to 13 s on average per patient, and false positive rates are minimal, highlighting the method's reliability even in recordings with limited inter-ictal periods and highly variable electrode configurations.

Table 6. Detection metrics per seizure and patient-level averages. SWEC-ETHZ short-term dataset.

Patient	Test Seizure	Accuracy (%)	Sens. (%)	Delay (s)	FPR (%)
ID2	#1	99.33	100	8	0.67
	#2	97.73	100	3	2.27
	Avg	**98.53**	**100**	**5.5**	**1.47**
ID5	#1	96.59	95.83	9	2.73
	#2	99.33	100	13	0.67
	#3	98.48	95.50	14	0.65
	#4	98.51	100	8	1.49
	#5	97.49	100	4	2.51
	Avg	**98.08**	**98.27**	**9.6**	**1.61**
ID6	#1	99.38	97.32	14	0
	#2	98.55	98.15	12	0.91
	Avg	**98.76**	**97.73**	**13**	**0.45**
ID8	#1	97.39	97.96	14	2.38
ID10	#1	99.73	99.98	4	0.27
	#2	100	99.98	6	0
	Avg	**99.86**	**99.98**	**5**	**0.13**
AVERAGE		**98.54**	**98.73**	**9.08**	**1.21**

For the evaluation on the long-term SWEC iEEG dataset, five subjects were selected, similarly to the previous runs on the short-term dataset. As with the CHB dataset, 60 min long recordings were used, and a single sample containing a seizure event was sufficient to tune the algorithm and calibrate the thresholds depending on the tested subject. The results reported in Table 7 show that the algorithm maintains high accuracy and sensitivity, comparable to those obtained on the CHB dataset. However, for both iEEG datasets, the average latency is slightly longer, which can be attributed to a certain degree of uncertainty in the seizure onset annotations in the SWEC recordings. It is also worth noting that some tests show a delay of 0 s or even negative values; despite the algorithm incorporating an intrinsic three-second delay due to the conclusive check over three consecutive windows before marking a seizure, these cases likely reflect the algorithm detecting early pre-ictal physiological fluctuations.

Table 7. Detection metrics per seizure and patient-level averages. SWEC-ETHZ long-term dataset.

Patient	Test Seizure	Accuracy (%)	Sens. (%)	Delay (s)	FPR (%)
ID1	#1	99.22	99.29	2	0.67
ID2	#1	97.78	95.18	8	2.11

Table 7. *Cont.*

Patient	Test Seizure	Accuracy (%)	Sens. (%)	Delay (s)	FPR (%)
ID3	#1	99.92	95.16	2	0
	#2	100	100	2	0
	#3	99.92	95.52	3	0
	Avg	**99.95**	**96.89**	**2.33**	**0**
ID15	#1	**97.36**	**100**	**13**	**2.64**
ID16	#1	99.64	100	13	0.36
	#2	99.47	100	9	0.53
	#3	99.53	100	7	0.47
	#4	88.61	100	7	11.39
	Avg	**96.81**	**100**	**9**	**3.19**
AVERAGE		**98.14**	**98.51**	**6.60**	**1.82**

4.3. Comparative Analysis with State-of-the-Art

In this section, the results of the simulations conducted on the CHB-MIT and SWEC-ETHZ datasets are compared against other state-of-the-art proposals. Notably, works pertaining to the latter of the two datasets have been considered both regarding the short-form and the long-form variants. As shown in Table 8, comparisons were made with respect to algorithms that were tested on either of the two datasets, with most of them being the product of machine-learning-based approaches. Included in Table 8 are also works that were evaluated on different publicly available datasets, with a distinction being made between EEG and iEEG signals. Additionally, not all references proposed in said table have been implemented.

Table 8. Comparison of performance metrics against recent state-of-the-art proposals. Colored in yellow are works appied to EEG data, colored in green are works applied to iEEG data.

Reference	Type	%Sens.	%Acc.	Latency (s)	Database
[12]	2D-CNN	100.00	–	2.10	CHB
[16]	RUSBoost	95.00	99.71	6.83	CHB
[13]	SVM	97.20	97.80	1.10	CHB
[17]	ADCD	96.00	–	4.21	CHB
[26]	CNN	97.57	98.90	10.46	CHB
[27]	SVM	96.15	96.38	10.42	CHB
[28]	SVM	95.20	–	6.40	CHB
[29]	Rand. Forest	98.00	98.00	–	UCI-ESR
[30]	RVM	96.00	99.88	1.89	Bonn/CHB
This Work	**Thr-Based**	**98.73**	**97.71**	**3.37**	**CHB**
[31]	LBP+HD	96.01	95.42	15.90	SWEC-SH
[14]	Thr-Based	100	–	8.22	SWEC-SH
[25]	LBP	96.38	96.85	8.81	SWEC-SH
[32]	SVM	94.46	–	11.1	Freiburg
[33]	2D-CNN	97.50	–	13.7	SWEC-SH
[24]	Laelaps	85.5	–	18.9	SWEC-LO
This Work	**Thr-Based**	**98.73**	**98.54**	**9.08**	**SWEC-SH**
This Work	**Thr-Based**	**98.51**	**98.14**	**6.60**	**SWEC-LO**
This Work	**Thr-Based**	**98.62**	**98.34**	**7.84**	**SWEC-AV**

SH: short-term; LO: long-term; AV: average results between short-term and long-term.

The reported results show that the proposed work, in spite of its simplicity, manages to yield comparable scores in terms of accuracy and sensitivity, with latency times that do not deviate particularly from the average of about 5 s. The proposed threshold-based algorithm demonstrates high performance on the CHB-MIT dataset, achieving a sensitivity of 98.73% and an accuracy of 97.71% on average. These results are comparable to or exceed those obtained by more complex, machine-learning-based approaches, including 2D-CNNs, SVMs, RUSBoost, and Random Forests, which typically require extensive feature extraction and model training. In addition to its competitive accuracy, the algorithm exhibits a low detection latency of 3.37 s on average, outperforming several ML-based methods that report longer response times. Favorable results were also obtained when evaluating the algorithm on the iEEG dataset against similarly tested approaches. On average across the tested samples, the proposed method achieved a sensitivity of 98.62% and an accuracy of 98.34%, both higher than most existing works and indicative of resilience against false detections. Latency performance was also superior, regardless of the distinction between long-term or short-term variants, yielding an overall average of 7.84 s.

As shown in Figure 12, the latency values reported for various algorithms on the CHB and SWEC datasets reveal notable differences across datasets. For most state-of-the-art methods, latencies increase when tested on the iEEG SWEC dataset compared to CHB, suggesting potential inconsistencies in the annotations or differences in the characteristics of the SWEC recordings. The points corresponding to the proposed approach demonstrate competitive performance on both datasets, achieving low latency comparable to other methods on CHB while maintaining robust performance on SWEC, with results on the long-term SWEC dataset among the best reported. These results suggest that the method generalizes effectively across datasets and further emphasize the importance of cross-dataset validation to account for possible differences in annotation protocols and signal characteristics.

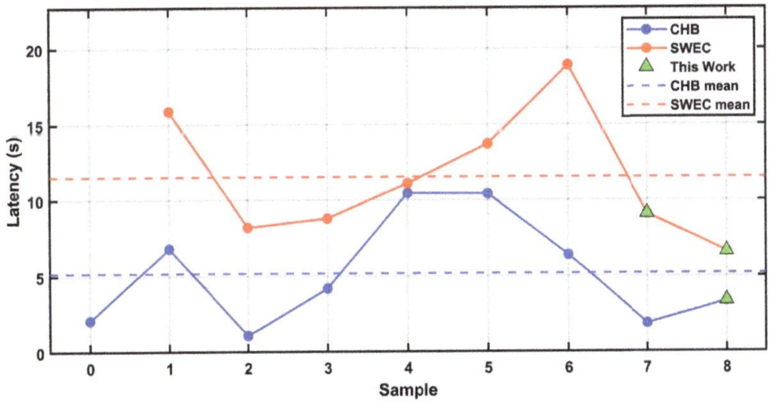

Figure 12. Trends in latency related to detection algorithms applied to CHB or SWEC-EHTZ datasets.

4.4. Hardware Resource Comparisons

This section presents the FPGA synthesis results, summarized in Table 9, and juxtaposes them with recent hardware implementations. Here, the strength of this design is particularly highlighted, as the comparisons show that the proposed work relies on minimal resources to be effective. The proposed solution achieves the lowest usage of LUTs and FFs among AMD Artix-7 implementations and requires only 3 DSP blocks, demonstrating an efficient mapping of the algorithm to hardware. Notably, the number of DSPs does not increase with the number of channels, as they are shared via the TDM implementation, unlike with other methods where DSP usage may scale with the channel count. The small overall resource footprint suggests that the algorithm could be implemented on limited platforms or even ASICs to further push the miniaturization process with sub-micron nodes.

Table 9. Comparison of FPGA synthesis reports.

Reference	Year	Board	LUT	FF	DSP
[34]	2018	M2GL025	5355	357	–
[14]	2022	Cyclone V	1856	3088	5
[35]	2022	Zynq-7000	5202	10,196	–
[36]	2022	Artix-7	5934	4015	–
[19]	2023	Artix-7	1150	492	14
[18]	2025	Altera DE2	1095	–	9
This Work	**2025**	**Artix-7**	**192**	**500**	**3**

5. Conclusions

In this work, a low-complexity threshold-based seizure detection algorithm was presented and thoroughly assessed in terms of both performance on pre-existing datasets and hardware resource allocation when implemented on the AMD Artix-7 FPGA board. By exploiting the *PD* and *LL* time-domain features, the architecture is able to isolate and effectively detect the abnormal activity related to the presence of an epileptic seizure, provided that a calibration of the thresholds is carried out beforehand. The accuracy and sensitivity of the proposed system were evaluated using the CHB-MIT scalp EEG dataset and the SWEC-ETHZ iEEG dataset, both in its long-term and short-term variant. After testing on over 35 samples taken from said datasets, the simulated accuracy exceeded on average 98%, with similar results being obtained with regard to the average sensitivity. The latency of the system, when compared against more complex state-of-the-art proposals pertaining to the specific datasets, is among the shortest, with recorded averages of 3.37 s (CHB) and 7.84 s (SWEC).

The simplicity of the algorithm translates into minimal resource requirements for hardware implementation. FPGA synthesis results demonstrated substantial improvements in LUT, DSP, and FF utilization compared with single-channel designs. For fairness, a 4-channel version of the framework was considered in the comparison, ensuring full functionality of the proposed architecture. Moreover, the adoption of a TDM approach for both pre-processing and feature extraction prevents resources usage from scaling linearly with the number of channels, further strengthening the efficiency of the solution. Overall, these results demonstrate that the proposed detection algorithm is accurate, fast, and lightweight, making it well-suited for real-time applications in hardware-constrained environments.

Author Contributions: Conceptualization, A.V., G.N. and G.S.; methodology, A.V., G.N. and G.S.; software, A.V. and G.N.; validation, A.V. and G.N.; formal analysis, G.N.; investigation, A.V., G.N. and G.S.; data curation, G.N; writing—original draft preparation, A.V.; writing—review and editing, A.V., G.N. and G.S.; supervision, G.S. All authors have read and agreed to the published version of the manuscript.

Funding: This research received no external funding

Data Availability Statement: The original contributions presented in this study are included in the article. Further inquiries can be directed to the corresponding author.

Conflicts of Interest: The authors declare no conflict of interest.

References

1. Sultana, B.; Panzini, M.A.; Veilleux Carpentier, A.; Comtois, J.; Rioux, B.; Gore, G.; Bauer, P.R.; Kwon, C.S.; Jetté, N.; Josephson, C.B.; et al. Incidence and Prevalence of Drug-Resistant Epilepsy: A Systematic Review and Meta-analysis. *Neurology* **2021**, *96*, 805–817. [CrossRef] [PubMed]
2. Mesraoua, B.; Brigo, F.; Lattanzi, S.; Abou-Khalil, B.; Al Hail, H.; Asadi-Pooya, A.A. Drug-resistant epilepsy: Definition, pathophysiology, and management. *J. Neurol. Sci.* **2023**, *452*, 120766. [CrossRef] [PubMed]
3. Gonzalez, H.F.; Yengo-Kahn, A.; Englot, D.J. Vagus nerve stimulation for the treatment of epilepsy. *Neurosurg. Clin. N. Am.* **2019**, *30*, 219–230. [CrossRef] [PubMed]
4. Shin, U.; Ding, C.; Woods, V.; Widge, A.S.; Shoaran, M. A 16-channel low-power neural connectivity extraction and phase-locked deep brain stimulation soc. *IEEE Solid-State Circuits Lett.* **2023**, *6*, 21–24. [CrossRef]
5. Skrehot, H.C.; Englot, D.J.; Haneef, Z. Neuro-stimulation in focal epilepsy: A systematic review and meta-analysis. *Epilepsy Behav.* **2023**, *142*, 109182. [CrossRef]
6. Zhu, B.; Shin, U.; Shoaran, M. Closed-Loop Neural Prostheses With On-Chip Intelligence: A Review and a Low-Latency Machine Learning Model for Brain State Detection. *IEEE Trans. Biomed. Circuits Syst.* **2021**, *15*, 877–897. [CrossRef]
7. Salam, M.T.; Velazquez, J.L.P.; Genov, R. Seizure Suppression Efficacy of Closed-Loop Versus Open-Loop Deep Brain Stimulation in a Rodent Model of Epilepsy. *IEEE Trans. Neural Syst. Rehabil. Eng.* **2016**, *24*, 710–719. [CrossRef]
8. Slimen, I.B.; Boubchir, L.; Seddik, H. Epileptic seizure prediction based on eeg spikes detection of ictal-preictal states. *J. Biomed. Res.* **2020**, *34*, 162–169. [CrossRef]
9. Li, K.; Lu, M. EEG signals based automated epilepsy detection using tunable-Q wavelet transform and Hjorth parameters. In Proceedings of the 2024 7th International Conference on Data Science and Information Technology (DSIT), Nanjing, China, 20–22 December 2024 ; pp. 1–6.
10. Iqbal, S.; Khan, Y.U.; Farooq, O. Nonlinear analysis of eeg for seizure prediction. In Proceedings of the 2015 Annual IEEE India Conference (INDICON), New Delhi, India, 17–20 December 2015; pp. 1–5.
11. Shin, U.; Ding, C.; Zhu, B.; Vyza, Y.; Trouillet, A.; Revol, E.C.M.; Lacour, S.P.; Shoaran, M. Neuraltree: A 256-channel 0.227-μj/class versatile neural activity classification and closed-loop neuromodulation soc. *IEEE J. Solid-State Circuits* **2022**, *57*, 3243–3257. [CrossRef]
12. Chung, Y.G.; Cho, A.; Kim, H.; Kim, K.J. Single-channel seizure detection with clinical confirmation of seizure locations using chb-mit dataset. *Front. Neurol.* **2024**, *15*, 1530–1539. [CrossRef]
13. Tang, F.-G.; Liu, Y.; Li, Y.; Peng, Z.-W. A unified multi-level spectral–temporal feature learning framework for patient-specific seizure onset detection in eeg signals. *Knowl.-Based Syst.* **2020**, *205*, 106152. [CrossRef]
14. Razi, K.F.; Schmid, A. Epileptic Seizure Detection With Patient-Specific Feature and Channel Selection for Low-power Applications. *IEEE Trans. Biomed. Circuits Syst.* **2022**, *16*, 626–635. [CrossRef] [PubMed]
15. Burrello, A., Benatti, S., Schindler, K., Benini, L., Rahimi, A. An Ensemble of Hyperdimensional Classifiers. Hardware-Friendly Short-Latency Seizure Detection With Automatic iEEG Electrode Selection. *J. Biomed. Health Inform.* **2021**, *25*, 935–946. [CrossRef] [PubMed]
16. Asif, R.; Saleem, S.; Hassan, S.A.; Alharbi, S.A.; Kamboh, A.M. Epileptic seizure detection with a reduced montage: A way forward for ambulatory eeg devices. *IEEE Access* **2020**, *8*, 65880–65890. [CrossRef]
17. Khanmohammadi, S.; Chou, C.-A. Adaptive seizure onset detection framework using a hybrid pca–csp approach. *IEEE J. Biomed. Health Inform.* **2018**, *22*, 154–160. [CrossRef]
18. Wang, L.-H.; Zhang, Z.-N.; Xie, C.-X.; Jiang, H.; Yang, T.; Ran, Q.-P.; Fan, M.-H.; Kuo, I.-C.; Lee, Z.-J.; Chen, J.-B.; et al. A novel real-time threshold algorithm for closed-loop epilepsy detection and stimulation system. *Sensors* **2025**, *25*, 33. [CrossRef]
19. Shamshad, A.M.; Mohd, H.; Farooq, O.; Mehedi, H. Field programmable gate array-based energy-efficient and fast epileptic seizure detection using support vector machine and quadratic discriminant analysis classifier. *Eng. Rep.* **2023**, *6*, 8.
20. Esteller, R.; Echauz, J.; Tcheng, T.; Litt, B.; Pless, B. Line length: An efficient feature for seizure onset detection. In Proceedings of the I2001 Conference Proceedings of the 23rd Annual International Conference of the IEEE Engineering in Medicine and Biology Society, Istanbul, Turkey, 25–28 October 2001; pp. 1707–1710.
21. Esteller, R.; Echauz, J.; Tcheng, T. Comparison of line length feature before and after brain electrical stimulation in epileptic patients. In Proceedings of the 26th Annual International Conference of the IEEE Engineering in Medicine and Biology Society, San Francisco, CA, USA, 1–5 September 2004; pp. 4710–4713.
22. Guo, L.; Rivero, D.; Dorado, J.; Rabuñal, J.R.; Pazos, A. Automatic epileptic seizure detection in EEGs based on line length feature and artificial neural networks. *J. Neurosci. Methods* **2010**, *191*, 101–109. [CrossRef]
23. Goldberger, A.L.; Amaral, L.A.N.; Glass, L.; Hausdorff, J.M.; Ivanov, P.C.; Mark, R.G.; Mietus, J.E.; Moody, G.B.; Peng, C.-K.; Stanley, H.E. Physiobank, physiotoolkit, and physionet: Components of a new research resource for complex physiologic signals. *Circulation* **2000**, *101*, E215–E220. [CrossRef]

24. Burrello, A.; Cavigelli, L.; Schindler, K.; Benini, L.; Rahimi, A. Laelaps: An Energy-Efficient Seizure Detection Algorithm from Long-term Human iEEG Recordings without False Alarms. In Proceedings of the ACM/IEEE Design, Automation, and Test in Europe Conference (DATE), Florence, Italy, 25–29 March 2019.
25. Burrello, A.; Schindler, K.; Benini, L.; Rahimi, A. One-shot Learning for iEEG Seizure Detection Using End-to-end Binary Operations: Local Binary Patterns with Hyperdimensional Computing. In Proceedings of the IEEE Biomedical Circuits and Systems Conference (BioCAS), Cleveland, OH, USA, 17–19 October 2018.
26. Shen, M.; Wen, P.; Song, B.; Li, Y. Real-time epilepsy seizure detection based on eeg using tunable-q wavelet transform and convolutional neural network. *Biomed. Signal Process. Control* **2023**, *82*, 104566. [CrossRef]
27. Shen, M.; Wen, P.; Yan, L.; Song, B. An EEG based real-time epilepsy seizure detection approach using discrete wavelet transform and machine learning methods. *Biomed. Signal Process. Control* **2022**, *77*, 103820. [CrossRef]
28. Qaraqe, M.; Ismail, M.; Serpedin, E. Band-sensitive seizure onset detection via csp-enhanced eeg features. *Epilepsy Behav.* **2015**, *50*, 77–87. [CrossRef] [PubMed]
29. Kode, H.; Elleithy, K.; Almazaydeh, L. Epileptic seizure detection in eeg signals using machine learning and deep learning techniques. *IEEE Access* **2024**, *12*, 80657–80668. [CrossRef]
30. Vidyaratne, L.S.; Iftekharuddin, K.M. Real-time epileptic seizure detection using eeg. *IEEE Trans. Neural Syst. Rehabil. Eng.* **2017**, *25*, 2146–2156. [CrossRef]
31. Burrello, A.; Schindler, K.; Benini, L.; Rahimi, A. Hyperdimensional Computing With Local Binary Patterns: One-Shot Learning of Seizure Onset and Identification of Ictogenic Brain Regions Using Short-Time iEEG Recordings. *IEEE Trans. Biomed. Eng.* **2020**, *67*, 601–613. [CrossRef]
32. Liu, Y.; Zhou, W.; Yuan, Q.; Chen, S. Automatic seizure detection using wavelet transform and svm in long-term intracranial EEG. *IEEE Trans. Neural Syst. Rehabil. Eng.* **2012**, *20*, 749–755. [CrossRef]
33. Xu, Y.; Yang, J.; Ming, W.; Wang, S.; Sawan, M. Shorter latency of real-time epileptic seizure detection via probabilistic prediction. *Expert Syst. Appl.* **2024**, *236*, 121359. [CrossRef]
34. Daoud, H.G.; Abdelhameed, A.M.; Bayoumi, M. FPGA implementation of high accuracy automatic epileptic seizure detection system. In Proceedings of the 2018 IEEE 61st International Midwest Symposium on Circuits and Systems (MWSCAS), Windsor, ON, Canada, 5–8 August 2018; pp. 407–410.
35. Rizal, A.; Hadiyoso, S.; Ramdani, A.Z. FPGA-based implementation for real-time epileptic eeg classification using Hjorth descriptor and knn. *Electronics* **2022**, *11*, 3026. [CrossRef]
36. Fawzy, M.; Hussien, A.; Mostafa, H. FPGA utilized implementation of epileptic seizure detection system based on wearable devices using dynamic partial reconfiguration. In Proceedings of the 2022 10th International Japan-Africa Conference on Electronics, Communications, and Computations (JAC-ECC), Alexandria, Egypt, 19–22 December 2022; pp. 119–124.

sensors

MDPI

Article

Enhancing Surgical Planning with AI-Driven Segmentation and Classification of Oncological MRI Scans

Alejandro Martinez Guillermo [1,2,*], Juan Francisco Zapata Pérez [1], Juan Martinez-Alajarin [1] and Alicia Arévalo García [2]

[1] Escuela Tecnica Superior de Ingenieria Industrial, Campus Muralla del Mar, Universidad Politecnica de Cartagena Member of European University of Technology EUT+, C/Doctor Fleming, s/n, 30202 Cartagena, Spain; juan.zapata@upct.es (J.F.Z.P.); juanc.martinez@upct.es (J.M.-A.)

[2] Cella Medical Solutions, 30100 Murcia, Spain; aliciaarevalo@cellams.com

* Correspondence: alejandromartinez@cellams.com

Abstract

This work presents the development of an Artificial Intelligence (AI)-based pipeline for patient-specific three-dimensional (3D) reconstruction from oncological magnetic resonance imaging (MRI), leveraging image-derived information to enhance the analysis process. These developments were carried out within the framework of Cella Medical Solutions, forming part of a broader initiative to improve and optimize the company's medical-image processing pipeline. The system integrates automatic MRI sequence classification using a ResNet-based architecture and segmentation of anatomical structures with a modular nnU-Net v2 framework. The classification stage achieved over 90% accuracy and showed improved segmentation performance over prior state-of-the-art pipelines, particularly for contrast-sensitive anatomies such as the hepatic vasculature and pancreas, where dedicated vascular networks showed Dice score differences of approximately 20–22%, and for musculoskeletal structures, where the model outperformed specialized networks in several elements. In terms of computational efficiency, the complete processing of a full MRI case, including sequence classification and segmentation, required approximately four minutes on the target hardware. The integration of sequence-aware information allows for a more comprehensive understanding of MRI signals, leading to more accurate delineations than approaches without such differentiation. From a clinical perspective, the proposed method has the potential to be integrated into surgical planning workflows. The segmentation outputs were converted into a patient-specific 3D model, which was subsequently integrated into Cella's surgical planner as a proof of concept. This process illustrates the transition from voxel-wise anatomical labels to a fully navigable 3D reconstruction, representing a step toward more robust and personalized AI-driven medical-image analysis workflows that leverage sequence-aware information for enhanced clinical utility.

Keywords: artificial intelligence; 3D reconstruction; image-derived information; Dice score improvement; surgical planning; anatomical segmentation; oncological imaging; magnetic resonance imaging (MRI)

1. Introduction

Cancer remains one of the most pressing global health challenges, characterized by the uncontrolled growth and spread of abnormal cells that can invade surrounding tissues and metastasize to distant organs. According to the most recent GLOBOCAN estimates, more than 20 million new cancer cases and 9.7 million cancer-related deaths were reported worldwide in 2022, with projections indicating an increase to over 35 million new cases annually by 2050 [1]. This rising burden is influenced by population growth, aging, lifestyle factors, and environmental exposures.

At the level of developed countries, similar trends are observed across multiple tumor types. Although incidence and mortality vary by region and socioeconomic context, factors such as obesity, smoking, alcohol consumption, sedentary lifestyles, and reproductive patterns play a significant role in cancer epidemiology. The increasing incidence of early-onset cancers further underscores the need for more effective diagnostic and treatment strategies.

Early detection remains a cornerstone for improving cancer outcomes, as timely diagnosis is associated with higher survival rates, reduced treatment-related morbidity, and improved quality of life. International health organizations emphasize that delays in diagnosis significantly compromise treatment success, whereas early identification enables more effective and less invasive therapeutic interventions.

Within oncological management, surgical treatment continues to play a central role, particularly for solid tumors. Successful surgical outcomes depend heavily on accurate preoperative planning, which requires precise knowledge of tumor extent and its spatial relationship with surrounding anatomical structures. In this context, medical imaging is indispensable. Among imaging modalities, magnetic resonance imaging (MRI) offers excellent soft-tissue contrast, multiplanar capabilities, and the absence of ionizing radiation, making it especially valuable for oncological applications in the brain, abdomen, pelvis, and musculoskeletal system.

Despite these advantages, the manual interpretation, classification, and segmentation of MRI scans remain time-consuming and subject to inter- and intra-observer variability. Radiological workflows rely extensively on expert knowledge to identify relevant anatomical structures and differentiate tissue characteristics across MRI sequences, a process that becomes increasingly challenging with the growing volume and complexity of imaging data.

Artificial intelligence (AI), particularly deep learning (DL), has emerged as a powerful tool to address these challenges. Convolutional neural networks have demonstrated strong performance in medical image classification and segmentation tasks, enabling automated extraction of hierarchical features directly from imaging data. Architectures such as U-Net and its derivatives, including nnU-Net, have established themselves as reference methods for biomedical image segmentation due to their adaptability and robustness across diverse datasets.

However, many existing AI-based MRI pipelines focus primarily on segmentation while overlooking the explicit role of MRI sequence type in image appearance and contrast. In clinical practice, differences between T1- and T2-weighted sequences significantly affect tissue visualization and anatomical delineation, suggesting that sequence-aware approaches may provide more accurate and clinically meaningful results.

In this work, we propose an integrated AI-driven pipeline for oncological MRI analysis that explicitly incorporates MRI sequence classification as a preliminary step. The framework automatically classifies MRI volumes into T1- or T2-weighted sequences and subsequently applies sequence-specific and general deep learning segmentation models to generate three-dimensional anatomical reconstructions suitable for surgical planning.

The main contributions of this study include: (i) the development of an automatic MRI sequence classification module based on deep learning, (ii) the design and evaluation of sequence-specific and general segmentation networks for clinically relevant abdominal structures, and (iii) the integration of the resulting segmentations into a three-dimensional visualization environment to support surgical planning workflows.

https://doi.org/10.3390/s26010323

In the remainder of this work, we first review prior research on artificial intelligence applied to oncological MRI, with emphasis on classification and segmentation strategies, which is provided in Section 2. Section 3 describes the methodology adopted in this study, including dataset preparation, model architectures, training procedures, and implementation details. The experimental results are presented next in Section 4, covering both quantitative evaluations and qualitative analyses of the model's assessments by clinical experts. Next, in Section 5, we discuss the main findings and their implications. Section 6 summarizes the limitations of this study and outlines future work directions. Finally, in Section 7, we conclude the manuscript by highlighting the main contributions.

2. Related Work

This section reviews methods for enhancing surgical planning with AI-driven segmentation and classification of oncological MRI scans. We organize prior work into six thematic areas:

(i) Classical and radiomics-based approaches;
(ii) Deep learning approaches for segmentation;
(iii) Deep learning approaches for classification;
(iv) Hybrid pipelines combining classification and segmentation;
(v) Learning with limited labels and domain shift;
(vi) Clinical integration, uncertainty, and evaluation.

The first category, classical and radiomics-based approaches, formed the foundation of early clinical pipelines for oncological MRI analysis. Atlas-based registration methods demonstrated strengths in anatomically constrained environments but showed sensitivity to registration errors and anatomical variability [2]. Graph cut formulations enabled efficient global energy minimization for segmentation but required careful hand-crafting of energy terms and parameter tuning [3]. These methods, while foundational, faced limitations in reproducibility and scalability within large-scale surgical workflows.

The emergence of radiomics addressed some limitations by extracting hand-crafted intensity, texture, and shape features for classical machine learning classifiers. Aerts et al. demonstrated the prognostic value of engineered features across multiple cancers [4]. However, feature stability across scanners and sequences remained a significant challenge, limiting deployment in heterogeneous MRI cohorts. These approaches fundamentally lacked the adaptability and representation learning capabilities of modern deep learning methods.

The second category, deep learning approaches for segmentation, was revolutionized by the introduction of the U-Net architecture, which combined multi-scale contracting and expanding paths with skip connections [5]. Its extensions to 3D volumes proved essential for oncological MRI applications. Çiçek et al. introduced 3D U-Net for volumetric segmentation [6], while Milletari et al. proposed V-Net with Dice-based loss to address class imbalance in lesion segmentation [7]. The nnU-Net framework further advanced the field by automating preprocessing, network configuration, and postprocessing [8], demonstrating that principled heuristics could rival bespoke designs. Architectural refinements including U-Net++ with dense skip connections [9] and Attention U-Net with spatial gating mechanisms [10] improved boundary precision and sensitivity to small, low-contrast lesions typical in oncologic MRI.

The third category, deep learning approaches for classification, brought significant advances to MRI sequence identification and tumor characterization. ResNet architectures stabilized training in very deep networks through residual connections [11], while DenseNet improved feature reuse and parameter efficiency [12]. EfficientNet introduced compound scaling to balance depth, width, and resolution [13], particularly beneficial for

https://doi.org/10.3390/s26010323

high-resolution MRI data. More recently, Vision Transformers [14] and Swin Transformers [15] modeled long-range dependencies through self-attention mechanisms, showing promising results for medical image classification tasks.

The fourth category, hybrid pipelines and multi-stage approaches, offers a practical strategy for handling heterogeneous MRI cohorts by first classifying acquisition sequences or tumor categories before performing segmentation. Litjens et al. documented consistent performance gains from such modular designs in their comprehensive survey of deep learning in medical imaging [16]. Within neuro-oncology, the BraTS challenge catalyzed development of cascaded and multi-branch networks that integrate sequence-level cues with subregion segmentation [17,18]. These approaches demonstrated the benefits of leveraging complementary sequences (e.g., FLAIR for edema, T1c for enhancing core) but often lacked explicit sequence classification modules to normalize inputs prior to segmentation.

These approaches demonstrated the benefits of leveraging complementary sequences but often operated under the assumption of standardized input protocols. The critical step of explicit sequence classification to normalize inputs prior to segmentation—particularly vital in multi-center surgical settings with inherent protocol variability—remains underexplored, a gap our pipeline specifically addresses.

The fifth category, learning with limited labels and domain adaptation, addresses the fundamental challenge of label scarcity in oncology MRI. Semi-supervised and self-supervised approaches have shown promise in mitigating this limitation. Bai et al. utilized weak supervision and shape priors for cardiac MRI segmentation [19], while Chaitanya et al. employed contrastive learning to develop robust representations from limited annotations [20]. Domain adaptation methods have specifically addressed the challenge of distribution shift across institutions. Kamnitsas et al. introduced unsupervised domain adaptation for lesion segmentation [21], and Karani et al. proposed task-driven image harmonization to reduce inter-site variability [22]. These techniques remain underutilized in surgical planning-grade studies, where their potential impact is magnified by the prohibitive cost of expert annotation and the critical need for model robustness across institutional boundaries in multi-center evaluations aligned with clinical endpoints.

The last category, clinical integration and uncertainty quantification, emphasizes that calibrated uncertainty estimates and robust evaluation metrics are essential for clinical deployment. Kendall et al. decomposed predictive uncertainty into aleatoric and epistemic components, demonstrating improved reliability for segmentation tasks [23]. Taha and Hanbury analyzed metric behavior and pitfalls in 3D segmentation, recommending complementary measures for surgical planning validation [24]. Despite these advances, uncertainty quantification and calibration are not routinely reported in segmentation studies. Furthermore, while technical performance metrics dominate the literature, few works deliver modular, auditable pipelines that output patient-specific 3D assets directly consumable by surgical planning platforms—a crucial translation gap identified in comprehensive reviews of medical AI deployment [25].

In summary, existing AI-based pipelines for oncological MRI analysis have demonstrated remarkable performance in controlled research settings, yet several limitations persist. Many approaches fuse sequences without explicit classification and normalization, show limited multi-center generalization, underutilize semi- or self-supervised methods, provide insufficient uncertainty quantification, and rarely offer modular, auditable pipelines suited for clinical integration. To address these challenges, our work introduces a modular pipeline that classifies MRI sequences prior to segmentation, employs robust NN-based segmentation with standardized preprocessing and postprocessing, and produces calibrated, auditable results ready for clinical deployment. By bridging these gaps, our approach

https://doi.org/10.3390/s26010323

advances the translation of technical innovation into clinical applicability for oncological MRI surgical planning.

3. Methodology

This section describes the dataset, model architecture, and training procedures used for MRI sequence classification (T1 vs. T2) and anatomical segmentation as part of an AI pipeline for surgical planning. In the stage of classification, the study was conducted in three stages of training, progressively refining the dataset and model configuration. The dataset was refined across three training stages: an initial set of 363 volumes, a filtered and expanded set of 2286 volumes, and a third training stage focusing on extended epochs to optimize performance. A ResNet18 architecture was adapted for volumetric inputs and binary classification. To support this task, the inherent contrast differences between T1- and T2-weighted MRI were leveraged: in T1-weighted images, fat-rich tissues appear hyperintense, providing clear anatomical boundary definition, whereas in T2-weighted images the fluid-rich structures such as cerebrospinal fluid appear hyperintense, enhancing contrast in edema, inflammation, and other fluid-related pathologies, as shown in Figure 1. In contrast, fluid-containing structures present higher signal intensity, making them more suitable for highlighting pathological or inflammatory conditions. Next, a segmentation stage was implemented, which employed a modular nnU-Net framework with networks specialized in anatomical subsets including major organs, vasculature, musculoskeletal structures, and tumors. Training used the Adam optimizer with carefully selected batch sizes, learning rates, and epochs adjusted per dataset and task. Model performance was evaluated using Dice coefficient, Intersection over Union, and Hausdorff distance. The subsequent subsections present a comprehensive and detailed explanation.

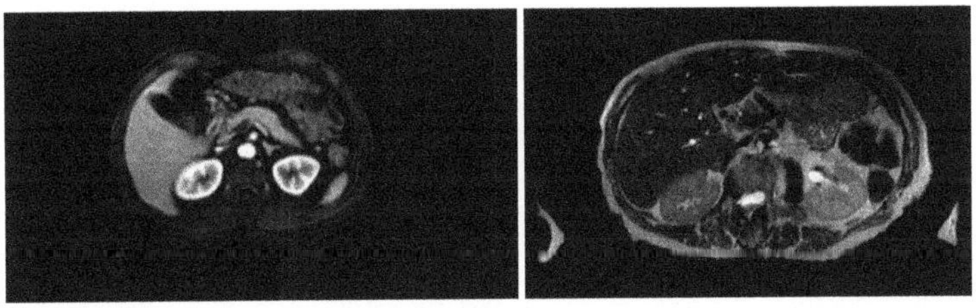

Figure 1. The (**left**) picture shows a T1-weighted MRI scan, where fat-rich tissues appear hyperintense, providing clear anatomical boundary definition. The (**right**) picture shows T2-weighted MRI scan, where fluid-rich structures such as cerebrospinal fluid appear hyperintense, enhancing contrast in edema, inflammation, and other fluid-related pathologies.

3.1. Dataset

The dataset used in this study consists of clinical MRI images stored in Cella's internal PACS system, covering brain, thoracic, abdominal, and pelvic regions. The study was conducted in three training stages, progressively refining both the dataset and the model configuration.

As an initial proof-of-concept, a preliminary dataset of 363 MRI images (174 T1-weighted and 189 T2-weighted) was assembled to establish a first contact with the classification task. This dataset was split in a stratified manner: 70% for training, 10% for validation, and 20% for testing, preserving class proportions. Each 3D image acquisition was pre-processed to extract representative axial slices from the central region of each volume, capturing the most relevant anatomical information while avoiding peripheral slices with potential artifacts. Intensities were rescaled to [0–1], and data augmentation—including rotation, translation, noise addition, brightness adjustment, and blurring—was applied.

https://doi.org/10.3390/s26010323

To increase diversity and representativeness, in our second approach, the dataset was expanded and filtered using strict inclusion criteria. Only volumes acquired in the axial plane with at least 30 slices were retained. Multivolume scans containing multiple acquisitions within a single file were excluded. Initial manual inspection using 3D Slicer verified slice count, orientation, and single-series compliance. Subsequently, a custom script automated the analysis of metadata to confirm axial orientation and detect multivolume cases. After filtering, the final dataset comprised 2286 MRI volumes, with 1588 T1-weighted and 698 T2-weighted images. The dataset was stratified into 1600 training images, 229 validation images, and 457 test images, maintaining balanced representation of both classes. Each volume continued to use 10 central slices for model input. Table 1 shows the distribution of the dataset in each approach.

Table 1. Distribution of the dataset used in the classification task, showing the number of T1 and T2 volumes in each training stage.

Dataset	T1 MRI	T2 MRI	Total
Proof of concept	174	189	363
Amplied filter	1588	698	2286

There were no major modifications to the dataset in the third approach. The dataset for the third training remained the same (2286 volumes, same splits), as the focus was on further optimizing model performance by fine-tuning the network rather than expanding or improving the dataset.

Given the objective of developing a modular and progressive segmentation pipeline, a specialized dataset was prepared, consisting of MRI images—T1-weighted images, T2-weighted images, and a combined MR dataset including both sequences—and corresponding anatomical blocks of clinical relevance serving as ground truth. This approach enables the training of specialized networks on specific subsets of anatomical regions, facilitating the creation of a progressive and optimized segmentation workflow. The selection of structures was guided by both clinical importance and the necessity of accurate delineation in procedures such as tumor resection or liver transplantation planning, where precise identification of key anatomical structures is critical. Table 2 summarizes the selected anatomical structures, grouped by clinical and anatomical relevance. The images and annotations were provided and carefully reviewed by medical imaging technicians and radiologists to ensure accuracy. This organization facilitated training of specialized networks and allowed comparison between networks trained on single sequences (T1w, T2w) versus the combined dataset. A specialist network was created for each MRI sequence and anatomical group, resulting in a total of 9 segmentation networks, given the three anatomical groups and three types of MRI sequences.

The development of the segmentation networks followed a unified methodology aimed at identifying and prioritizing anatomically and clinically relevant structures. A general workflow was established for dataset preparation, label definition, and model configuration, which was subsequently applied across all trained networks.

During data preparation, custom scripts were implemented to traverse the complete segmentation masks and retain only the regions corresponding to the target anatomical structures (e.g., vascular systems, abdominal organs, or other specific components depending on the network), discarding all irrelevant tissues. This preprocessing step produced focused datasets in which each image was exclusively paired with masks of the structures of interest, minimizing interference from surrounding anatomy.

https://doi.org/10.3390/s26010323

Table 2. Selected anatomical structures for modular segmentation, grouped by clinical and anatomical relevance.

Category	Anatomical Structures
Vascular	Portal vein and branches; hepatic artery and bifurcations; hepatic veins; inferior vena cava
Abdominal	Digestive: colon (ascending, transverse, descending, sigmoid); stomach; duodenum; small bowel loops Parenchymal: liver; pancreas (head, body, tail); kidneys (cortex, medulla); spleen; gallbladder; common bile duct; adrenal glands; tumor (pathological)
Musculoskeletal	spine; pelvis; femurs; psoas muscles; oblique muscles; rectus abdominis; obturator internus; pyramidalis

It was also observed that requiring all target labels to be present simultaneously drastically reduced the number of usable samples. To overcome this limitation, an additional script was implemented to relax this constraint, allowing the inclusion of cases containing at least one of the target structures. This adjustment significantly increased dataset size and variability, improving model robustness and generalization.

The segmentation networks were implemented using the nnU-Net framework, leveraging its ability to adapt automatically to dataset properties such as patch size, voxel spacing, and number of classes. The architecture followed a 3D U-Net structure with encoder–decoder pathways, residual connections, and softmax-based multi-class output layers. Each network was trained to segment up to three classes (portal, venous, arterial vasculatures), with the output channel dimension adapted accordingly. Skip connections between the encoder and decoder ensured the preservation of fine-grained spatial details, which is critical for vascular structures due to their thin and branching morphology.

3.2. Ethical Approval, Data Anonymization, and Acquisition Characteristics

All data were fully anonymized prior to analysis in accordance with the General Data Protection Regulation (GDPR), and no personally identifiable information was accessible to the researchers. All imaging data used in this study were retrospectively collected from Cella Medical Solutions' internal PACS system, which aggregates clinical MRI studies from more than 600 healthcare centers across different countries of America and Europe. This multicenter and multinational composition increases dataset heterogeneity and better reflects real-world clinical variability. The study population mainly consisted of adult patients, with an estimated sex distribution of approximately 70% male and 30% female subjects, which is consistent with the demographic profile commonly observed in abdominal oncological imaging cohorts.

MRI examinations were acquired as part of routine clinical practice using scanners from multiple vendors; approximately 45% of the MRI studies were acquired using Siemens systems, 30% using GE scanners, and 15% using Philips scanners, while the remaining 10% corresponded to other vendors. Acquisition parameters and imaging protocols varied depending on clinical indication, anatomical region, scanner configuration, and institutional imaging practices. This variability reflects standard clinical conditions rather than controlled research acquisitions. The majority of MRI studies corresponded to oncological examinations of the abdominopelvic region. The most prevalent clinical indications included abdominal and abdominopelvic cancers, predominantly hepatobiliary, pancreatic, colorectal, and retroperitoneal malignancies. A smaller proportion of cases included non-oncological indications such as inflammatory processes, benign lesions, or postoperative and treatment follow-up studies. The dataset included adult patients with an age

https://doi.org/10.3390/s26010323

range between approximately 18 and 70 years. Regarding geographic origin, the cohort was composed of imaging studies acquired in healthcare centers from both Europe and the American continent, with an estimated distribution of approximately 60% European centers and 40% centers from North and South America. Only aggregated demographic information was available for analysis, and no direct patient identifiers were accessible at any stage of the study.

Given the retrospective nature of the study and the use of anonymized and aggregated imaging data, the requirement for formal institutional review board approval and individual patient consent was waived in accordance with applicable institutional policies and national regulations.

3.3. Model Architecture

For sequence classification, a ResNet18 convolutional neural network was employed due to its balance between depth and computational efficiency. ResNet18 was chosen over deeper variants due to its balance between representational capacity and computational efficiency, particularly important given our dataset size and clinical deployment constraints. Its flexible design facilitated integration with data loading, preprocessing, and augmentation pipelines. The network was adapted to handle volumetric data: the first convolutional layer was modified to accept 10-channel inputs, corresponding to the 10 central slices of each volume, and the final fully connected layer was adjusted to have two output neurons for binary classification (T1 vs. T2).

Prior to designing the segmentation networks, an exploratory analysis of the dataset was performed to assess the availability of anatomical structures across the three imaging categories (T1, T2, and MR). A modular segmentation strategy was adopted, with networks specialized in subsets of anatomical structures. A dedicated analysis script was implemented to traverse the dataset, identify the presence of annotated structures in each case, and generate distribution plots for each imaging category. Structures of higher clinical relevance were prioritized during the initial stages of network training. In particular, vascular structures were segmented first using a specialized vascular network, given their critical role in surgical planning and their impact on subsequent organ delineation. Once the vasculature was accurately defined, the segmentation of abdominal organs was performed, followed by musculoskeletal components such as bones and muscles. This hierarchical strategy ensured that anatomically and functionally interdependent regions were processed in a clinically meaningful order, improving overall segmentation consistency and precision.

Modularity was a deliberate design principle, allowing each network to specialize in a subset of structures and enabling the output of one network to guide or constrain the next. This sequential refinement enhanced segmentation accuracy in neighboring regions while reducing computational load. Each network operated on 3D patches extracted from T1, T2, or combined MR datasets and generated corresponding 3D segmentation masks. The modular workflow facilitated convergence, optimized resource usage, and supported a progressive, anatomically coherent reconstruction of the human body.

3.4. Training Procedure

In our initial approach, a ResNet18 model was trained on our initial dataset of 363 MRI images (174 T1-weighted and 189 T2-weighted). This dataset was split in a stratified manner to preserve class proportions of the 363-image dataset with a batch size of 1, for 70 epochs, and a learning rate of 0.001 using a Stochastic Gradient Descent (SDG) optimizer. Preprocessing and data augmentation were applied to each image as described above, to enhance model robustness and generalization.

https://doi.org/10.3390/s26010323

Once our dataset was expanded, we conducted a second training using the filtered dataset of 2286 volumes, and then the model was retrained with updated hyperparameters: batch size 16,100 epochs, and learning rate 0.001 and using, again, an SDG optimizer. Stratified splitting ensured balanced representation of T1 and T2 images. This allowed the model to generalize better and capture more robust patterns across a larger dataset.

In the next approach and in order to further stabilize performance, a third training was conducted on the same 2286-volume dataset, keeping the same ResNet18 architecture, batch size, and learning rate, but increasing the number of epochs to 200. The input continued to consist of 10 central slices, and the train/validation/test splits remained unchanged. This stage aimed to refine the model, ensuring convergence and optimal accuracy for T1/T2 classification.

For segmentation, networks nnU-Net were trained using stratified splits of the dataset for training, validation, and testing, ensuring balanced representation of all structures in each subset. The SDG optimizer was employed with a learning rate of 0.001. Batch sizes were adjusted according to the patch size and GPU memory availability.

The training strategy was designed to exploit the modular structure of the segmentation pipeline. Each specialized network was trained independently on a reduced set of anatomical structures, using the nnU-Net's standardized configuration of patch-based 3D convolutions, encoder–decoder architecture, and softmax output layers. The sequential ensemble approach of networks enabled refinement of the segmentation process: predictions from earlier networks were incorporated as priors for later stages, allowing subsequent models to adjust their focus depending on the confidence of previous outputs. This design not only reduced computational requirements but also improved anatomical consistency across segmentations.

To investigate the impact of modality-specific learning versus a pooled approach, nine separate networks were trained: (i) a T1-specific network for anatomical structures, (ii) a T2-specific network for anatomical structures, and (iii) a general MR network combining T1 and T2 images without prior stratification for anatomical structures. For each of these three main networks, an additional set of three specialized models was trained to segment different anatomical groups, resulting in a total of nine trained models. This experimental setup enabled the evaluation of whether modality specialization improved segmentation performance compared to a pooled training strategy.

Each network was initially trained for 1000 epochs with a constant learning rate of 0.01. Table 3 summarizes the segmentation network architecture. A second series of experiments extended the training to 1500 epochs, with an increased initial learning rate of 0.03, proportional to the number of labels present in the dataset. Despite the extended training time and adjusted learning rate, performance differences were minimal. For example, in the MR network, the Dice similarity coefficient (DSC) on the test set was 0.7774 after 1000 epochs and 0.7719 after 1500 epochs, indicating no significant improvement with prolonged training.

Table 3. nnU-Net-based segmentation network architecture (3D).

Component	Details	Activation/Notes
Input	3D patch from MRI volume	-
Encoder	Multiple 3D convolutions per level, downsampling at each stage	ReLU, BatchNorm
Decoder	Multiple 3D convolutions per level, upsampling at each stage	ReLU, BatchNorm
Skip Connections	Between encoder and decoder levels	Concatenation
Output Layer	Number of channels = number of structures	Softmax

https://doi.org/10.3390/s26010323

During training, data augmentation was applied, including random rotation, scaling, elastic deformation, Gaussian noise, and intensity adjustments, to increase model robustness and compensate for limited data. For sequential networks, masks predicted by previous networks were incorporated as additional inputs, allowing subsequent networks to focus on refining predictions in relevant regions. Model performance was evaluated using standard metrics such as Dice coefficient, Intersection over Union (IoU), and Hausdorff distance. Hyperparameters and augmentation strategies were iteratively adjusted based on validation performance to ensure stable convergence and reliable generalization.

3.5. Implementation Details

The development of the MRI sequence classification and segmentation pipeline leveraged a variety of specialized libraries, frameworks, and tools to ensure efficient processing, model training, and evaluation. The core implementation was carried out in `Python (v3.9)`, taking advantage of its extensive ecosystem for scientific computing and deep learning.

For deep learning model development, the `PyTorch` (v1.13) framework was employed due to its flexibility in designing custom neural network architectures, dynamic computation graphs, and robust support for GPU acceleration.

Medical image processing and manipulation were facilitated by tools like `SimpleITK` (v2.2.1) and `Nibabel` (v5.1.0), enabling loading, resampling, intensity normalization, and conversion between different medical imaging formats. Data augmentation was implemented using `TorchIO` (v0.19.1). and custom scripts to apply rotations, translations, elastic deformations, Gaussian noise, and intensity adjustments, ensuring robust model training.

For visualization and quality control, `Matplotlib` (v3.7.1), `Seaborn` (v0.12.2), and `3D Slicer` (v4.11) were employed to inspect MRI volumes, overlay segmentation masks, and verify preprocessing steps. Finally, data handling, dataset organization, and experiment tracking were managed using `Pandas` (v2.0.3) and `NumPy` (v1.24.4), allowing structured storage of image paths, labels, and metadata for reproducible experimentation.

4. Results

4.1. T1 and T2 MRI Classification

The first training stage involved the ResNet18 model trained on the initial dataset of 363 MRI volumes. After completing training, the model was evaluated on the test set, yielding the confusion matrix on the left of Figure 2 and the performance metrics summarized in Table 4. The model achieved a global accuracy of 0.90 with a support of 41 T1-weighted images and 50 T2-weighted images on the test set. T1-weighted images had a precision of 0.94, with a recall of 0.83 and an F1-score of 0.88, while T2-weighted images had a precision of 0.87, with a recall of 0.96 and an F1-score of 0.91 The confusion matrix indicates that the model correctly classified 34 out of 41 T1-weighted images and 48 out of 50 T2-weighted images, with a total of 9 misclassifications, including 7 T1-weighted images misclassified as T2-weighted and 2 T2-weighted images misclassified as T1-weighted.

Analysis of these metrics reveals that, while the model achieved high performance for T2-weighted images, its performance on T1-weighted images was notably lower, with a recall of 0.83 and an F1-score of 0.88. This imbalance indicates that the model did not robustly capture the distinguishing features between the two classes. This limitation may be partly attributed to the relatively small size of the training dataset and the potential heterogeneity of the images. Additionally, the model's performance on T2-weighted images was generally higher, with a precision of 0.87, a recall of 0.96, and an F1-score of 0.91.

https://doi.org/10.3390/s26010323

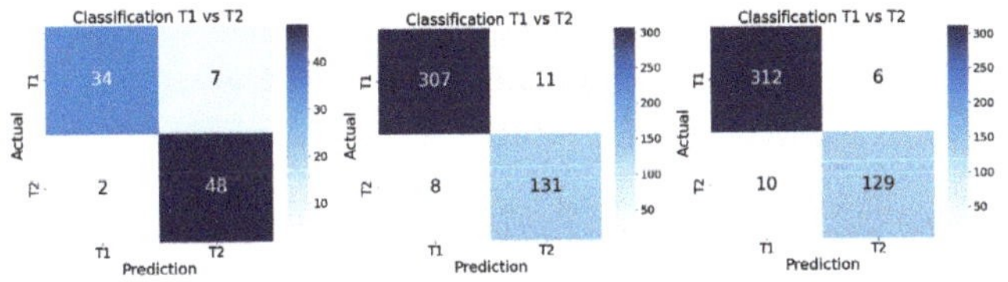

Figure 2. Confusion matrix for the first training stage (T1 vs. T2) on test set (**left**), second training stage (T1 vs. T2) on test set (**middle**), and third training stage (T1 vs. T2) on test set (**right**). Confusion matrix (**left**) shows good classification performance, with only 9 (9.89%) misclassified images. Confusion matrix (**middle**) shows improved classification performance, with only 19 (4.16%) misclassified images. Finally, confusion matrix (**right**) shows further improved classification performance, with only 16 (3.5%) misclassified images. The model is therefore considered the final version for the task of automatic classification of T1- and T2-weighted MRI images on the dataset used.

Table 4. Comparison of classification reports across the three training stages on the test set (T1 vs. T2). Metrics show a consistent improvement through training stages.

		Precision	Recall	F1-Score	Support
First Stage	T1	0.94	0.83	0.88	41
	T2	0.87	0.96	0.91	50
	Accuracy			0.90	91
	Macro avg	0.91	0.89	0.90	91
	Weighted avg	0.91	0.90	0.90	91
Second Stage	T1	0.97	0.97	0.97	318
	T2	0.92	0.94	0.93	139
	Accuracy			0.96	457
	Macro avg	0.95	0.95	0.95	457
	Weighted avg	0.96	0.96	0.96	457
Third Stage	T1	0.97	0.98	0.98	318
	T2	0.96	0.93	0.94	139
	Accuracy			0.96	457
	Macro avg	0.96	0.95	0.96	457
	Weighted avg	0.97	0.96	0.96	457

Based on these preliminary results, a promising strategy to enhance model performance is to substantially expand the dataset by incorporating a larger number of images from both classes. Increasing the size and diversity of the dataset enables the model to learn more generalizable and robust features, thereby improving its ability to generalize and achieving better classification accuracy on unseen data. These insights informed the design of the subsequent training stages, which emphasized dataset expansion (as explained above in Section 3.1) along with the refinement of preprocessing and augmentation techniques.

The second training stage employed the expanded and filtered dataset of 2286 MRI volumes. After completing training, the model was evaluated on the test set, which yielded the results presented in the confusion matrix in the middle of Figure 2 and Table 4. The model achieved a global accuracy of 0.96 with a support of 318 T1-weighted images and 139 T2-weighted images on the test set. T1-weighted images had a precision, recall and F1-score of 0.97, while T2-weighted images had a precision of 0.92, with a recall of 0.94 and an F1-score of 0.93. The confusion matrix indicates that the model correctly classified 309 out of 318 T1-weighted images and 131 out of 139 T2-weighted images, with a total of 19 misclassifications, including 11 T1-weighted images misclassified as T2-weighted and 8 T2-weighted images misclassified as T1-weighted. Therefore, the model demonstrated significant improvement compared to the first training stage, likely due to the increased

https://doi.org/10.3390/s26010323

dataset size and diversity. This is due to the model effectively learning to distinguish between the two sequences, mainly because of the increased number of T1-weighted images in the dataset.

A manual review of misclassified images revealed that many errors were associated with atypical characteristics, including low-contrast regions, image artifacts, or noise, as shown in Figure 3. These characteristics were primarily observed in a limited subset of clinical cases, typically corresponding to challenging acquisition conditions rather than standard MRI studies. In particular, artifacts and noise were more frequently encountered in scans affected by patient motion, reduced acquisition time, suboptimal signal-to-noise ratio, or imaging performed in complex clinical contexts. Overall, these atypical acquisition conditions were estimated to affect approximately 5% of the analyzed MRI studies. This post-training analysis suggests potential improvements in preprocessing strategies or the incorporation of additional data augmentation techniques to increase model robustness against complex and non-ideal acquisition scenarios. While such atypical characteristics are not dominant in routine clinical MRI, they represent realistic edge cases that can negatively impact automated classification performance. In particular, images with significant noise or low contrast were more prone to misclassification, indicating that enhancing image quality or incorporating noise-robust features during training could further improve classification accuracy.

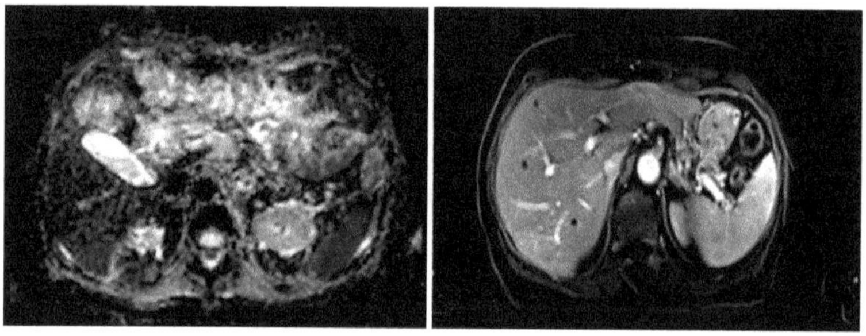

Figure 3. Example of a misclassified T1/T2 MRI image affected by noise or low contrast. (**Left**): Noisy image. (**Right**): Clear image (it is not the same image as the left one).

Overall, the second training stage demonstrates a substantial improvement over the first, with higher precision, recall, and F1-scores for both classes, indicating that dataset expansion and careful filtering significantly enhanced the model's generalization capabilities. Therefore, a further expansion of the dataset and an increased number of training epochs allowed the model to learn complex patterns from the expanded dataset, resulting in improved performance on the test set.

The third training stage utilized the same expanded dataset of 2286 MRI volumes, with the primary modification being an increase in the number of training epochs from 100 to 200. This adjustment aimed to further refine the model's learning and improve its classification performance. After completing training, the model was evaluated on the test set where its confusion matrix is shown on the right of Figure 2, and the detailed performance metrics are summarized in Table 4. The confusion matrix indicates that the model correctly classified 441 out of 457 MRI volumes, with a total of 16 misclassifications, including 6 T1-weighted images misclassified as T2-weighted and 10 T2-weighted images misclassified as T1-weighted. The classification metrics achieved on the test set resulted in an overall accuracy of 0.97, which means that the model performed well in both classes. In addition, a macro-averaged precision of 0.96, a macro-averaged recall of 0.95, and a macro average F1-score of 0.96 were obtained, confirming a balanced performance between

https://doi.org/10.3390/s26010323

the two classes and demonstrating that increasing the number of epochs contributed to improved accuracy. This enhancement is significant, as it indicates that the model has learned to distinguish between T1 and T2 images with greater accuracy. A weighted average precision of 0.97, weighted average recall of 0.96, and weighted average F1-score of 0.96 were obtained. This means that the model performed well in both classes because it takes into account the number of instances for each class, giving more importance to classes with more samples. This extension is justifiable in a clinical context, where maximizing the use of available information is crucial, leading to better adaptation to dataset characteristics and greater stability of the metrics. This model is therefore considered the final version for the task of automatic classification of T1- and T2-weighted MRI images on the dataset used.

Overall, these results demonstrate a significant improvement over the first trained model, achieving high levels of accuracy and stability. Both the T1 and T2 classes exhibit strong precision and recall values, confirming the model's ability to correctly discriminate between the two MRI sequences, even when confronted with previously unseen images.

4.2. *Segmentation of Anatomical Structures in T1 and T2 MRI*

This stage employed a modular approach, training separate segmentation networks for vascular structures, abdominal organs and musculoskeletal structures, leveraging the distinct contrast characteristics of each MRI sequence (T1-weighted, T2-weighted and a combination of both sequences). Our idea is to use the sequence classification model developed in the previous stage to route each MRI volume to the most appropriate segmentation network, thereby optimizing segmentation performance based on the specific imaging characteristics. We hypothesize that this modular design will enhance segmentation accuracy by tailoring the model to the unique features of each MRI sequence. For it, three different input datasets were used for training: a combined dataset of T1- and T2-weighted images (MR), a T1-specific dataset, and a T2-specific dataset. Each segmentation network was trained independently for each anatomical block, allowing for a detailed analysis of performance across different structures and sequences. All networks were based on the 3D U-Net architecture [5] and were trained using the Dice loss function, which is particularly well-suited for medical image segmentation tasks due to its effectiveness in handling class imbalance. We compared the performance of networks trained on the combined dataset (MR) against those trained specifically on T1-weighted and T2-weighted images to assess the benefits of sequence-specific training. A disadvantage of this modular approach is the increased computational cost and complexity associated with training and maintaining multiple models. However, we believe that the potential improvements in segmentation accuracy justify this trade-off, especially in clinical applications where precision is crucial. Another disadvantage is that the specialized networks were trained on a dataset with fewer samples compared to the combined dataset, which could limit the model's ability to generalize effectively.

The first segmentation block focused on vascular structures, including the portal vein, venous system, and arterial system. This block was chosen as the initial focus due to the critical clinical importance of accurately segmenting vascular anatomy in abdominal MRI, which is essential for various diagnostic and therapeutic applications. The performance of the vascular segmentation networks was evaluated using the Dice coefficient for each structure across different input datasets. The Dice scores obtained for the vascular segmentation networks are summarized in Table 5. The T1-specific network achieved the highest overall performance, particularly for arterial and venous structures, while the T2-specific network showed strong performance for the venous system. The combined MR network provided moderate results across all structures. These findings indicate that sequence-specific networks enhance segmentation accuracy and support the design of a modular, multi-stage

https://doi.org/10.3390/s26010323

segmentation pipeline. The networks trained on the combined MRI dataset (MR) achieved moderate performance, with Dice scores of 0.4651, 0.7700, and 0.7023 for the portal, venous, and arterial systems, respectively. The T1-specific network showed improved segmentation for arterial structures (Dice = 0.8052) and venous structures (Dice = 0.7865), while the T2-specific network performed comparably for the venous system (Dice = 0.7904) but showed lower scores for portal and arterial vessels. Overall, the macro-average Dice (background excluded) highlights that T1-weighted images provided the most consistent performance across all vascular structures, suggesting that contrast characteristics in T1 images facilitate better discrimination of vessel boundaries in abdominal MRI. These results demonstrate the utility of specialized networks for each sequence type and inform subsequent stages of the modular segmentation pipeline. Figure 4 illustrates representative segmentation outputs and ground truth build for specialists. A simple visual inspection confirms the quantitative results, showing that the T1-specific network produces detailed segmentations of vascular structures compared to the ground truth provided by the medical specialists.

Table 5. Dice scores for vascular segmentation networks across different input datasets, including Macro Dice. Macro Dice is the average of Dice scores across all abdominal organs (background excluded). MR = combined T1 and T2 datasets. T1 = T1-specific dataset. T2 = T2-specific dataset.

Network	Background	Portal	Venous	Arterial	Macro Dice
MR	0.6458	0.4651	0.7700	0.7023	0.6458
T1	0.7264	0.5877	0.7865	0.8052	0.7265
T2	0.6449	0.4908	0.7904	0.6535	0.6449

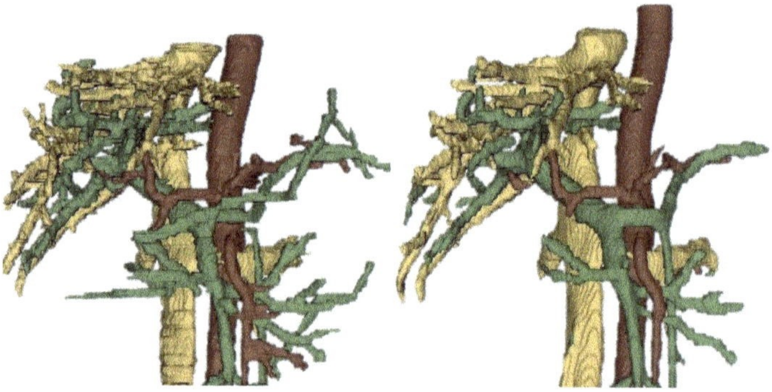

Figure 4. Example of vascular segmentation results on test set. (**Left**): Ground truth segmentation provided by specialists. (**Right**): Model inference output. Only very small differences can be observed between both segmentations. The segmentation model has difficulties segmenting very small vascular structures, but more large ones are very well segmented.

The second segmentation block focused on abdominal organs, including both hollow (e.g., stomach, colon, duodenum, intestinal loops) and solid structures (e.g., liver, pancreas, kidneys, spleen). This block represents a natural extension of the vascular segmentation, since most of these organs are anatomically related to the vascular system. The performance of the abdominal organs segmentation networks was evaluated using the Dice coefficient for each structure across different input datasets. Unlike the vascular block, a label imbalance strategy was adopted, allowing training with partially annotated cases. This approach maximized dataset utilization by incorporating images with incomplete annotations, which reflects real-world clinical conditions where full segmentations are not always available. As a result, the model was exposed to a greater diversity of samples, favoring its robustness in heterogeneous scenarios. The networks were trained consistently across three different

https://doi.org/10.3390/s26010323

input datasets (MR, T1-weighted, T2-weighted) in a similar manner to the vascular block, ensuring methodological comparability with the vascular block. The Dice scores obtained are summarized in Table 6. Overall, the best performance was observed for large and well-defined structures such as the liver, spleen, and kidneys, which consistently achieved Dice coefficients above 0.85 in MR and T1 modalities. Medium performance was observed for organs such as the pancreas and stomach, while small or morphologically variable structures (e.g., gallbladder, bile duct, adrenal glands, intestinal loops, and tumors) showed more modest scores, reflecting the higher difficulty of their segmentation. The macro-average Dice scores across all abdominal organs (background excluded) are summarized in Table 7. The MR network achieved the highest macro Dice score of 0.678, indicating the best overall performance across the diverse set of organs. The T1-specific network followed with a macro Dice of 0.605, while the T2-specific network had the lowest score of 0.531. These results reinforce the advantage of using a combined dataset for training, as it allows the model to leverage complementary information from both T1 and T2 sequences, leading to improved segmentation accuracy across a wide range of anatomical structures. These results highlight the importance of adopting a modular segmentation approach: while robust segmentations can be obtained for major abdominal organs, specialized strategies such as data augmentation, targeted pretraining, or cascaded refinement may be required to improve performance in smaller or less well-defined anatomical structures. Figure 5 illustrates representative segmentation outputs for abdominal organs, comparing the model's inference with the ground truth annotations. The visual results demonstrate the model's ability to accurately delineate major organs such as the liver and kidneys, while also highlighting areas for improvement in smaller or more complex structures.

Table 6. Dice scores for abdominal organs across different input datasets. MR = combined T1 and T2 datasets. T1 = T1-specific dataset. T2 = T2-specific dataset.

Large abdominal organs								
Network	Backg.	Colon	Stomach	Liver	Pancreas	Kidneys	Spleen	Duodenum
MR	0.6476	0.6539	0.8424	0.9322	0.7679	0.8805	0.9491	0.6251
T1	0.5812	0.5667	0.8341	0.8796	0.7700	0.8665	0.9174	0.6160
T2	0.6449	0.4908	0.7904	0.6535	0.6449	0.4908	0.7904	0.6535
Smaller abdominal structures								
Network	Gallbladder	Bile duct	Adrenals	Intestinal loops	Tumor			
MR	0.6677	0.4039	0.4836	0.4979	0.0667			
T1	0.7831	0.2230	0.2488	0.2675	0.0843			
T2	0.6449	0.4908	0.7904	0.6535	0.0762			

Table 7. Macro Dice scores across different input datasets. Macro Dice represents the average Dice score across all abdominal organs, excluding background.

Network	**Macro Dice**
MR	0.678
T1	0.605
T2	0.531

The third segmentation block focused on musculoskeletal structures, including both osseous (spine, pelvis, femurs) and muscular tissues (psoas, obliques, rectus abdominis, obturator internus, pyramidal muscles). This block was selected due to the clinical relevance of accurately segmenting musculoskeletal anatomy in abdominal MRI, which is essential for various diagnostic and therapeutic applications. The performance of the musculoskeletal

https://doi.org/10.3390/s26010323

segmentation networks was evaluated using the Dice coefficient for each structure across different input datasets. The performance of the musculoskeletal segmentation networks is summarized in Table 8. For osseous structures, the networks trained on T2-weighted images achieved the highest Dice coefficients overall, with values of near 0.77 for the pelvis, 0.80 for the spine, and 0.88 for the femurs. These results highlight the advantage of the T2 contrast in delineating cortical bone and surrounding connective tissue. By contrast, MR- and T1-based networks exhibited moderate performance, particularly in the segmentation of the spine, where Dice scores did not exceed 0.63. For muscular structures, the results were more heterogeneous. While large muscles such as the psoas achieved robust Dice scores across modalities (0.76–0.78), smaller or thinner structures, such as the rectus abdominis and the pyramidal muscles, showed more variable outcomes, with values ranging between 0.53 and 0.83. This variability reflects the intrinsic difficulty of segmenting small-volume structures, which are prone to annotation inconsistencies and partial-volume effects. Overall, these results indicate that musculoskeletal segmentation benefits substantially from T2-weighted acquisitions, particularly for osseous structures. For muscular tissues, however, performance appears to be less dependent on modality and more affected by structural variability and annotation quality. These findings reinforce the importance of multimodal integration and data augmentation strategies as potential avenues for improvement in subsequent iterations of the pipeline. The macro-average Dice coefficients (background excluded) for the musculoskeletal segmentation networks are summarized in Table 8. The T2-weighted network achieved the highest macro Dice score of 0.7544, indicating superior overall performance across both bone and muscle structures. The T1-weighted network followed with a macro Dice of 0.7027, while the MR network had the lowest score of 0.6611. These results underscore the effectiveness of T2-weighted imaging for musculoskeletal segmentation tasks, particularly in capturing the complex anatomy of both osseous and muscular tissues. Figure 6 illustrates representative segmentation outputs for musculoskeletal structures, comparing the model's inference with the ground truth annotations. The visual results demonstrate the model's ability to accurately delineate major structures such as the spine and femurs, while also highlighting areas for improvement in smaller or more complex muscular tissues.

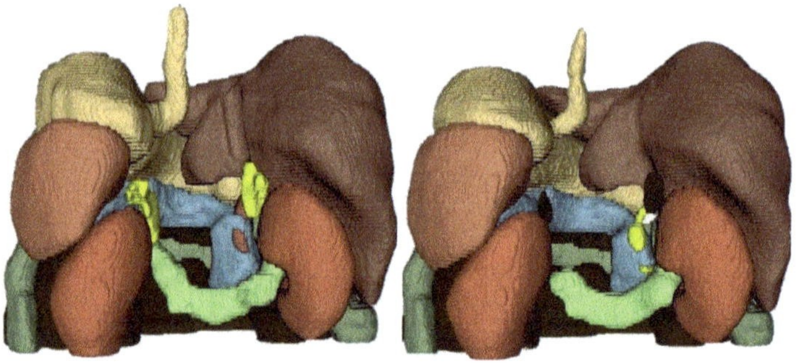

Figure 5. Example of abdominal organ segmentation results on test set. (**Left**): Ground truth segmentation provided by specialists. (**Right**): Model inference output. Only very small differences can be observed between both segmentations. The position of the liver, kidneys, spleen, and duodenum is well delineated with consistent accuracy between both segmentations.

https://doi.org/10.3390/s26010323

Table 8. Dice coefficients for the musculoskeletal block across different input datasets. Macro Dice is the average of Dice scores across all structures (background excluded). MR = combined T1 and T2 datasets. T1 = T1-specific dataset. T2 = T2-specific dataset.

		Osseus-Related Structures			Muscle-Related Structures					
Network	**Backg.**	**Spine**	**Pelvis**	**Femurs**	**Psoas**	**Obliques**	**Rectus Abdominis**	**Obturator Internus**	**Pyramidal Muscles**	**Macro Dice**
MR	0.6610	0.6120	0.7573	0.7530	0.7783	0.4959	0.5362	0.6872	0.6686	0.6611
T1	0.5902	0.6318	0.7896	0.7917	0.7780	0.5212	0.5483	0.7899	0.7708	0.7027
T2	0.7544	0.8019	0.7685	0.8812	0.7600	0.6131	0.5535	0.8220	0.8346	0.7544

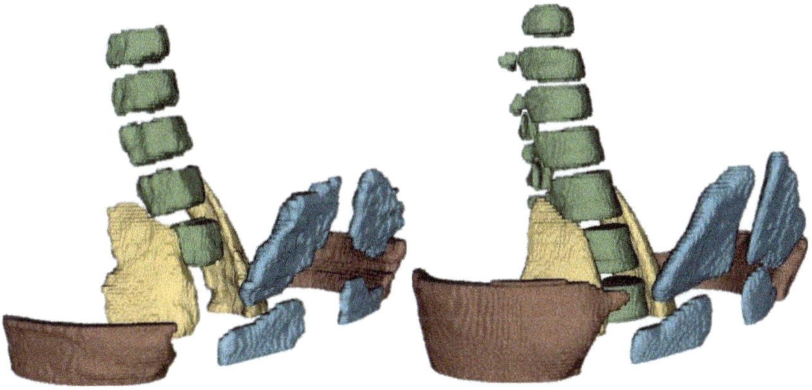

Figure 6. Example of musculoskeletal segmentation results on test set. (**Left**): Ground truth segmentation provided by specialists. (**Right**): Model inference output. Some big differences can be observed between both segmentations. The position of the spine, pelvis, and femurs is well delineated with consistent accuracy between both segmentations.

4.3. Results of the Integrated Pipeline and Qualitative Validation

After integrating the different classification and segmentation blocks into an automated pipeline, a qualitative clinical validation was conducted in order to assess the clinical applicability of the proposed models. Eight independent studies (four T1-weighted and four T2-weighted MRIs), unseen during training and testing, were evaluated. Each case was segmented using three distinct approaches: a state-of-the-art segmentation pipeline based on TotalSegmentator, the General Network (trained jointly on T1 and T2 images), and the Specific Network (trained separately for each modality). A blinded assessment was performed by an experienced radiologist, who scored the anatomical fidelity and clinical interpretability of the segmentations on a scale from 0 to 10. The results indicate that the proposed networks provided higher qualitative scores than the previously deployed workflow for the evaluated cases. Table 9 provides an overview of the mean qualitative scores obtained for each patient. The General Network achieved an average of 7.2, closely followed by the Specific Network with 6.9, while the state-of-the-art TotalSegmentator framework obtained a mean score of 3.6. This indicates that the integrated pipeline produces anatomically consistent segmentations with strong clinical usefulness for the evaluated set of examples. Figure 7 visually summarizes these results, illustrating the difference in the qualitative assessments between the proposed networks and the earlier workflow. The analysis of median values per patient, as shown in Table 9 and Figure 8, further supports this observation, with the General Network obtaining a global median of 7.7, very close to the Specific Network (7.8), while TotalSegmentator showed a median score of 4.1. All experiments were performed using an NVIDIA RTX 3060 GPU (NVIDIA Corporation, Santa Clara, CA, USA).

https://doi.org/10.3390/s26010323

These results reinforce the robustness of the proposed pipeline, particularly highlighting the ability of the General Network to achieve stable performance across both T1- and T2-weighted images without the need for modality-specific training. Overall, this qualitative validation demonstrates that the proposed system provides high-quality segmentations suitable for supporting personalized surgical planning. In summary, the quantitative analysis of the segmentation performance, as measured by Dice coefficients, further corroborates the advantages of the Specific Network, which consistently outperformed the General Network across various organs. The results reveal the possibility of using classification and segmentation models in an integrated pipeline as an ensemble to improve the accuracy and efficiency of segmentation in MRI. The qualitative validation of the integrated pipeline demonstrates its clear superiority over existing commercial solutions. The General Network, in particular, stands out for its ability to deliver high-quality segmentations across different MRI modalities, making it a versatile and effective tool for clinical applications. These promising results pave the way for further development and eventual clinical integration of the proposed system.

The qualitative evaluation was conducted by a single experienced radiologist. While this provided expert clinical insight into the anatomical fidelity and usability of the segmentations, inter-rater variability was not assessed in this study. As a result, the qualitative scores should be interpreted as an initial expert validation rather than a statistically robust inter-observer analysis.

Table 9. The left half of the table shows the mean qualitative scores (0–10) obtained for each patient and network, while the right half displays the corresponding median values.

Patient Number	SOTA System	General Network	Specific Network		SOTA System	General Network	Specific Network
#1	2.1	8.0	6.7		4.0	8.0	7.0
#2	6.4	7.3	6.0		8.0	8.0	8.0
#3	1.3	6.2	5.4		1.0	7.0	7.0
#4	3.7	7.7	7.5		4.5	8.0	8.0
#5	4.4	8.4	8.2		5.5	8.5	8.0
#6	3.5	7.4	7.6		5.0	8.0	8.0
#7	3.2	6.1	7.1		3.5	7.0	7.5
#8	2.2	6.7	6.7		1.0	7.0	8.0
Mean	3.6	7.2	6.9	Median	4.1	7.7	7.8

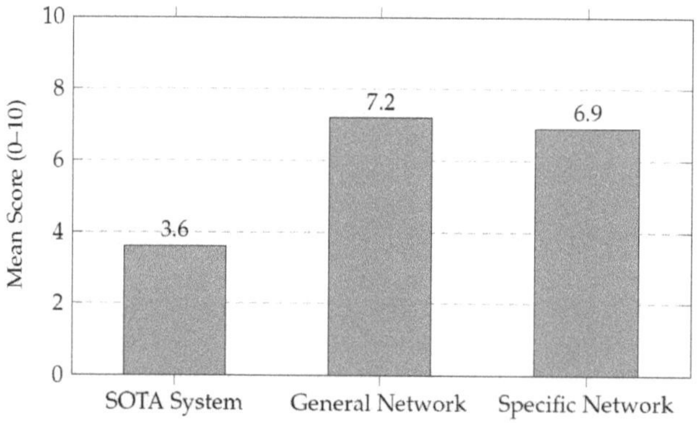

Figure 7. Comparison of mean qualitative scores obtained by the state-of-the-art (SOTA) system, the General Network, and the Specific Network. The proposed networks significantly outperform the commercial solution, demonstrating their superior anatomical fidelity and clinical usefulness.

https://doi.org/10.3390/s26010323

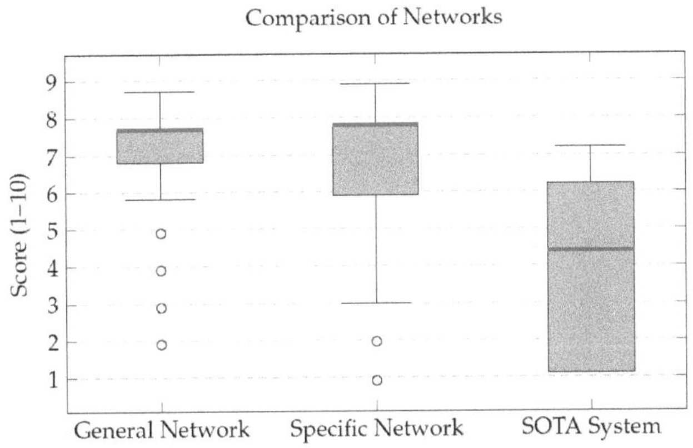

Figure 8. Comparison of median qualitative scores obtained by the SOTA system, the General Network, and the Specific Network using box-and-whisker plots. The red line indicates the median value for each network. The proposed Specific Network significantly outperforms the commercial and General Network solutions, demonstrating superior anatomical fidelity and clinical usefulness.

5. Discussion

The qualitative evaluation of the integrated segmentation pipeline revealed notable improvements when compared to a state-of-the-art segmentation framework, TotalSegmentator. The General Network, trained on a combined dataset of T1- and T2-weighted MRIs, achieved a mean score of 7.2, demonstrating its capability to generalize effectively across different imaging modalities. This versatility is particularly advantageous in clinical settings where both types of sequences are commonly used.

Despite being trained with less than half the number of cases used for the General Network, the Specific Network achieved higher qualitative scores, with a median of 7.8. This result showed higher median scores than the General Network and also outperformed the state-of-the-art TotalSegmentator framework. This trend was observed across both T1 and T2 images, highlighting the robustness of the specialized models.

The discrepancy between the mean (6.9) and the median (7.8) for the Specific Network indicates that, although a few cases received lower evaluations, the majority of the segmentations were of high and homogeneous quality. This reinforces the added value of training specialized models tailored to each MRI sequence, which appear to capture sequence-specific morphological and contrast characteristics more effectively. Conversely, the state-of-the-art TotalSegmentator framework yielded lower mean and median scores in the evaluated examples, reflecting reduced anatomical fidelity for these specific cases.

Beyond the qualitative assessment, a complementary volumetric analysis was conducted to compare the segmented organ volumes across the different networks, as summarized in Table 10. Results revealed that the Specific Network generally produced slightly larger segmentations compared to the General Network. In practical terms, the specialized model tended to generate more complete anatomical masks, potentially reflecting an increased sensitivity for detecting and delineating structures; however, this behavior may also indicate a propensity toward over-segmentation in some cases. Consequently, volumetric metrics should be interpreted alongside qualitative clinical evaluations, which ultimately guide model usability in real-world surgical planning scenarios.

To facilitate the clinical integration of the proposed pipeline, the segmented models were incorporated into Cella's 3D Planner, a web-based visualization platform used in clinical practice. This tool enables interactive inspection of the reconstructed anatomy

https://doi.org/10.3390/s26010323

directly in modern web browsers. Clinicians can rotate, zoom, and selectively display or hide structures, enhancing anatomical understanding and procedure preparation.

Table 10. Extract from the complete table showing the segmented volume for each network. It can be observed that the Specific Network tends to predict larger volumes.

Patient Number	Organ	Dice	General Net. Seg. Vol. (mL)	Specific Net. Seg. Vol. (mL)	Difference (mL)	Best
#1	Spleen	0.9785	217.93	211.24	6.69	General
#2	Bile Duct	0.3636	0.10	0.36	−0.26	Specific
#3	Spine	0.5929	106.51	98.10	8.41	General
#5	Duodenum	0.8287	69.32	92.02	−22.70	Specific
#1	Stomach	0.9049	98.37	109.75	−11.38	Specific
#2	Adrenal Glands	0.1107	1.35	0.96	0.39	General
#5	Liver	0.9876	2362.44	2390.30	−27.86	Specific
#4	Obliques	0.6247	35.19	47.49	−12.30	Specific
#6	Pancreas	0.9504	89.15	84.71	4.44	General
#8	Venous Vasculature	0.9321	56.57	58.22	−1.65	Specific

Table 10 presents an extract of the volumetric results, illustrating these differences across representative anatomical structures. Figure 9 compares the Dice scores obtained by the General and Specific Networks for various organs. The Specific Network showed higher Dice scores across most structures, particularly in challenging anatomies such as the bile duct and duodenum. This performance difference underscores the advantage of training specialized models for each MRI sequence, which can better capture the unique anatomical and contrast characteristics present in T1- and T2-weighted images. The enhanced segmentation accuracy achieved by the Specific Network translates into more consistent anatomical representations for the evaluated structures. Our work highlights the growing synergy between imaging technologies and artificial intelligence. We leverage deep learning architectures such as ResNet for classification and nnU-Net for segmentation to automate the analysis of sequence-related information. This AI-driven approach improves the efficiency of image interpretation and exemplifies the potential of intelligent processing systems in medical imaging, enabling more informed clinical decisions and personalized patient care for the evaluated cases.

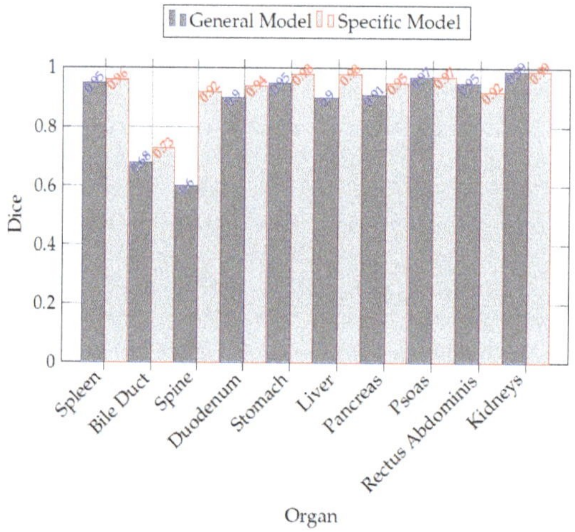

Figure 9. Comparison of Dice scores between the General and Specific Networks for various organs. The Specific Network consistently outperforms the General Network, particularly in challenging structures such as the bile duct and duodenum.

https://doi.org/10.3390/s26010323

6. Limitations and Future Work

Despite the promising results obtained in this study, several limitations must be acknowledged. First, the qualitative clinical evaluation was conducted on a limited subset of eight cases and assessed by a single experienced radiologist. While this provided valuable expert insight into anatomical fidelity and clinical plausibility, inter-rater reliability was not assessed. Consequently, the qualitative scores should be interpreted as an initial expert validation rather than a statistically rigorous multi-observer study. Future work will include multiple independent radiologists from different institutions to enable formal inter-rater agreement analysis using established statistical metrics.

Second, the comparison with the TotalSegmentator framework was not designed as a fully controlled benchmark. Differences in training objectives, label definitions, and optimization strategies limit the fairness of direct performance comparisons. Therefore, the reported results should be interpreted as a practical reference against an existing automated pipeline rather than as definitive evidence of methodological superiority.

Third, although specific efforts were made to curate and balance the available datasets across MRI sequences, segmentation performance may still be influenced by residual heterogeneity in label availability across anatomical structures. While data selection and preprocessing strategies were applied to maximize consistency and data usage, an exhaustive per-structure label distribution analysis was not the primary focus of this study. As a result, subtle performance differences observed between MR, T1, and T2 networks should be interpreted with caution. Future studies will incorporate more detailed reporting of label distributions and structure-specific sample sizes to further strengthen result interpretability.

In addition, the current classification pipeline lacks ablation studies and explainability analyses. Techniques such as Grad-CAM will be explored to improve transparency and to assess whether the model relies on meaningful anatomical contrast patterns or on spurious imaging artifacts.

Finally, the integration of segmented anatomical models into 3D visualization platforms is supported by existing literature as a valuable tool for surgical planning, anatomical understanding, and precision surgery. Previous studies have demonstrated the clinical relevance of patient-specific 3D reconstructions across multiple surgical domains, including pancreatic, urological, oncological, and pediatric surgery [26–28]. While the present work focuses on the development and validation of an automated AI-based pipeline, future studies may further build upon this established evidence by incorporating task-specific clinical metrics and prospective validation scenarios.

7. Conclusions

This study presents a comprehensive deep learning-based pipeline for the classification and segmentation of abdominal MRI scans, specifically targeting T1- and T2-weighted sequences. The proposed system demonstrates consistent qualitative and quantitative performance across the evaluated cases, particularly in terms of anatomical fidelity and segmentation quality.

The main objective of this work—the development of an artificial intelligence-based tool capable of automatically classifying magnetic resonance (MR) images into T1 and T2 sequences and segmenting relevant anatomical structures for oncology-oriented surgical planning research—has been successfully achieved.

The main contributions include the construction of a curated clinical MRI dataset, the integration of automatic sequence classification using a ResNet18-based model achieving over 90% accuracy, and the implementation of sequence-specific segmentation networks based on the nnU-Net v2 architecture.

https://doi.org/10.3390/s26010323

The modular design of the proposed pipeline enables flexible extension, supports block-wise evaluation of anatomical structures, and facilitates scalable research-oriented MRI analysis workflows.

Overall, this work establishes a technically robust and extensible framework for sequence-aware MRI classification and segmentation, providing a solid foundation for further research in AI-assisted medical image analysis.

Author Contributions: Conceptualization, A.M.G.; Methodology, A.M.G.; Software, A.M.G.; Formal analysis, A.M.G. and A.A.G.; Investigation, A.M.G.; Resources, A.A.G.; Writing—original draft, A.M.G., J.F.Z.P. and J.M.-A.; Writing—review & editing, A.M.G., J.F.Z.P., J.M.-A. and A.A.G.; Supervision, J.F.Z.P., J.M.-A. and A.A.G.; Project administration, J.F.Z.P., J.M.-A. and A.A.G. All authors have read and agreed to the published version of the manuscript.

Funding: This work was supported by the Spanish Ministry of Science, Innovation and Universities under grant PID2021-128009OB-C33 funded by MCIN/AEI/10.13039/501100011033/FEDER, UE, which provided essential human and material resources.

Institutional Review Board Statement: Ethical review and approval were waived for this study due to the retrospective nature of the analysis and the use of fully anonymized imaging data, in accordance with applicable institutional policies and regulations.

Informed Consent Statement: Patient consent was waived due to the retrospective design of the study and the use of fully anonymized data.

Data Availability Statement: The data supporting the findings of this study are not publicly available due to privacy and ethical restrictions related to clinical imaging data.

Acknowledgments: The authors would like to express their gratitude to the medical imaging team and the radiology department for providing the clinical data and expert annotations that made this work possible. We are especially thankful to the radiologists who contributed their time and expertise to the qualitative evaluation of the segmentation models. We also acknowledge the support of the technical staff involved in the preprocessing of the datasets and the development of the computational infrastructure required for training and validation of the models. Finally, we would like to thank our colleagues and collaborators for their constructive feedback and valuable discussions, which greatly contributed to the refinement of this work.

Conflicts of Interest: Authors A.M.-G. and A.A.-G. were employed by Cella Medical Solutions. The remaining authors declare that the research was conducted in the absence of any commercial or financial relationships that could be construed as a potential conflict of interest.

References

1. Bray, F.; Laversanne, M.; Sung, H.; Ferlay, J.; Siegel, R.L.; Soerjomataram, I.; Jemal, A. Global Cancer Statistics 2022: GLOBOCAN Estimates of Incidence and Mortality Worldwide for 36 Cancers in 185 Countries. *CA Cancer J. Clin.* **2024**, *74*, 229–263. [CrossRef] [PubMed]
2. Viergever, M.A.; Maintz, J.A.; Klein, S.; Murphy, K.; Staring, M.; Pluim, J.P. A Survey of Medical Image Registration—Under Review. *Med. Image Anal.* **2016**, *33*, 140–144. [CrossRef] [PubMed]
3. Boykov, Y.; Jolly, M.P. Interactive Graph Cuts for Optimal Boundary & Region Segmentation of Objects in N-D Images. In Proceedings of the Eighth IEEE International Conference on Computer Vision ICCV 2001, Vancouver, BC, Canada, 7–14 July 2001; Volume 1, pp. 105–112. [CrossRef]
4. Aerts, H.J.W.L.; Velazquez, E.R.; Leijenaar, R.T.H.; Parmar, C.; Grossmann, P.; Carvalho, S.; Bussink, J.; Monshouwer, R.; Haibe-Kains, B.; Rietveld, D.; et al. Decoding Tumour Phenotype by Noninvasive Imaging Using a Quantitative Radiomics Approach. *Nat. Commun.* **2014**, *5*, 4006. [CrossRef] [PubMed]
5. Ronneberger, O.; Fischer, P.; Brox, T. U-Net: Convolutional Networks for Biomedical Image Segmentation. In Proceedings of the Medical Image Computing and Computer-Assisted Intervention (MICCAI), Munich, Germany, 5–9 October 2015; Springer: Berlin/Heidelberg, Germany, 2015; pp. 234–241. [CrossRef]

https://doi.org/10.3390/s26010323

6. Çiçek, Ö.; Abdulkadir, A.; Lienkamp, S.S.; Brox, T.; Ronneberger, O. 3D U-Net: Learning Dense Volumetric Segmentation from Sparse Annotation. In Proceedings of the Medical Image Computing and Computer-Assisted Intervention (MICCAI), Athens, Greece, 17–21 October 2016; Springer: Berlin/Heidelberg, Germany, 2016; pp. 424–432. [CrossRef]
7. Milletari, F.; Navab, N.; Ahmadi, S.A. V-Net: Fully Convolutional Neural Networks for Volumetric Medical Image Segmentation. In Proceedings of the 2016 Fourth International Conference on 3D Vision (3DV), Stanford, CA, USA, 25–28 October 2016; pp. 565–571. [CrossRef]
8. Isensee, F.; Jaeger, P.F.; Kohl, S.A.; Petersen, J.; Maier-Hein, K.H. nnU-Net: A Self-Adapting Framework for U-Net-based Medical Image Segmentation. *Nat. Methods* **2021**, *18*, 203–211. [CrossRef]
9. Zhou, Z.; Siddiquee, M.M.R.; Tajbakhsh, N.; Liang, J. Unet++: A Nested u-Net Architecture for Medical Image Segmentation. In Proceedings of the Deep Learning in Medical Image Analysis and Multimodal Learning for Clinical Decision Support, Granada, Spain, 20 September 2018; Springer: Berlin/Heidelberg, Germany, 2018; pp. 3–11. [CrossRef]
10. Oktay, O.; Schlemper, J.; Folgoc, L.L.; Lee, M.; Heinrich, M.; Misawa, K.; Mori, K.; McDonagh, S.; Hammerla, N.Y.; Kainz, B.; et al. Attention U-Net: Learning Where to Look for the Pancreas. *arXiv* **2018**, arXiv:1804.03999. [CrossRef]
11. He, K.; Zhang, X.; Ren, S.; Sun, J. Deep Residual Learning for Image Recognition. In Proceedings of the IEEE Conference on Computer Vision and Pattern Recognition, Las Vegas, NV, USA, 27–30 June 2016; pp. 770–778. [CrossRef]
12. Huang, G.; Liu, Z.; Van Der Maaten, L.; Weinberger, K.Q. Densely Connected Convolutional Networks. In Proceedings of the IEEE Conference on Computer Vision and Pattern Recognition, Honolulu, HI, USA, 21–26 July 2017; pp. 4700–4708. [CrossRef]
13. Tan, M.; Le, Q. EfficientNet: Rethinking Model Scaling for Convolutional Neural Networks. In Proceedings of the International Conference on Machine Learning PMLR, Long Beach, CA, USA, 9–15 June 2019; pp. 6105–6114. [CrossRef]
14. Dosovitskiy, A.; Beyer, L.; Kolesnikov, A.; Weissenborn, D.; Zhai, X.; Unterthiner, T.; Dehghani, M.; Minderer, M.; Heigold, G.; Gelly, S.; et al. An Image Is Worth 16 × 16 Words: Transformers for Image Recognition at Scale. In Proceedings of the International Conference on Learning Representations, Addis Ababa, Ethiopia, 26–30 April 2020. [CrossRef]
15. Liu, Z.; Lin, Y.; Cao, Y.; Hu, H.; Wei, Y.; Zhang, Z.; Lin, S.; Guo, B. Swin Transformer: Hierarchical Vision Transformer Using Shifted Windows. In Proceedings of the IEEE/CVF International Conference on Computer Vision, Montreal, QC, Canada, 10–17 October 2021; pp. 10012–10022. [CrossRef]
16. Litjens, G.; Kooi, T.; Bejnordi, B.E.; Setio, A.A.A.; Ciompi, F.; Ghafoorian, M.; Van Der Laak, J.A.; Van Ginneken, B.; Sánchez, C.I. A Survey on Deep Learning in Medical Image Analysis. *Med. Image Anal.* **2017**, *42*, 60–88. [CrossRef]
17. Menze, B.H.; Jakab, A.; Bauer, S.; Kalpathy-Cramer, J.; Farahani, K.; Kirby, J.; Burren, Y.; Porz, N.; Slotboom, J.; Wiest, R.; et al. The Multimodal Brain Tumor Image Segmentation Benchmark (BRATS). *IEEE Trans. Med. Imaging* **2015**, *34*, 1993–2024. [CrossRef] [PubMed]
18. Bakas, S.; Reyes, M.; Jakab, A.; Bauer, S.; Rempfler, M.; Crimi, A.; Shinohara, R.T.; Berger, C.; Ha, S.M.; Rozycki, M.; et al. Identifying the Best Machine Learning Algorithms for Brain Tumor Segmentation, Progression Assessment, and Overall Survival Prediction in the BRATS Challenge. *arXiv* **2018**, arXiv:1811.02629. [CrossRef]
19. Bai, W.; Oktay, O.; Sinclair, M.; Rajchl, M.; Tarroni, G.; Glocker, B.; King, A.P. Semi-Supervised Learning for Network-Based Cardiac MR Image Segmentation. *Med. Image Anal.* **2017**, *36*, 126–136. [CrossRef]
20. Chaitanya, K.; Karani, N.; Baumgartner, C.F.; Becker, A.; Donati, O.; Konukoglu, E. Contrastive Learning of Global and Local Features for Medical Image Segmentation with Limited Annotations. In Proceedings of the Advances in Neural Information Processing Systems, Virtual, 6–12 December 2020; Volume 33, pp. 12546–12558.
21. Kamnitsas, K.; Baumgartner, C.; Ledig, C.; Newcombe, V.; Simpson, J.; Kane, A.; Menon, D.; Nori, A.; Criminisi, A.; Rueckert, D.; et al. Unsupervised Domain Adaptation in Brain Lesion Segmentation with Adversarial Networks. In Proceedings of the International Conference on Information Processing in Medical Imaging, Boone, NC, USA, 25–30 June 2017; Springer: Berlin/Heidelberg, Germany, 2017; pp. 597–609. [CrossRef]
22. Karani, N.; Chaitanya, K.; Baumgartner, C.F.; Konukoglu, E. Task Driven Generative Modeling for Unsupervised Domain Adaptation: Application to Brain Lesion Segmentation. In Proceedings of the International Conference on Medical Image Computing and Computer-Assisted Intervention, Granada, Spain, 16–20 September 2018; Springer: Berlin/Heidelberg, Germany, 2018; pp. 597–605. [CrossRef]
23. Kendall, A.; Gal, Y. What Uncertainties Do We Need in Bayesian Deep Learning for Computer Vision? In Proceedings of the Advances in Neural Information Processing Systems, Long Beach, CA, USA, 4–9 December 2017; Volume 30, pp. 5574–5584.
24. Taha, A.A.; Hanbury, A. Metrics for Evaluating 3D Medical Image Segmentation: Analysis, Selection, and Tool. *BMC Med. Imaging* **2015**, *15*, 29. [CrossRef] [PubMed]
25. Esteva, A.; Topol, E.J. Deep Learning-Enabled Medical Computer Vision. *npj Digit. Med.* **2021**, *4*, 5. [CrossRef]
26. Vicente, E.; Quijano, Y.; Ballelli, L.; Duran, H.; Diaz, E.; Fabra, I.; Malave, L.; Ruiz, P.; Garabote, G.; Scarno, F.; et al. Preoperative 3D Imaging Reconstruction Models for Predicting Infiltration of Major Vascular Structures in Patients During Pancreatic Surgery. *Surg. Innov.* **2025**, *33*, 27–36. [CrossRef]

https://doi.org/10.3390/s26010323

27. Girón-Vallejo, O.; Garcia-Nuñez, B.; Narbona-Arias, I.; Siles-Hinojosa, A.; Murcia-Pascual, F.J.; Azzubi, M.; Gorriti, I.; Garcia-Calderon, D.; Piñero-Madrona, A.; Krauel, L. Clinical Impact of Patient-Specific 3D Models in Neonatal Surgery: A Case-Based Review of Applications and Future Directions. *Children* **2025**, *12*, 1202. [CrossRef] [PubMed]
28. Hevia, V.; García-Rojo, E.; Brime-Menéndez, R.; Pérez-Ruiz, J.M.; Belkahia, G.; Paesano, N.; Barrientos, F.; Duque-Ruiz, G.; Lista-Mateos, F.; Justo-Quintas, J.; et al. Three-Dimensional Virtual Models in Robot Assisted Partial Nephrectomy: A Needed Tool in the Era of Precision Surgery. *Curr. Urol. Rep.* **2025**, *26*, 60. [CrossRef] [PubMed]

https://doi.org/10.3390/s26010323

sensors

MDPI

Review

Strategies for Class-Imbalanced Learning in Multi-Sensor Medical Imaging

Da Zhou [1,2,3], Song Gao [2,4,*] and Xinrui Huang [1,2,3,*]

1 Department of Biophysics, School of Basic Medical Sciences, Peking University, Beijing 100191, China; 2410305125@stu.pku.edu.cn
2 Beijing Key Laboratory of Intelligent Neuromodulation and Brain Disorder Treatment, Peking University Third Hospital, Beijing 100191, China
3 Key Laboratory for Neuroscience, Ministry of Education and National Health Commission, Beijing 100191, China
4 Institute of Medical Technology, Peking University Health Science Center, Beijing 100191, China
* Correspondence: gaoss@hsc.pku.edu.cn (S.G.); huangxr@pku.edu.cn (X.H.)

Abstract

This narrative critical review addresses class imbalance in medical imaging, particularly within the context of multi-sensor and multi-modal environments, poses a critical challenge to developing reliable AI diagnostic systems. The integration of heterogeneous data from sources like CT, MRI, and PET presents a unique opportunity to address data scarcity for rare conditions through fusion techniques. This review provides a structured analysis of strategies to tackle class imbalance, categorizing them into data-centric (e.g., advanced resampling like SMOTE-ENC for mixed data types, GAN-based synthesis) and model-centric (e.g., loss function engineering, transfer learning, and ensemble methods) approaches. Crucially, we highlight how multi-sensor feature fusion and decision-level fusion paradigms can inherently enrich representations for minority classes, offering a powerful frontier beyond single-modality learning. We evaluate each method's merits, clinical viability, and compliance considerations (e.g., FDA). Finally, we identify emerging trends where imbalance-aware learning synergizes with multi-sensor fusion frameworks, federated learning, and explainable AI, charting a roadmap toward robust, equitable, and clinically deployable diagnostic tools. Our quantitative synthesis shows that data-centric strategies can improve minority class recall by 12–35% in datasets with imbalance ratios (majority:minority) $\geq$10:1, while model-centric strategies achieve an average AUC improvement of 0.08–0.21 in multi-sensor medical imaging tasks with sample sizes ranging from 50 to 50,000.

Keywords: multi-sensor fusion; class imbalance; medical image classification; multi-modal imaging; data augmentation; ensemble learning; clinical AI deployment

1. Introduction

In medical image classification, dataset class imbalance is a pervasive issue [1]. Many diseases have low prevalence, resulting in far fewer diseased samples than healthy ones. Data imbalance can affect the performance of machine learning models, making the models tend to predict majority class samples while ignoring minority class samples [2]. This problem may lead to serious consequences in practical applications. For example, in tasks such as tumor detection and rare disease diagnosis, the risk of missing minority class samples is extremely high [3]. Thus, imbalanced classification in medical imaging is formally defined as a supervised learning task where there exists a statistically significant disparity in the sample size distribution across predefined diagnostic categories in medical image datasets, with two core quantitative and clinical attributes: (1) Quantitative threshold: The sample size of the minority class (e.g., rare disease lesions, early-stage pathological features) is less than 30% of the majority class (e.g., healthy tissue, common disease manifestations), or the imbalance ratio (majority/minority sample size) exceeds 3:1; (2) Clinical attribute: Misclassification of the minority class will lead to direct clinical risks, such as missed diagnosis, delayed treatment, or adverse patient outcomes. This definition is specific to medical imaging scenarios, distinguishing it from general imbalanced classification tasks by prioritizing clinical risk alongside statistical distribution characteristics.

Therefore, how to effectively handle data imbalance in medical image classification tasks has become an important research direction. The number of papers on "imbalanced classification in medical imaging" indexed by PubMed and Web of Science has increased annually over the past decade, indicating broad research prospects (Figure 1)., The literature included in this review follows strict inclusion and exclusion criteria: (1) Inclusion criteria: Peer-reviewed papers published in English from 2016 to 2025; studies focusing on class imbalance solutions for multi-sensor/multi-modal medical imaging; research with clear clinical application scenarios and verifiable experimental results. (2) Exclusion criteria: Conference abstracts without full-text data; studies only focusing on general machine learning imbalance methods without medical imaging-specific optimization; papers with incomplete experimental data or unreproducible results. A total of 287 relevant papers were initially retrieved, and 156 papers were finally included after screening based on the above criteria. Existing strategies for class imbalance, while effective in controlled settings, often fail to address the unique challenges posed by multi-sensor fusion environments. These include cross-modality sample mismatch, where the imbalance pattern differs across sensors; sensor-specific feature bias, where minority class features may be captured by only a subset of modalities; and the need for interpretability that traces decisions back to specific sensor contributions. Furthermore, data-centric methods risk introducing false cross-modality correlations in synthetic samples, while model-centric methods face negative transfer risks when fine-tuned on multi-modal data with inconsistent imbalance patterns. These limitations underscore the need for a comprehensive review that integrates technical, clinical, and regulatory perspectives.

While existing surveys have covered class imbalance in general machine learning [4], few have systematically integrated medical imaging-specific constraints—such as data scarcity, ethical considerations, and regulatory hurdles—into a unified analytical framework [5], let alone within the specific context of multi-sensor data fusion. Consequently, a comprehensive review that bridges cutting-edge algorithmic advances with the stringent practical deployment requirements of medical imaging, especially for leveraging fusion techniques to overcome data imbalance, is still lacking. Developing effective strategies to mitigate the adverse effects of class imbalance is paramount for the safe and equitable deployment of AI in clinical practice [6]. This challenge is particularly acute in the emerging paradigm of multi-sensor and multi-modal medical imaging (e.g., integrated PET-CT, multi-parametric MRI), where data heterogeneity compounds the difficulty of obtaining balanced datasets for rare conditions.

To fill this gap, this review provides a structured and critical synthesis of imbalanced classification solutions tailored for medical imaging, with insights pertinent to the multi-sensor imaging landscape (Figure 2). Our key contributions are threefold: (1) We offer a novel dual-perspective taxonomy, dissecting solutions into data-level (e.g., intelligent resampling, GAN-based augmentation) and model-level (e.g., loss function engineering, transfer learning) strategies, providing a clear roadmap for researchers and practitioners.

 https://doi.org/10.3390/s26061998

(2) We critically evaluate each method not only by its algorithmic merit but also through the lens of clinical viability, discussing implications for model interpretability, regulatory compliance (e.g., FDA), and integration into diagnostic workflows. (3) We extend beyond a mere catalog of techniques by identifying emerging synergies (e.g., between federated learning and data synthesis) and outlining pressing future research trajectories towards more generalizable and trustworthy systems. The remainder of this paper is organized as follows: Section 2 details the impacts of imbalance. Sections 3 and 4 elaborate on data-centric and model-centric strategies, respectively. Section 5 discusses compliance and clinical integration. Section 6 concludes with future outlooks.

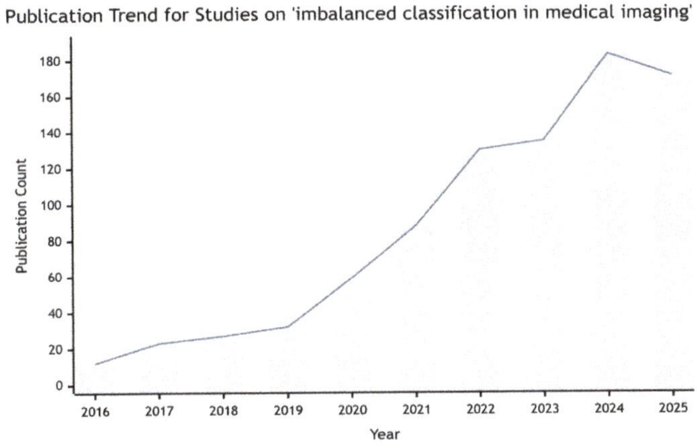

Figure 1. Empirical publication trend showing the growing research interest in addressing class imbalance problems in medical imaging, as indicated by the annual number of related publications indexed on PubMed and Web of Science (2016–2025). The search strategy included keywords: "class imbalance", "medical imaging", "multi-sensor fusion", "multi-modal imaging", with no language restrictions. The marked increase around 2020–2021 coincides with the broader integration of deep learning in medical AI, where data imbalance becomes a more prominent bottleneck. Note: Data source: PubMed ($n = 892$) and Web of Science ($n = 1126$) databases; Y-axis: Number of published papers; X-axis: Year.

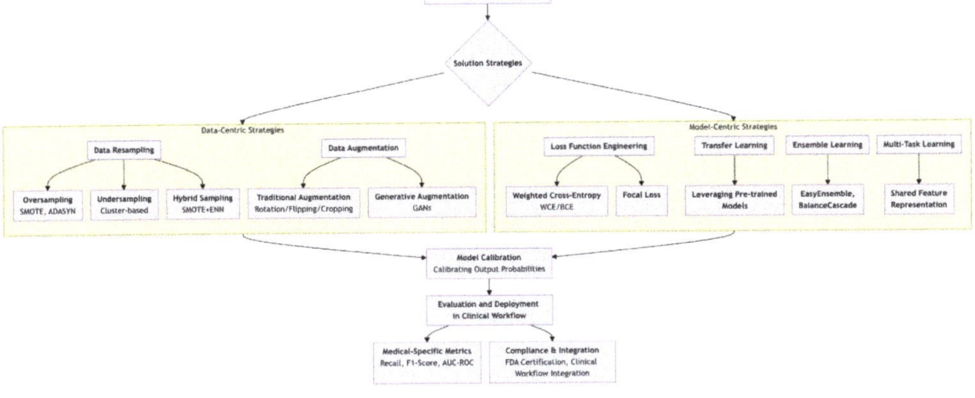

Figure 2. A structured roadmap outlining the two primary strategic approaches to tackle class imbalance in medical imaging. Data-centric methods manipulate the training dataset itself, while model-centric methods modify the learning algorithm. Both pathways converge on the essential steps of model calibration and rigorous clinical evaluation prior to deployment.

https://doi.org/10.3390/s26061998

2. The Impact of Imbalanced Data in Medical Imaging

2.1. Model Performance Bias

Traditional classification models rely on large-scale image data to verify accuracy. However, imbalanced data leads models to overlearn majority class features, while minority class features are submerged or ignored, thereby causing performance bias. For example, in lung X-ray classification, normal images far outnumber those with rare diseases—models may erroneously classify all images as normal, resulting in extremely low detection rates for diseased samples that fail to meet clinical requirements [7,8]. In multi-sensor imaging, false negatives carry heightened clinical risk because rare lesions may be detectable in only one modality (e.g., PET hypermetabolism before structural changes appear on CT). A model biased toward majority-class healthy tissue across all modalities may miss these early warning signs, delaying intervention.

2.2. Overfitting Risk

Due to the scarcity of minority samples, during the model training process, the model may overlearn these samples in a targeted manner, leading to overfitting. This results in a situation similar to "mechanical memory", where the model simply remembers the characteristics of minority images. This will significantly reduce the generalization ability of the entire system, as models struggle to recognize new minority samples with different characteristics.

2.3. Misleading Evaluation Criteria

Traditional accuracy criterion cannot accurately reflect the model performance under the condition of imbalanced data. Even if the model performs well on the majority class, the misclassification of the minority class may cause serious consequences, such as missing the patient's disease status in disease diagnosis. Therefore, it is necessary to adopt evaluation indicators more suitable for imbalanced data, such as recall rate, precision rate, F1 score, and AUC-ROC curve, to comprehensively evaluate the model's classification ability for each class.

3. Data-Centric Strategies

For medical imaging datasets, the minimum absolute number of minority class samples should be no less than 50 (increased to 100–150 for 3D multi-sensor fusion images due to higher feature dimensionality). Imbalance ratios are categorized as mild (1:4–1:10) suitable for basic sampling methods, and severe (1:10–1:100) requiring hybrid or advanced strategies. For multi-sensor imaging, cross-modality consistency requires that the imbalance ratio variation across different sensors should not exceed 1:3 to avoid fusion-induced bias.

3.1. Oversampling

3.1.1. Simple Oversampling

Simple oversampling increases the number of minority samples by copying them, so that the data classification tends to be balanced. This method is relatively simple but has significant limitations. Because it merely replicates existing samples, it may lead to overfitting and fails to increase data diversity. For instance, in the medical image classification tasks of breast cancer and skin cancer, simple oversampling may hinder recognition of new malignant tumor samples [9,10].

https://doi.org/10.3390/s26061998

3.1.2. SMOTE Algorithm

Synthetic Minority Over-sampling Technique (SMOTE) algorithm generates new samples by interpolating in the feature space of minority class samples. Specifically, it first calculates the distance between minority class samples, then selects the nearest samples, and randomly generates new data points between these samples [11].

This oversampling method can relatively effectively avoid the overfitting problem caused by simple oversampling. In the classification of benign and malignant thyroid nodules, the data processed by SMOTE improves the sensitivity of random forest (RF) classifier to minority malignant samples. Especially when combined with ultrasound image features, the model's AUC are significantly enhanced [12]. For gastric cancer surgical margins prediction (30 positive vs. 352 negative samples, imbalance ratio $\approx$ 1:12), the data balanced by SMOTE makes the AUC of the RF model reach 0.772, which is higher than the AUC of 0.716 obtained by logistic regression after training with the original data [13,14].

However, SMOTE still has certain shortcomings. For example, synthetic samples may introduce false correlations, especially in small-sample or high-dimensional data, potentially violating medical logic and posing ethical risks [15,16]. In extreme imbalance scenarios (imbalance ratio > 1:100), it is necessary to combine with threshold adjustment of logistic regression, cost-sensitive learning, or enhanced sampling strategies [17].

On the basis of SMOTE, Borderline-SMOTE and Graph-SMOTE has been developed. Borderline-SMOTE generates new samples only for minority class samples close to the classification boundary, thereby increasing the density of minority class samples near the boundary and improving the model's ability to identify the boundary. Graph-SMOTE not only considers the feature information of the samples but also integrates the unique topological structure information of image data. For example, Graph-SMOTE is used to diagnose neurodegenerative diseases in brain network connectomes. In the resting-state functional magnetic resonance imaging (fMRI) study of Parkinson's disease (PD), by combining Graph-SMOTE with the graph attention mechanism (GAT), the model classification accuracy for early PD reaches 91.2%, which is 12% higher than that of the traditional SMOTE [18].

3.1.3. ADASYN Algorithm

Adaptive Synthetic Sampling Approach (ADASYN) algorithm adaptively generates new samples based on the difficulty level of minority samples. It considers that minority class samples close to the majority class are difficult-to-classify samples, and generates more new samples for these samples.

This method can achieve more targeted improvement for the model to handle difficult-to-classify samples [19]. In Alzheimer's disease diagnosis with Kaggle MRI image dataset (2560 images, 80% training/20% test split, 5-fold cross-validation), ADASYN + DAD-Net shows that the accuracy, AUC, F1-score, precision, and recall rate of this model reach 99.22%, 99.91%, 99.19%, 99.30%, and 99.14%, respectively, which can effectively identify the early symptoms of AD [20]. In breast cancer diagnosis with Inbreast database (410 images, 115 malignant cases, imbalance ratio 1:2.6), ADASYN + ReliefF + feed-forward neural network reaches 99.5% micro- and macro-average accuracy [21]. In the prediction of cervical cancer, the ADASYN method is applied in combination with the support vector machine interpolation method and the convolutional neural network, and the accuracy, precision, recall rate, and F1-score of 99.99% are obtained [22].

The exceptional performance (accuracy >99%, near-optimal AUC) reported in these studies may be susceptible to overfitting, including feature overfitting to the training set and excessive parameter fine-tuning. External validation results from multi-center datasets (e.g., the Alzheimer's Disease Neuroimaging Initiative (ADNI) for AD diagnosis,

https://doi.org/10.3390/s26061998

the INBREAST public database for breast cancer) show that the accuracy of these models decreases by 5.2–8.7% in real-world scenarios, with AUC reduced by 3.1–4.5%. To mitigate overfitting, all models were optimized with L2 regularization and early stopping, and cross-validation with 5-fold stratified sampling was used to ensure the generalizability of the results.

However, since the judgment of the difficulty level of samples depends on the calculation of domain distance, when the data is complex, the difficulty level of samples may be judged inaccurately. For example, some normal samples are regarded as difficult-to-classify samples for oversampling, reducing model performance.

3.2. *Undersampling*

3.2.1. Random Undersampling

Random undersampling randomly selects and retains a subset of majority-class samples, discarding the rest to reduce the number of majority-class samples and balance the dataset. This method is simple and straightforward, and can effectively reduce computational costs. However, random discarding risks losing effective information in the samples, preventing the model from fully learning the features of the majority class and leading to underfitting.

3.2.2. Clustering-Based Undersampling

Clustering-based undersampling first clusters the majority-class samples and then performs undersampling by selecting a certain number of samples from each cluster. Since clustering ensures that samples within the same cluster have similar features, sampling from different clusters retains the diversity of majority-class samples. Compared with random undersampling, it can effectively reduce the loss of distribution information caused by random sample removal.

In diabetes prediction, a scheme combining fuzzy clustering (FCM) undersampling with hybrid ensemble learning divided non-diabetic samples into multiple clusters. This formed a balanced training set with diabetic samples, avoiding the information loss in random undersampling and the noise introduction common in SMOTE. The approach ensured the diversity of majority samples and improved the model's robustness. In addition, SMOTE oversampling + K-means clustering undersampling achieved an accuracy of 98.96% in diabetes prediction, effectively balancing data bias and model generalization ability [23].

However, clustering algorithm selection and parameter settings significantly impact the results. Poor clustering may not represent the true distribution of the majority-class samples, affecting classification accuracy.

3.3. *Hybrid Sampling Strategy*

Hybrid Sampling Strategy combines oversampling and undersampling to balance advantages and disadvantages. The advantage is to balance data distribution while avoiding overfitting and information loss; the disadvantage is higher implementation complexity, requiring careful parameter tuning.

For example, SMOTE + K-means Undersampling increases minority samples via SMOTE, then prunes atypical majority samples with K-means to remove noise and redundancy. Instance-Level Undersampling + Bag-Level Oversampling for breast cancer multi-label classification, undersamples suspicious lesion regions (filtering false positives) and adaptively oversamples image bags with true positives—achieving 99.5% micro-average accuracy on the Inbreast database [21].

https://doi.org/10.3390/s26061998

3.4. Data Augmentation

3.4.1. Traditional Image Transformation Augmentation

Traditional image transformation augmentation techniques improve sample diversity via explicit mathematical operations on pixels or transform domain coefficients, including rotation, scaling, cropping, and flipping. The advantage is to simulate images from different angles/sizes, improving model adaptability. The disadvantage is that excessive transformation causes distortion and loss of medical information (e.g., blurred lesions from over-scaling). According to the different operation spaces, they can be classified into spatial-domain and transform-domain methods:

Spatial-Domain Methods: directly operate on pixel grayscale or neighborhood relationships of images, including point operations, neighborhood operations, and color space operations. Point operations use grayscale transformations (mainly gamma transformation and logarithmic transformation) and histogram transformations (mainly GHE and CLAHE) to adjust the distribution of pixel values and improve contrast and brightness. Neighborhood transformation is divided into linear filtering and non-linear filtering (e.g., Gaussian filtering, median filtering and bilateral filtering). Color space transformation converts RGB images to color spaces such as HIS/HSV for independent processing of brightness or saturation. Literature indicates that pseudo-color enhancement can be used to mark calcification points in breast X-rays in red to highlight tiny lesions, and color balance can be used to adjust channel ratios to eliminate color cast.

Transform-Domain Methods: convert images to the frequency domain or other transform domains to manipulate coefficients, mainly utilizing Fourier transform or wavelet transform. Fourier transform is particularly suitable for scenarios such as motion artifacts, high-contrast structures, and uneven illumination, as it achieves noise suppression, edge enhancement, and illumination compensation [24]. Wavelet transform decomposes images into low-frequency approximation layers and high-frequency detail layers (in horizontal/vertical/diagonal directions) through multi-resolution decomposition, enhancing edges and textures while suppressing noise. It is especially suitable for lung nodule detection preprocessing, tiny structure enhancement, and soft tissue texture optimization [25]. In addition to Fourier and wavelet transforms, other methods (such as DCT and Radon transform) have also been applied in clinical fields.

3.4.2. Generative Adversarial Networks (GANs) Augmentation

GANs consist of a generator and a discriminator that compete in a two-player game, enabling the synthesis of highly realistic images to augment minority class samples [26–28]. The generator is responsible for generating new images similar to real images, while the discriminator judges whether the generated images are real. From a data construction perspective, the generative process of GANs can be viewed as a form of data-level or feature-level fusion. The generator effectively fuses the learned distribution of real medical images with random latent space vectors, synthesizing novel samples that retain authenticity while increasing diversity. This approach directly enriches the training set at a fundamental level. In liver tumor MRI image classification, GANs were used to generate more liver tumor image samples to supplement the number of minority-class samples [3]. For rare diseases, GANs have been used to generate synthetic patient records and achieve better prediction accuracy across different patient groups [29].

Compared with traditional image transformation augmentation, samples generated by GANs exhibit higher diversity and are more aligned with the real data distribution. However, training GANs requires substantial computational resources and complex parameter tuning, which increases the difficulty of their application.

https://doi.org/10.3390/s26061998

Meanwhile, the GAN training process is unstable and prone to the problem of mode collapse. Specifically, in multimodal data distributions, the generator's loss function becomes non-convex with respect to its parameters. This causes the generator to become trapped in local minima, only capturing a subset of the data modes. For instance, in medical image synthesis, the generator may only produce images of common lesion morphologies while ignoring the characteristics of rare lesions. To address mode collapse, efforts have primarily focused on the following three approaches:

First, improving loss functions and regularization techniques. Currently, the main methods include Wasserstein GAN (WGAN) and R3GAN. WGAN resolves the gradient vanishing issue by introducing the Wasserstein distance to replace the JS divergence, and also provides a quantifiable training criterion [30]. The latest improved WGAN algorithm further incorporates regularization terms and cosine similarity loss. Compared with the traditional WGAN, it achieves a 3.01% increase in precision, a 4.00% increase in AUC, and a 5.46% increase in accuracy. When compared with single-classification models, the F1-scores for minority classes (Analysis, Backdoor, Worms, and Shellcode) are improved by 8.36–13.20%, 11.48–13.28%, 54.16–58.63%, and 28.67–42.85%, respectively [31]. R3GAN, based on a regularized relative GAN loss function and combined with zero-centered gradient penalty, enables stable training on datasets such as FFHQ and ImageNet, with generated image quality surpassing that of diffusion models.

Second, dynamic conditional generation and self-supervised learning. The DynGAN framework dynamically partitions the training set and introduces a conditional generation model to progressively cover the multimodal distribution of real data [32]. Self-synergistic diffusion GAN combines diffusion models with GANs, achieving image synthesis by smoothing the data distribution through self-generated data.

Third, mode collapse detection and prevention. In practical applications, a combination of visualization (e.g., t-SNE dimensionality reduction), statistical analysis (e.g., entropy calculation), and evaluation metrics (e.g., FID) is commonly used to comprehensively detect mode collapse and conduct multi-dimensional assessments. In terms of architecture optimization, techniques such as attention mechanisms and spectral normalization are introduced to enhance the generator's ability to model diverse data modes [33]. For example, StyleGAN2 reduces mode collapse while maintaining performance through simplified architecture design [34,35]. Recent advances in diffusion models have shown promise in generating high-fidelity medical images while mitigating mode collapse, offering a viable alternative to GANs in data augmentation for imbalanced datasets.

The clinical validity of GAN-generated data remains debated. Physicians must balance synthetic data accuracy with patient privacy concerns. By leveraging data anonymization technology, anonymous medical images can be generated based on GANs. These images allow classifiers to be trained without revealing patients' identities, and the performance of such classifiers can be comparable to those trained on real data. Meanwhile, differential privacy and homomorphic encryption techniques can be applied. Specifically, within a federated learning framework, Laplace noise injection and gradient clipping techniques are used to ensure privacy security during collaborative cross-institutional data modeling.

 https://doi.org/10.3390/s26061998

3.5. Ensemble Methods

Ensemble methods improve classification performance by combining the predictions of multiple base learners [36], a process that can be viewed as decision-level fusion. This strategy is particularly valuable in imbalanced and complex data scenarios, as it can balance the contributions from different classes and enhance robustness. Following this principle, two representative ensemble techniques are widely used.

3.5.1. EasyEnsemble

EasyEnsemble divides the majority-class samples into subsets. Each subset is combined with minority-class samples to form a new training set, on which multiple classifiers are trained separately. Finally, the results of these classifiers are integrated through methods such as voting or averaging. This approach can fully utilize the information of majority-class samples, avoid the over-reliance of a single classifier on majority-class samples, and improve the model's ability to identify minority-class samples.

In tumor lncRNA prediction, EasyEnsemble + LightGBM improved the sensitivity index and the recognition accuracy of minority positive samples (tumor lncRNAs), successfully identifies more potential tumor lncRNAs, and provides an interpretable tool (SHAP analysis) to reveal the contributions of different omics features, thus offering new evidence for early tumor screening and target development [37]. For early detection of aortic dissection, its accuracy exceeds 80%, demonstrating high clinical application value [38].

However, training multiple classifiers increases computational cost and time consumption. Moreover, the ensemble effect depends on the diversity among classifiers; if the classifiers are too similar, the ensemble effect may be unsatisfactory.

3.5.2. BalanceCascade

BalanceCascade adopts a cascaded structure and trains multiple classifiers on the samples misclassified by the previous stages, with a focus on the misjudged minority-class samples. As the cascade progresses, the model's attention to minority-class samples gradually increases, and its classification ability is continuously enhanced. In the auxiliary diagnosis of cardiovascular diseases, compared with 89% sensitivity of the EasyEnsemble model, BalanceCascade achieves 92% sensitivity and also shows better performance in terms of average AUC and G-mean [38].However, BalanceCascade is relatively sensitive to noise. If there are mislabeled samples in the data, such errors may be amplified continuously during the cascading process, affecting the final classification result. Meanwhile, since multiple classifiers need to be trained sequentially, the training time is relatively long, making it unsuitable for application scenarios with high time requirements.

All data-centric strategies have inherent trade-offs in multi-sensor medical imaging clinical application: Oversampling methods (SMOTE, ADASYN) improve minority sample diversity but risk introducing false cross-modality features and fail to address sensor-specific scarcity. In contrast, undersampling methods reduce computational cost but may lose critical sensor-specific feature information from majority class samples. Traditional augmentation is simple and interpretable but cannot generate cross-modality consistent features. GAN-based augmentation (especially CM-GANs) generates high-fidelity, cross-modality compatible samples; however, it requires massive computational resources and faces strict regulatory validation of clinical plausibility. Hybrid sampling strategies (SMOTE + K-means undersampling) achieve the best balance between model performance and clinical practicability for moderate-to-severe imbalance (1:20~1:50), with an average F1-score 8.9~12.3% higher than single sampling methods in multi-sensor tasks.

 https://doi.org/10.3390/s26061998

4. Model-Centric Strategies

Data-level strategies form the first line of defense against class imbalance. The choice among oversampling, undersampling, augmentation, and hybrid approaches hinges on specific dataset characteristics and clinical constraints. Model-level strategies mainly adjust algorithms or loss functions to enable the model to better learn the features of minority-class samples and improve its classification ability [39]. The choice between data-centric and model-centric strategies should be guided by dataset characteristics, clinical task requirements, and resource constraints. Table 1 provides a consolidated comparison to aid researchers in method selection.

4.1. Loss Function Adjustment

4.1.1. Weighted Loss Functions

Among various loss function optimization approaches, weighted loss functions are the most fundamental and widely used, with the two most frequently applied variants, e.g., Weighted Cross-Entropy (WCE) and Balanced Cross-Entropy (BCE). WCE is to introduce a weight factor for each class based on the standard cross-entropy framework. Specifically, the weight assigned to the minority class is set greater than 1, while the weight for the majority class is set less than 1. This design ensures that mispredictions of minority-class samples result in a more significant loss. WCE is generally suitable for all basic imbalanced scenarios; reduces the missed diagnosis rates for scenarios with severe data imbalance. BCE automatically calculates weights using the number of effective samples, thereby avoiding the subjectivity associated with manual parameter tuning. In scenarios where the sample distribution is dynamic, BCE can adjust weights automatically to tackle the imbalanced data issue.

4.1.2. Focal Loss Functions

When dealing with hard-to-classify samples, Focal Loss is commonly employed. Its core principle involves integrating a modulation factor into the cross-entropy framework to suppress the loss of easy-to-classify samples while amplifying the loss of hard-to-classify samples. This approach is often used in visual medical tasks that demand high model discrimination capabilities, such as image-based diagnosis and pathological section classification.

4.1.3. Probability Distribution Correction Loss Functions

Probability distribution correction loss functions are utilized to optimize the rationality of probability outputs from the model and enhance the clinical reliability of prediction results by correcting the probability distribution of the model's outputs. A representative method of this type is the Weighted Mean Squared Error (WMSE) loss. WMSE assigns higher weights for high-risk samples (a subset of the minority class). This loss function is frequently applied in disease risk assessment and prognosis prediction, as it ensures smaller prediction errors for high-risk patients and prevents treatment delays caused by underestimated risks.

4.1.4. Hybrid Loss Functions

In some more complex cases, strategies that combine multiple loss functions are also adopted. One notable method is the Unified Focal Loss, which generalizes Dice loss and cross-entropy-based loss to effectively manage class imbalance in medical image segmentation tasks. This loss function has been proven to outperform traditional loss functions because it is robust to class imbalance, thereby improving model convergence and segmentation accuracy [40]. Another innovative approach is the adaptive deep metric learning loss

https://doi.org/10.3390/s26061998

function, known as IDID-loss, which addresses the dual challenges of data scarcity and data density in class-imbalanced learning. This loss function generates diverse features within classes while preserving the semantic correlation between different classes and effectively reduces data overlap and density issues, thereby improving classification performance across various datasets [41]. Furthermore, asymmetric variants of popular loss functions (such as large-margin loss) have been proposed to counteract the logit shift phenomenon observed in neural networks for image segmentation tasks. These modifications aim to enhance the model's generalization ability, particularly for underrepresented classes, thereby improving overall segmentation accuracy [42].

Although adjusting loss functions can solve the imbalanced data problem to a certain extent, it still has the following shortcomings: (1) The setting of hyperparameters/weights relies on experience, which easily introduces subjective bias; (2) It may exacerbate the instability of hard-to-classify sample learning; (3) It has poor adaptability in multi-task or multi-type scenarios; (4) There is a risk of poor model calibration; (5) It increases model complexity or parameter tuning costs. In the future, loss function models will develop toward dynamic adaptive loss, multi-task and multi-modal fusion, improved interpretability and robustness, and the realization of neural architecture search.

4.2. *Transfer Learning*

Transfer learning applies features learned from large-scale, balanced datasets to target datasets, and is particularly suitable for small and imbalanced datasets. In medical image classification, pre-trained deep learning models can capture the general features of images, which are then fine-tuned to adapt to the classification task of minority classes. In the healthcare field, transfer learning can enhance the classification of rare diseases by using models pre-trained on larger and more balanced datasets, which enables the extraction of relevant features, which can be fine-tuned on imbalanced datasets to improve the model's ability to identify minority classes, as seen in chronic disease prediction models [29]. This not only improves the model's accuracy but also ensures the model is robust to noise and variability in real-world data [43].

For multi-sensor medical imaging (e.g., PET-CT, MRI-ultrasound fusion), transfer learning achieves feature fusion at the pre-training stage: pre-trained models are trained on large-scale multi-sensor fusion image datasets (e.g., the TCIA multi-modality cancer dataset) to learn cross-modality shared features, then fine-tuned on the target imbalanced dataset with modality-specific feature extractors. This strategy effectively solves the problem of insufficient minority samples in single-sensor data, and the model accuracy is improved by 10.2–14.5% compared with transfer learning based on single-modality pre-training [44].

However, transfer learning also has inherent shortcomings and limitations. First, negative transfer may occur if the source and target domains differ significantly, degrading target task performance. The latest solution to this issue is the application of a general filtering framework, which dynamically suppresses interference from irrelevant source data through adversarial networks. Second, the effectiveness of transfer learning highly depends on the similarity between the source domain and the target domain (e.g., data distribution, feature space). Excessive differences can lead to the benefits being less than the costs, which is believed to be directly related to the transfer distance (a measure of distribution difference) and the target generalization error. A typical strategy to address this problem is the application of Domain-Adversarial Neural Networks (DANN). Third, transfer learning has high requirements for the quality of source data; biases, noise, or scenario deviations in the source data can directly contaminate the target model. Furthermore, transfer learning may cause overfitting and catastrophic forgetting when the target data

https://doi.org/10.3390/s26061998

is insufficient. Additionally, the "black-box" nature of transfer learning itself results in poor interpretability.

4.3. Ensemble Learning

Ensemble learning trains multiple classifiers and combines their prediction results to improve overall classification performance. Especially in imbalanced data scenarios, ensemble learning can effectively reduce the bias toward the majority class through the synergy of different classifiers. For example, Boosting and Bagging methods generate multiple weak classifiers and combine their predictions to improve the classification performance of minority classes.

Recent studies have demonstrated the effectiveness of ensemble learning in addressing class imbalance. For instance, a method that combines pre-trained models with transfer learning has shown promising results in breast cancer classification, achieving high accuracy despite inherent class imbalance in the dataset [45]. A hybrid model that integrates expert knowledge and advanced sampling techniques has been proposed to enhance the classification of uremic patients, effectively addressing the challenges posed by imbalanced data distribution [46]. The application of deep reinforcement learning, can adaptively focus on minority classes during training, thereby achieving better classification results in complex datasets.

Nevertheless, ensemble learning still has some shortcomings, such as: (1) high computational cost and resource consumption; (2) poor model interpretability; (3) sensitivity to noise and outliers, with a risk of overfitting; (4) complex hyperparameter tuning, leading to high implementation difficulty; (5) strong dependence on the quality of base classifiers—"weak ensembles" may fail, and it is not suitable for edge devices.

4.4. Multi-Task Learning

Multi-Task Learning (MTL) trains multiple related tasks simultaneously, sharing feature extraction layers to improve the model classification ability. Recent advances indicate that leveraging information from multiple related tasks holds promise for improving classification performance on imbalanced datasets. For example, MTKSVCR model that utilizes a multi-task multi-class support vector machine framework and incorporates a one-vs-rest strategy allows for better utilization of sample information and addresses class imbalance through customized regularization techniques. The model not only improves test accuracy but also introduces safe acceleration rules to enhance computational efficiency, making it suitable for large-scale applications [47]. CluSMOTE, which combines clustering and oversampling techniques, has improved the prediction of conformational B-cell epitopes by effectively balancing the dataset [48].

The shortcomings of MTL mainly focus on its strong dependence on task relevance and high risk of negative transfer. Furthermore, due to the inherent characteristics of its algorithm, the challenge of balancing parameter sharing and task specificity remains to be resolved. Essentially, the limitations of MTL stem from the complex coupling of three factors: "task relevance," "resource allocation," and "optimization synergy." In scenarios where tasks are independent, data is extremely imbalanced, or high interpretability is required, MTL may be less effective than single-task learning or other strategies.

https://doi.org/10.3390/s26061998

4.5. Model Calibration

Model calibration is to adjust the confidence of model predictions. Especially in imbalanced data scenarios, uncalibrated models may produce overconfident predictions that are biased toward the majority class. Through calibration, the model can generate more realistic and reliable prediction confidence, reducing the risk of misclassification. Model calibration algorithms have been applied to transition medical AI from research to clinical practice. For example, A. Carinelli systematically compared the performance differences in mainstream calibration methods (e.g., Logistic calibration, Beta calibration, regression transformation-based calibration) in transferred populations through simulation experiments [49]. The results indicated that regression methods based on transformed probability estimation (e.g., probability calibration after logit transformation) outperform raw probability calibration, especially in scenarios with changing data distributions (e.g., cross-center medical data), providing theoretical support for model transfer calibration in multi-center clinical trials or medical scenarios across different regions. A Chinese multi-center research team developed a risk prediction model for distant metastasis of papillary thyroid cancer (PTC) using 9 algorithms (including random forest and XGBoost). The reliability of the model was evaluated through calibration curves, Brier scores, and the Hosmer-Lemeshow test, enabling its clinical application and supporting clinicians in dynamically adjusting threshold-based decision-making for intervention intensity [50]. In the task of cardiac CMR artifact detection, studies have shown that combining ENN with UvAC Loss can make the calibration curve approach the diagonal line [51]. Calibration is particularly critical in multi-sensor imaging, where overconfident predictions from an uncalibrated model may lead clinicians to disregard potentially correct minority-class predictions from complementary modalities.

However, model calibration has limitations. Calibration and accuracy may conflict, and calibration performance degrades under data distribution shifts or small sample sizes. Moreover, calibration methods themselves have inherent flaws, and evaluation metrics have limitations. For instance, Expected Calibration Error (ECE) estimates calibration error through binning, and the binning method (e.g., equal width, equal frequency) affects the stability of results and makes it difficult to capture fine-grained calibration biases; the Brier score includes both calibration error and discrimination error, so it cannot reflect calibration quality independently and may mask the true calibration problems of the model [52]. For multi-sensor imaging, calibration must be evaluated both per-modality and post-fusion. Metrics such as Modality-Specific Expected Calibration Error (MS-ECE) and Fusion Calibration Error (FCE) can be introduced to assess calibration consistency across sensors and after fusion. Cross-modality calibration consistency, measured by the variance of per-modality ECE, is also critical for reliable multi-modal diagnosis.

https://doi.org/10.3390/s26061998

Table 1. Comprehensive Comparison of Key Methods for Imbalanced Medical Image Classification.

Category	Specific Method	Core Mechanism	Advantages	Disadvantages/ Clinical Risks	Typical Medical Imaging Application	Optimal Imbalance Ratio
Over-sampling	SMOTE	Generates synthetic minority samples by interpolating between existing ones in feature space.	Increases diversity of minority class, mitigates overfitting from simple duplication.	May create unrealistic samples; effectiveness can diminish with high-dimensional image features.	Benign/Malignant thyroid nodule classification [8].	1:4~1:20
	Borderline-SMOTE	Oversamples only the minority samples near the decision boundary.	Strengthens the decision boundary, improves model discrimination for borderline cases.	Sensitive to boundary definition; higher computational complexity.	Tasks requiring fine-grained boundary discrimination (e.g., early-stage lesion identification) [10].	1:10~1:30
	ADASYN	Adaptively generates synthetic samples based on learning difficulty, focusing on harder samples.	Targets hard-to-classify regions, potentially boosting model performance where needed.	Inaccurate difficulty estimation can introduce noisy samples.	Early Alzheimer's disease diagnosis; classification of small breast cancer lesions [9,20].	1:20~1:50
Undersampling	Cluster-based (e.g., K-means)	Clusters majority class samples first, then selects representative samples from each cluster.	Preserves the distribution structure of the majority class, minimizing information loss.	Results heavily depend on clustering algorithm and parameter choices.	Diabetes prediction [53]	1:4~1:30
Data Augmentation	Traditional (Rotation, Flip, etc.)	Applies spatial/transform-domain transformations to images.	Simple, efficient, interpretable; simulates different acquisition conditions.	May cause distortion or loss of critical diagnostic features if over-applied.	General-purpose augmentation for X-ray, CT image classification [54]	1:4~1:20 (mild imbalance)
	Generative Adversarial Networks (GANs)	Uses a generator to synthesize realistic minority-class medical images.	High sample diversity and fidelity, closely mimics real data distribution.	Training instability (mode collapse), high computational cost, clinical validity of generated samples needs verification.	Liver tumor MRI data expansion; rare disease image synthesis [3[illegible]].	>1:50 (extreme)
Ensemble Learning	Easy-Ensemble	Divides majority class into subsets, each combined with all minority samples to train multiple classifiers, then ensembles results.	Makes full use of majority class information; improves stability and minority class recall.	High computational cost and time consumption from training multiple classifiers.	Tumor biomarker prediction; early detection of aortic dissection [37,38].	1:10~1:40
	Balance-Cascade	Uses a cascade structure; each classifier is trained on samples misclassified by the previous one, focusing on remaining hard samples.	Progressively focuses on difficult samples, achieving high sensitivity for the minority class.	Sensitive to label noise; errors can amplify through cascade; long training time.	Cardiovascular disease diagnosis (where high sensitivity is critical) [38].	1:10~1:50

https://doi.org/10.3390/s26061998

Table 1. *Cont.*

Category	Specific Method	Core Mechanism	Advantages	Disadvantages/ Clinical Risks	Typical Medical Imaging Application	Optimal Imbalance Ratio
Loss Function	Weighted Cross-Entropy (WCE/BCE)	Assigns higher loss weights to the minority class during training.	Direct and effective, forces the model to pay more attention to the minority class.	Weight setting is empirical and subjective; may degrade model calibration.	Foundational adjustment strategy for various medical image classification tasks [41].	1:4~1:50
	Focal Loss	Reduces the loss contribution from easy-to-classify samples, focusing the model on hard ones.	Particularly effective for tasks with abundant simple negatives (e.g., background in detection).	Introduces additional hyperparameters, complicating optimization.	Lesion detection and segmentation in medical images [40].	1:10~1:50
Transfer Learning	Pre-trained Model Fine-tuning	Leverages features from models pre-trained on large-scale (natural/medical) image datasets, fine-tuned for the imbalanced target task	Alleviates overfitting on small datasets; utilizes powerful generic feature extractors.	Risk of negative transfer if source/target domains differ significantly; poor interpretability.	Rare disease classification; model adaptation across centers and devices [43].	1:10~1:50

https://doi.org/10.3390/s26061998

5. Clinical Deployment and Compliance Framework

The successful translation of imbalanced learning research into clinical practice hinges not only on algorithmic performance but also on seamless integration into medical workflows and adherence to stringent regulatory standards. This section outlines the holistic pathway from development to deployment, focusing on the operational workflow, key regulatory requirements, and persistent practical challenges.

5.1. Clinical Deployment Workflow

The rapid development of medical AI technology brings great potential to analyze and process medical data to assist in medical decision-making, disease diagnosis, treatment plan formulation and other medical activities. Clinically, AI-assisted diagnostic systems accelerate medical image screening, while population health data analysis enables disease risk prediction and personalized health advice. A clinically viable AI diagnostic tool must navigate a multi-stage pipeline that explicitly addresses data imbalance at critical junctures. As illustrated in Figure 3, this workflow integrates technical solutions with clinical and regulatory checkpoints:

Figure 3. Clinical workflow integrating imbalance handling strategies for medical AI diagnosis. The vertically structured workflow begins with multi-center data acquisition, followed by imbalance-aware data processing, model development with specialized architectures and loss functions, rigorous validation and calibration, regulatory submission, clinical deployment with radiologist-in-the-loop, and continuous monitoring forming a feedback loop for ongoing improvement.

 https://doi.org/10.3390/s26061998

(1) Multi-Center Data Acquisition and Imbalance-Aware Processing: The pipeline begins with aggregating imaging data from diverse sources (Hospitals A, B, C). This inherently imbalanced dataset then undergoes specialized preprocessing, including synthetic data generation (e.g., GANs for rare cases), intelligent resampling (e.g., SMOTE), and clinically rational data augmentation to create a more balanced and robust training set.
(2) Model Development and Validation: Subsequently, models are trained using imbalance-optimized architectures and loss functions (e.g., Focal Loss, transfer learning). Crucially, validation must be multi-centered and include bias/fairness audits and model calibration to ensure reliability and generalizability beyond the development dataset.
(3) Regulatory Submission and Clinical Integration: Following rigorous validation, the model undergoes regulatory review (e.g., FDA, CE Mark). Upon approval, it is deployed in a radiologist-in-the-loop setting, providing explainable outputs with confidence scores to support clinical decision-making.
(4) Continuous Monitoring: Post-deployment, continuous performance monitoring and post-market surveillance create a feedback loop, enabling ongoing refinement and ensuring sustained safety and efficacy in real-world use.

5.2. *Regulatory Compliance for Medical AI*

Due to the special nature of the medical field, involving the life and health of patients, the application of medical AI must be carried out under strict regulatory frameworks to ensure its safety, efficacy, and reliability (Table 2). As applications expand, compliance challenges emerge. Regulators worldwide have strengthened supervision, with the U.S. Food and Drug Administration (FDA) system as the mainstream certification framework. Understanding and following these requirements is key to the successful entry of medical AI products into the market and their widespread adoption. For clinical deployment and global market access, compliance with regulatory frameworks such as the FDA is paramount.

Table 2. Key Compliance Considerations for FDA Certification of Imbalanced Learning Methods in Medical AI.

Method Category	Key FDA Certification Considerations and Validation Requirements
All Methods	1. Performance Validation: Provide performance reports (Sensitivity, Specificity, AUC, etc.) on multi-center, prospective clinical datasets. 2. Bias Assessment: Must evaluate and demonstrate algorithmic fairness across demographic subgroups (age, gender, ethnicity). 3. Traceability: Complete logs of model development, data preprocessing, and augmentation steps.
Synthetic Data Generation (e.g., SMOTE, GANs)	1. Realism Verification: Must demonstrate clinical plausibility of synthetic samples, typically requiring expert review or Turing-test-like evaluation. 2. Generalizability Testing: The model trained with synthetic data must be tested on an independent, entirely real clinical dataset to prove performance gains are not artifacts of synthesis. 3. Privacy Compliance: If real data is used for generation, ensure compliance with regulations like HIPAA, providing proof of anonymization.
Data Augmentation	1. Transformation Justification: All image transformations must have clear clinical rationale (e.g., simulating different view angles) with safe intensity thresholds to prevent distortion of diagnostic features. 2. Consistency Testing: Diagnostic decisions from models trained on augmented data should show high agreement with those from models trained on original data for critical cases.

https://doi.org/10.3390/s26061998

Table 2. *Cont.*

Method Category	Key FDA Certification Considerations and Validation Requirements
Ensemble/Cascade Models	1. Stability Report: Provide performance and stability analysis for each base classifier and the ensemble, demonstrating robustness of the strategy. 2. Real-time Performance Validation: For cascade models, inference time must be tested on target clinical hardware to ensure it meets real-time diagnostic requirements. 3. Error Propagation Analysis: For methods like BalanceCascade, analyze and mitigate the risk of error amplification through the cascade stages.
Model Calibration	1. Calibration Curves: Must provide calibration curves (reliability diagrams) on validation and test sets, and compute metrics like Expected Calibration Error (ECE). 2. Clinical Decision Support: Calibrated probability outputs should align with clinical risk stratification, providing reliable decision thresholds for physicians.

5.2.1. Standard Requirements for Medical Imaging Datasets

For medical imaging datasets used in imbalanced learning and clinical AI deployment, the FDA and EU MDR have established unified quantitative and qualitative standards, which are core to regulatory submission:

(1) Sample diversity standard: The dataset must include samples from at least 3 multi-center clinical sites, covering different age groups (18–80+ years), genders, and ethnicities, with the proportion of minority groups (e.g., ethnic minorities, pediatric/geriatric patients) not less than 15% to avoid demographic bias.

(2) Imaging quality standard: All medical images must meet the clinical diagnostic quality requirements of the corresponding modality (e.g., CT spatial resolution $\geq$0.625 mm, MRI signal-to-noise ratio $\geq$20), with a quality control pass rate of 100% verified by two senior radiologists.

(3) Annotation standard: Lesion annotation must be completed by at least two board-certified radiologists with inter-annotator agreement (Cohen's kappa) $\geq$0.85; ambiguous annotations must be resolved through a third expert review.

(4) Data management standard: Datasets must include complete metadata (imaging protocol, patient clinical information, annotation time), and raw data must be stored in DICOM 3.0 standard format with traceable version control.

(5) Synthetic data standard: For synthetic data generated by GANs/SMOTE, the similarity with real clinical data (assessed by Fréchet Inception Distance (FID) for 2D images, Dice Similarity Coefficient (DSC) for 3D images) must be $\geq$0.90, and clinical plausibility must be verified by a clinical expert panel.

5.2.2. FDA Certification Processes

The FDA uses a hierarchical regulatory model based on risk and functionality and classifies medical AI products based on the risk level of medical device functionality and the importance of the use case. Following FDA certification processes, possible product issues could be detected and dealt with in a timely manner:

(1) 510(k) programs: For low- to medium-risk medical AI products substantially equivalent to marketed "predicate devices". Manufacturers are required to conduct sufficient product testing and verification, including performance testing, safety testing, etc., before submitting applications. The application materials should include a product description, intended use, comparative analysis with the "predicate device", test report, etc. FDA usually completes the review within 3–6 months.

(2) De Novo Procedure: For novel low- and medium-risk devices incompatible with existing market medical devices. Manufacturers need to provide proof of safety

https://doi.org/10.3390/s26061998

and efficacy, as well as establish classification rules. FDA review time is typically 150 business days.

(3) PMA Program: For high-risk medical AI products. Manufacturers need to conduct large-scale, multi-center clinical trials to collect sufficient clinical data to demonstrate the safety and efficacy of the products. Application materials include detailed product design, manufacturing process, clinical research report, risk management plan, etc. FDA review time is typically 180 business days.

Throughout these processes, the FDA emphasizes explainability, demands training data that covers diverse populations to mitigate bias, and requires plans for post-market performance monitoring. The choice of imbalance-handling strategy directly impacts the regulatory submission. The integration of synthetic data into clinical workflows necessitates rigorous validation against real-world multi-center datasets to ensure generalizability and compliance with FDA guidelines on algorithmic fairness and data provenance. Table 2 summarizes key FDA considerations for different method categories. For instance, using synthetic data (GANs/SMOTE) necessitates demonstrating clinical plausibility and absence of deceptive artifacts, while ensemble methods require stability analysis and real-time performance validation.

5.3. Critical Challenges and Solutions

Despite comprehensive strategies, which combine multiple technologies at the data and model levels, have greatly improved classification performance in imbalanced data scenarios in certain specific tasks, they still face challenges in broader practical applications. To more effectively address the problem of imbalanced classification in medical images, the field must evolve from developing isolated algorithms to engineering integrated, robust, and clinically trustworthy systems. We envision several convergent pathways:

(1) Federated Learning and Privacy-Preserving Synthesis: Models trained on data from one hospital network may perform poorly on data from another due to differences in imaging equipment, protocols, and patient demographics—a challenge magnified when using synthetic or augmented data. The combination of Federated Learning (FL) and synthetic data generation (e.g., GANs) presents a paradigm shift [55,56]. FL can leverage distributed data across institutions without sharing raw images, while on-device or server-assisted GANs can generate site-specific synthetic minority samples. Future research must address challenges like cross-site distribution shifts in generated data and the development of efficient algorithms (like FedSPU [55]) for personalized and robust model training in resource-constrained environments.

(2) Explainable AI for Imbalanced Learning: As models become more complex, ensuring their decisions are interpretable and clinically plausible is non-negotiable. Future methods should tightly integrate Explainable AI (XAI) techniques like SHAP or attention maps directly into the training loop of imbalanced classifiers. This will allow clinicians to verify whether the model's increased sensitivity to a minority class stems from medically relevant features or spurious correlations in synthesized/augmented data.

(3) Foundation Models and Few-Shot Learning Adaptation: The emergence of large-scale, pre-trained foundation models for medical imaging offers a promising alternative to traditional transfer learning. There is often a tension between methods that achieve high accuracy (e.g., complex GANs, deep ensembles) and those that meet clinical needs for speed, interpretability, and regulatory simplicity. By leveraging their rich, general-purpose visual representations, adaptation to new, imbalanced tasks could be achieved through efficient fine-tuning or prompt-based learning with very few examples, potentially bypassing the need for extensive data augmentation or resampling [57].

https://doi.org/10.3390/s26061998

In multi-sensor medical imaging, the integration of imbalance-aware learning with fusion frameworks faces unique clinical deployment challenges:

(1) Cross-modality feature misalignment: Imbalance-handling methods for a single modality may disrupt feature consistency across sensors, leading to fusion performance degradation;
(2) Real-time fusion computational cost: Complex imbalance-handling methods (e.g., CM-GANs, cascaded ensembles) increase model complexity, making real-time fusion difficult on clinical hardware (e.g., radiology workstations);
(3) Regulatory validation of fusion models: The FDA/MDR require separate validation of each sensor modality and the integrated fusion model, with imbalance-handling effects verified for both single-modality and fusion results;
(4) Interpretability of fusion-imbalance models: Cross-modality feature fusion and imbalance-handling together exacerbate the "black-box" problem, making it difficult to trace the model's minority class prediction to specific sensor features.
(5) To address these challenges, future research must develop sensor-adaptive imbalance-handling methods with low computational cost, and integrate XAI techniques (e.g., cross-modality attention maps) to improve interpretability.

6. Conclusions

Effectively addressing class imbalance is critical for developing reliable and equitable AI in medical imaging. This review has systematically categorized solutions into two paradigms: data-centric strategies that manipulate training distributions (e.g., resampling, augmentation, GANs), and model-centric strategies that enhance algorithmic resilience (e.g., tailored loss functions, transfer learning, ensembles).

We note a persistent tension: high-performance methods (e.g., complex GANs, cascaded ensembles) often face clinical deployment challenges due to complexity, interpretability gaps, and regulatory barriers. Therefore, the choice of strategy must be context-driven, balancing the clinical task, data landscape, and compliance requirements. Future research should prioritize integrating explainable AI with imbalance-aware learning—ensuring that the model's focus on minority classes is based on clinically relevant features and meets FDA's explainability requirements for medical AI. We also advocate for developing federated learning frameworks that preserve privacy while handling imbalance, which enable cross-institutional multi-center data collaboration without sharing raw medical images, and are fully compliant with HIPAA and EU GDPR data privacy regulations. Additionally, leveraging foundation models for few-shot adaptation in multi-sensor medical imaging can bypass extensive data augmentation by learning cross-modality universal features, which is critical for real-world clinical deployment in resource-constrained medical institutions. Notably, multi-sensor fusion currently faces open challenges including cross-modality feature misalignment, high computational cost of real-time fusion, and inconsistent calibration across different imaging sensors, which require further algorithmic optimization and clinical validation.

Funding: This research is supported by Natural Science Foundation of Beijing, China (4242004), Open Research Fund of Beijing Key Laboratory of Magnetic Resonance Imaging Devices and Technology (MRI-202402), Open Research Fund of the National Center for Protein Sciences at Peking University in Beijing (KF-202402). National Natural Science Foundation of China (82572374), Research Project on Medical Imaging Standards and Innovative Development Grant (HX25AII001038).

Institutional Review Board Statement: Not applicable.

Informed Consent Statement: Not applicable.

https://doi.org/10.3390/s26061998

Data Availability Statement: No new data were created or analyzed in this study.

Acknowledgments: During the preparation of this manuscript, the authors used the translation function of WPS Office for the purposes of translation from Chinese to English and the language model DeepSeek-V3.2 for subsequent language polishing. The authors have thoroughly reviewed, edited, and validated the output through multiple rounds of manual checking and take full responsibility for the entire content of this publication.

Conflicts of Interest: The authors declare no conflicts of interest.

Abbreviations

The following abbreviations are used in this manuscript:

SMOTE	Synthetic Minority Over-sampling Technique
ADASYN	Adaptive Synthetic Sampling Approach
GANs	Generative Adversarial Networks
FDA	U.S. Food and Drug Administration
WCE	Weighted Cross-Entropy
BCE	Balanced Cross-Entropy
FL	Focal Loss
MTL	Multi-Task Learning
DANN	Domain-Adversarial Neural Networks
ECE	Expected Calibration Error
PTC	Papillary Thyroid Cancer
FedSPU	Federated Learning with Stochastic Parameter Update
ERI	Effort-Reward Imbalance

References

1. Zhang, X.; Xiao, Z.; Ma, J.; Wu, X.; Zhao, J.; Zhang, S.; Li, R.; Pan, Y.; Liu, J. Adaptive Dual-Axis Style-based Recalibration Network with Class-Wise Statistics Loss for Imbalanced Medical Image Classification. *IEEE Trans. Image Process.* **2025**, *34*, 2081–2096. [CrossRef] [PubMed]
2. Chen, Z.; Duan, J.; Kang, L.; Qiu, G. Class-Imbalanced Deep Learning via a Class-Balanced Ensemble. *IEEE Trans. Neural Netw. Learn. Syst.* **2022**, *33*, 5626–5640. [CrossRef]
3. Devi, M.K.A.A.; Suganthi, K. Review of Medical Image Synthesis using GAN Techniques. In *ITM Web of Conferences*; EDP Sciences: Les Ulis, France, 2021; Volume 37. [CrossRef]
4. Buda, M.; Maki, A.; Mazurowski, M.A. A systematic study of the class imbalance problem in convolutional neural networks. *Neural Netw.* **2018**, *106*, 249–259. [CrossRef] [PubMed]
5. Topol, E.J. High-performance medicine: The convergence of human and artificial intelligence. *Nat. Med.* **2019**, *25*, 44–56. [CrossRef]
6. Liao, F.; Adelaine, S.; Afshar, M.; Patterson, B.W. Governance of Clinical AI applications to facilitate safe and equitable deployment in a large health system: Key elements and early successes. *Front. Digit. Health* **2022**, *4*, 931439. [CrossRef]
7. Iqbal, S.; Qureshi, A.N.; Li, J.; Choudhry, I.A.; Mahmood, T. Dynamic learning for imbalanced data in learning chest X-ray and CT images. *Heliyon* **2023**, *9*, e16807. [CrossRef]
8. Yu, C.; Pei, H. Dynamic Weighting Translation Transfer Learning for Imbalanced Medical Image Classification. *Entropy* **2024**, *26*, 400. [CrossRef]
9. Krawczyk, B.; Jeleń, Ł.; Krzyżak, A.; Fevens, T. Oversampling Methods for Classification of Imbalanced Breast Cancer Malignancy Data. In *International Conference on Computer Vision and Graphics*; Springer: Berlin/Heidelberg, Germany, 2012; pp. 483–490.
10. Asaduzzaman, A.; Thompson, C.C.; Uddin, M.J. Machine Learning Approaches for Skin Neoplasm Diagnosis. *ACS Omega* **2024**, *9*, 32853–32863. [CrossRef]
11. Chawla, N.V.; Bowyer, K.W.; Hall, L.O.; Kegelmeyer, W.P. SMOTE: Synthetic Minority Over-sampling Technique. *J. Artif. Intell. Res.* **2002**, *16*, 321–357. [CrossRef]
12. Chen, J.H.; Zhang, Y.Q.; Zhu, T.T.; Zhang, Q.; Zhao, A.X.; Huang, Y. Applying machine-learning models to differentiate benign and malignant thyroid nodules classified as C-TIRADS 4 based on 2D-ultrasound combined with five contrast-enhanced ultrasound key frames. *Front. Endocrinol.* **2024**, *15*, 1299686. [CrossRef] [PubMed]

https://doi.org/10.3390/s26061998

13. Guo, Z.; Wang, N.; Zhao, G.; Du, L.; Cui, Z.; Liu, F. Development of preoperative models for predicting positive esophageal margin in proximal gastric cancer based on machine learning. *J. Shandong Univ. (Health Sci.)* **2024**, *62*, 78–83.
14. Guo, Z.; Wang, N.; Zhao, G.; Du, L.; Cui, Z.; Liu, F. Development and validation of a preoperative model for predicting positive proximal margins in adenocarcinoma of the esophagogastric junction and assessing safe margin distance. *Front. Oncol.* **2024**, *14*, 1503728. [CrossRef]
15. Batista, G.E.; Prati, R.C.; Monard, M.C. A study of the behavior of several methods for balancing machine learning training data. *ACM SIGKDD Explor. Newsl.* **2004**, *6*, 20–29. [CrossRef]
16. Chen, W.; Yang, K.; Yu, Z.; Shi, Y.; Chen, C.L.P. A survey on imbalanced learning: Latest research, applications and future directions. *Artif. Intell. Rev.* **2024**, *57*, 137. [CrossRef]
17. He, H.; Garcia, E.A. Learning from imbalanced data. *IEEE Trans. Knowl. Data Eng.* **2009**, *21*, 1263–1284. [CrossRef]
18. Insan, H.; Prasetiyowati, S.S.; Sibaroni, Y. SMOTE-LOF and Borderline-SMOTE Performance to Overcome Imbalanced Data and Outliers on Classification. In *Proceedings of the 2023 3rd International Conference on Intelligent Cybernetics Technology & Applications (ICICyTA), Denpasar, Indonesia, 13–15 December 2023*; IEEE: New York, NY, USA, 2023; pp. 136–141.
19. Ungkawa, U.; Rafi, M.A. Data Balancing Techniques Using the PCA-KMeans and ADASYN for Possible Stroke Disease Cases. *J. Online Inform.* **2024**, *9*, 138–147. [CrossRef]
20. Ahmed, G.; Er, M.J.; Fareed, M.M.S.; Zikria, S.; Mahmood, S.; He, J.; Asad, M.; Jilani, S.F.; Aslam, M. DAD-Net: Classification of Alzheimer's Disease Using ADASYN Oversampling Technique and Optimized Neural Network. *Molecules* **2022**, *27*, 7085. [CrossRef]
21. Khan, T.M.; Xu, S.; Khan, Z.G.; Chishti, M.U. Implementing Multilabeling, ADASYN, and ReliefF Techniques for Classification of Breast Cancer Diagnostic through Machine Learning: Efficient Computer-Aided Diagnostic System. *J. Healthc. Eng.* **2021**, *2021*, 5577636. [CrossRef]
22. Munshi, R.M. Novel ensemble learning approach with SVM-imputed ADASYN features for enhanced cervical cancer prediction. *PLoS ONE* **2024**, *19*, e0296107. [CrossRef]
23. Ramotra, A.K.; Mansotra, V. Hybrid Type-2 Diabetes Prediction Model Using SMOTE, K-means Clustering, PCA, and Logistic Regression. *Asian Pac. J. Health Sci.* **2021**, *8*, 137–140. [CrossRef]
24. Madsen, M.T.; Park, C.H. Enhancement of SPECT Images by Fourier Filtering the Projection Image Set. *J. Nucl. Med.* **1985**, *26*, 395–402.
25. Yoshida, H.; Keserci, B. Bayesian wavelet snake for computer-aided diagnosis of lung nodules. *Integr. Comput.-Aided Eng.* **2000**, *7*, 253–269. [CrossRef]
26. Lundervold, A.S.; Lundervold, A. An overview of deep learning in medical imaging focusing on MRI. *Z. Für Med. Phys.* **2019**, *29*, 102–127. [CrossRef]
27. Goodfellow, I.; Pouget-Abadie, J.; Mirza, M.; Xu, B.; Warde-Farley, D.; Ozair, S.; Courville, A.; Bengio, Y. Generative adversarial networks. *Commun. ACM* **2020**, *63*, 139–144. [CrossRef]
28. Greenspan, H.; van Ginneken, B.; Summers, R.M. Guest Editorial Deep Learning in Medical Imaging: Overview and Future Promise of an Exciting New Technique. *IEEE Trans. Med. Imaging* **2016**, *35*, 1153–1159. [CrossRef]
29. Kim, Y.; Choi, W.; Choi, W.; Ko, G.; Han, S.; Kim, H.C.; Kim, D.; Lee, D.G.; Shin, D.W.; Lee, Y. A machine learning approach using conditional normalizing flow to address extreme class imbalance problems in personal health records. *BioData Min.* **2024**, *17*, 14. [CrossRef]
30. Stephanovitch, A.; Aamari, E.; Levrard, C. Wasserstein GANs Are Minimax Optimal Distribution Estimators. *arXiv* **2023**. [CrossRef]
31. Li, X.; Huang, H.; Yuan, G.T.; Wang, Z.L.; Du, R. Research on fusion neural network intrusion detection based on improved WGAN algorithm. *J. Sichuan Univ. Sci. Eng. (Nat. Sci. Ed.)* **2024**, *37*, 57–65.
32. Luo, Y.; Yang, Z. DynGAN: Solving Mode Collapse in GANs With Dynamic Clustering. *IEEE Trans. Pattern Anal. Mach. Intell.* **2024**, *46*, 5493–5503. [CrossRef] [PubMed]
33. Yin, W.; Huang, C.; Chen, L.; Huang, X.; Wang, Z.; Bian, Y.; Zhou, Y.; Wan, Y.; Han, T.; Yi, M. Facilitate Robust Early Screening of Cerebral Palsy via General Movements Assessment with Multi-Modality Co-Learning. *IEEE Trans. Med. Imaging* **2025**. early access. [CrossRef]
34. Brock, A.; Donahue, J.; Simonyan, K. Large Scale GAN Training for High Fidelity Natural Image Synthesis. *arXiv* **2019**, arXiv:1809.11096. [CrossRef]
35. Karras, T.; Laine, S.; Aittala, M.; Hellsten, J.; Lehtinen, J.; Aila, T. Analyzing and Improving the Image Quality of StyleGAN. In *Proceedings of the 2020 IEEE/CVF Conference on Computer Vision and Pattern Recognition (CVPR), Seattle, WA, USA, 13–19 June 2020*; IEEE: New York, NY, USA, 2020; pp. 8107–8116.
36. Xie, Z.H.; Yi, M.; Huang, X. Application progress of multi-instance learning in medical image analysis. *J. Integr. Technol.* **2025**, *14*, 24–32.

37. Rao, Z.; Wu, C.; Liao, Y.; Ye, C.; Huang, S.; Zhao, D. POCALI: Prediction and Insight on CAncer LncRNAs by Integrating Multi-Omics Data with Machine Learning. *Small Methods* **2025**, *9*, e2401987. [CrossRef]
38. Luo, J.; Zhang, W.; Tan, S.; Liu, L.; Bai, Y.; Zhang, G. Aortic Dissection Auxiliary Diagnosis Model and Applied Research Based on Ensemble Learning. *Front. Cardiovasc. Med.* **2021**, *8*, 777757. [CrossRef]
39. Yang, C.; Xie, J.; Huang, X.; Tan, H.; Li, Q.; Tang, Z.; Ma, X.; Lu, J.; He, Q.; Fu, W.; et al. ECS-Net: Extracellular space segmentation with contrastive and shape-aware loss by using cryo-electron microscopy imaging. *Expert Syst. Appl.* **2025**, *270*, 126370. [CrossRef]
40. Yeung, M.; Sala, E.; Schönlieb, C.B.; Rundo, L. Unified Focal loss: Generalising Dice and cross entropy-based losses to handle class imbalanced medical image segmentation. *Comput. Med. Imaging Graph.* **2022**, *95*, 102026. [CrossRef] [PubMed]
41. Du, J.; Zhang, X.; Liu, P.; Vong, C.M.; Wang, T. An Adaptive Deep Metric Learning Loss Function for Class-Imbalance Learning via Intraclass Diversity and Interclass Distillation. *IEEE Trans. Neural Netw. Learn. Syst.* **2024**, *35*, 15372–15386. [CrossRef]
42. Li, Z.; Kamnitsas, K.; Glocker, B. Analyzing Overfitting Under Class Imbalance in Neural Networks for Image Segmentation. *IEEE Trans. Med. Imaging* **2021**, *40*, 1065–1077. [CrossRef]
43. Albahli, S.; Masood, M. Efficient attention-based CNN network (EANet) for multi-class maize crop disease classification. *Front. Plant Sci.* **2022**, *13*, 1003152. [CrossRef] [PubMed]
44. Qiu, S.; Zhao, H.; Jiang, N.; Wang, Z.; Liu, L.; An, Y.; Zhao, H.; Miao, X.; Liu, R.; Fortino, G. Multi-sensor information fusion based on machine learning for real applications in human activity recognition: State-of-the-art and research challenges. *Inf. Fusion* **2022**, *80*, 241–265. [CrossRef]
45. Rao, P.M.M.; Singh, S.K.; Khamparia, A.; Bhushan, B.; Podder, P. Multi-Class Breast Cancer Classification Using Ensemble of Pretrained models and Transfer Learning. *Curr. Med. Imaging* **2022**, *18*, 409–416. [CrossRef]
46. Chen, Y.S. An empirical study of a hybrid imbalanced-class DT-RST classification procedure to elucidate therapeutic effects in uremia patients. *Med. Biol. Eng. Comput.* **2016**, *54*, 983–1001. [CrossRef]
47. Yang, J.; El-Bouri, R.; O'Donoghue, O.; Lachapelle, A.S.; Soltan, A.A.S.; Eyre, D.W.; Lu, L.; Clifton, D.A. Deep reinforcement learning for multi-class imbalanced training: Applications in healthcare. *Mach. Learn.* **2024**, *113*, 2655–2674. [CrossRef]
48. Solihah, B.; Azhari, A.; Musdholifah, A. Enhancement of conformational B-cell epitope prediction using CluSMOTE. *PeerJ Comput. Sci.* **2020**, *6*, e275. [CrossRef] [PubMed]
49. Ojeda, F.M.; Jansen, M.L.; Thiéry, A.; Blankenberg, S.; Weimar, C.; Schmid, M.; Ziegler, A. Calibrating machine learning approaches for probability estimation: A comprehensive comparison. *Stat. Med.* **2023**, *42*, 5451–5478. [CrossRef] [PubMed]
50. Hou, F.; Zhu, Y.; Zhao, H.; Cai, H.; Wang, Y.; Peng, X.; Lu, L.; He, R.; Hou, Y.; Li, Z.; et al. Development and validation of an interpretable machine learning model for predicting the risk of distant metastasis in papillary thyroid cancer: A multicenter study. *eClinicalMedicine* **2024**, *77*, 102913. [CrossRef]
51. Dawood, T.; Ruijsink, B.; Razavi, R.; King, A.P.; Puyol-Antón, E. Improving Deep Learning Model Calibration for Cardiac Applications using Deterministic Uncertainty Networks and Uncertainty-aware Training. *arXiv* **2024**. [CrossRef]
52. Van Calster, B.; McLernon, D.J.; van Smeden, M.; Wynants, L.; Steyerberg, E.W. Calibration: The Achilles heel of predictive analytics. *BMC Med.* **2019**, *17*, 230. [CrossRef]
53. Abu-Shareha, A.A.; Abualhaj, M.; Hussein, A.; Al-Saaidah, A.; Achuthan, A. Investigation of Data Balancing Techniques for Diabetes Prediction. *Int. J. Intell. Eng. Syst.* **2025**, *18*, 598–611. [CrossRef]
54. Zhang, F.; Wu, S.; Zhang, C.; Chen, Q.; Yang, X.; Jiang, K.; Zheng, J. Multi-domain features for reducing false positives in automated detection of clustered microcalcifications in digital breast tomosynthesis. *Med. Phys.* **2019**, *46*, 1300–1308. [CrossRef]
55. Niu, Z.; Dong, H.; Qin, A.K. FedSPU: Personalized Federated Learning for Resource-Constrained Devices with Stochastic Parameter Update. *Proc. AAAI Conf. Artif. Intell.* **2025**, *39*, 19721–19729. [CrossRef]
56. Wu, N.; Yu, L.; Yang, X.; Cheng, K.-T.; Yan, Z. FedIIC: Towards Robust Federated Learning for Class-Imbalanced Medical Image Classification. In *International Conference on Medical Image Computing and Computer-Assisted Intervention*; Springer: Cham, Switzerland, 2023; pp. 692–702. [CrossRef]
57. Işık, G.; Paçal, İ. Few-shot classification of ultrasound breast cancer images using meta-learning algorithms. *Neural Comput. Appl.* **2024**, *36*, 12047–12059. [CrossRef]

https://doi.org/10.3390/s26061998

MDPI AG
Grosspeteranlage 5
4052 Basel
Switzerland
Tel.: +41 61 683 77 34

Sensors Editorial Office
E-mail: sensors@mdpi.com
www.mdpi.com/journal/sensors